Development and Application of Nonlinear Dissipative Device in Structural Vibration Control

Development and Application of Nonlinear Dissipative Device in Structural Vibration Control

Special Issue Editors

Zheng Lu
Tony Yang
Ying Zhou
Angeliki Papalou

MDPI • Basel • Beijing • Wuhan • Barcelona • Belgrade

MDPI

Special Issue Editors

Zheng Lu
Tongji University
China

Tony Yang
The University of British Columbia
Canada

Ying Zhou
State Key Laboratory of Disaster Reduction in Civil Engineering
China

Angeliki Papalou
Technological Educational Institute
Greece

Editorial Office
MDPI
St. Alban-Anlage 66
Basel, Switzerland

This edition is a reprint of the Special Issue published online in the open access journal *Applied Sciences* (ISSN 2076-3417) from 2017–2018 (available at: http://www.mdpi.com/journal/applsci/special_issues/Structural_Vibration_Control).

For citation purposes, cite each article independently as indicated on the article page online and as indicated below:

Lastname, F.M.; Lastname, F.M. Article title. *Journal Name* **Year**, *Article number*, page range.

ISBN 978-3-03897-037-8 (Pbk)
ISBN 978-3-03897-038-5 (PDF)

Cover image courtesy of Zheng Lu.

Contents

About the Special Issue Editors

Zheng Lu, Professor, College of Civil Engineering, Tongji University has more than 10 years' experience in structural sibration control, earthquake resistance of engineering structures, and particle damping technology. The studies he has carried out focus on developing the particle damping-based new dissipative devices and corresponding fundamental applications. He was the director of two projects subsidized by the Natural Science Foundation of China. He received the Shanghai "Chen Guang" project, which is a Talent Scheme for young scholars under the age of 30. He has published more than 50 first-authored or corresponding-authored SCI-indexed papers, which have received high acclaim. His H-index is 11 in Web of Science. He has published review papers pertaining to nonlinear dissipative dampers in two high-quality SCI indexed journals, namely, Structural Control and Health Monitoring and the Journal of Sound and Vibration.

Tony Yang, Associate Professor, Department of Civil Engineering, the University of British Columbia. Dr. Yang's research focuses on improving structural performance through advanced analytical simulation and experimental testing. He has developed next-generation, performance-based design guidelines (adopted by the Applied Technology Council, the ATC-58 research team) in the United States; furthermore, he has developed advanced experimental testing technologies, such as hybrid simulation and nonlinear control of shake table, to evaluate structural responses under extreme loading conditions.

Ying Zhou, Professor, College of Civil Engineering, Tongji University. Professor Zhou's research focuses on seismic performance of complex tall buildings, performance-based seismic design, structural performance of composite structures and hybrid structures, and methodology and technology for structural dynamic tests. In 2008, she received the Outstanding Paper Award for Young Experts at the 14th World Conference on Earthquake Engineering. She received the National Science Fund for Excellent Young Scholars (2014–2016), for which the project title was seismic resistance of high-rise buildings.

Angeliki Papalou, Associate Professor, Department of Civil Engineering, Technological Educational Institute (T.E.I.) of Western Greece. Dr. Papalou's research interests are focused on the seismic protection of structures, structural control, and health monitoring, in the seismic protection and restoration of historic structures and in the analysis and design of structures. She was the principal investigator of a funded research project and member of research groups for five other funded research projects. She is a reviewer for more than ten international scientific journals and a member of the American Society of Civil Engineers (ACSE), the Technical Chamber of Greece (TEE) and the American Mechanics Institute.

Preface to "Development and Application of Nonlinear Dissipative Device in Structural Vibration Control"

This book entitled Development and Application of Nonlinear Dissipative Device in Structural Vibration Control contains papers that focus on the development and application of innovative nonlinear dissipative systems that mitigate the potentially catastrophic effects of extreme loading by incorporating new materials or effective mechanical control technologies. Moreover, new nonlinear analytical methods for distinctive vibrating structures under different excitations are also highlighted in this book. It is notable that many research areas, especially in civil engineering, have attached much importance to nonlinear characteristics of both vibrating structures and dissipation devices. This is mainly because under strong excitations, such as severe earthquakes, vibrating structures tend to yield and generate excessive displacement, which leads to material and geometric nonlinearities, respectively. Both, in turn, exert a significant effect on the seismic performance of vibrating structures and dampen the effectiveness of dissipative devices. Additionally, nonlinear dampers present more superiorities in energy dissipation than linear dampers, such as a wide frequency band of vibration attenuation and high robustness. Therefore, these nonlinear dampers have been utilized in many different cases. For example, nonlinear fluid viscous dampers are applied to control the large maximum bearing displacement of isolation systems; pounding-tuned mass dampers are employed to alleviate the excessive vibration of the power transmission tower; self-powered magnetorheological dampers are used to suppress the undesirable vibration of long stay cables.

Therefore, the contents of this book cover a wide variety of topics, which can be mainly divided into three categories, namely, new nonlinear dissipative devices, new simulation tools for vibrating structures undergoing the nonlinear stage, and new design/optimum methods for dissipative devices and isolation systems. It is worth mentioning that to broaden the scope of nonlinearity, besides the nonlinear dissipative devices, the specific structures that contains nonlinear connections or express nonlinear behaviors, and the based-isolated structures whose isolators would yield under large displacements, are also the targets of this book. Moreover, to reinforce the point that linear dampers are capable of producing desirable damping performance under certain circumstances, the recent research pertaining to the linear dampers, including tuned mass damper and eddy current tuned mass damper (the damping force produced by the eddy currents is proportional to the relative velocity), are also contained in this book.

This book contains 13 very high-quality papers. The author groups represent currently active researchers in the structural vibration control area. The topics are not only current (cutting-edge research) but also of great academic (fundamental phase) and industrial (applied phase) interest. The readers will observe that compared to linear dissipative devices, the application of nonlinear dissipative devices in civil engineering is just beginning, and most of the research concentrates on theoretical study, numerical simulation, and experimental study. Hence, further efforts should be made regarding the applied phase of nonlinear dampers.

Zheng Lu, Tony Yang , Ying Zhou , Angeliki Papalou
Special Issue Editors

applied
sciences

MDPI

Editorial

Special Issue: Development and Application of Nonlinear Dissipative Device in Structural Vibration Control

Zheng Lu [1,*], Ying Zhou [1], Tony Yang [2] and Angeliki Papalou [3]

[1]   Research Institute of Structural Engineering and Disaster Reduction, College of Civil Engineering, Tongji University, Shanghai 200092, China; yingzhou@tongji.edu.cn

[2]   Department of Civil Engineering, University of British Columbia, Vancouver, BC V6T 1Z4, Canada; yang@civil.ubc.ca

[3]   Department of Civil Engineering, Technological Educational Institute (T.E.I.) of Western Greece, 26334 Patras, Greece; papalou@teiwest.gr

*   Correspondence: luzheng111@tongji.edu.cn; Tel.: +86-21-6598-6186

Received: 21 May 2018; Accepted: 22 May 2018; Published: 23 May 2018

This Special Issue (SI) of *Applied Sciences* on Development and Application of Nonlinear Dissipative Devices in Structural Vibration Control contains papers that focus on the development and application of innovative nonlinear dissipative systems that mitigate the potentially catastrophic effects of extreme loading by incorporating new materials or effective mechanical control technologies. Moreover, the new nonlinear analytical methods for distinctive vibrating structures under different excitations are also highlighted in this Special Issue. It is notable that many research areas, especially those related to civil engineering, have placed more importance on the nonlinear characteristics of both vibrating structures and dissipation devices. This is mainly because under strong excitations, such as severe earthquakes, the vibrating structures tend to yield and generate excessive displacement, which leads to material and geometric nonlinearities, respectively. Both of these nonlinearities have significant effects on the seismic performance of vibrating structures and the damping effectiveness of dissipative devices. Additionally, the nonlinear dampers present more advantages in energy dissipation than linear dampers, such as wide frequency bands of vibration attenuation and high robustness. Therefore, these nonlinear dampers have been utilized in many different cases. For example, nonlinear fluid viscous dampers are applied to control the large maximum bearing displacement of isolation systems; pounding tuned mass dampers are employed to alleviate the excessive vibration of power transmission towers; and self-powered magnetorheological dampers are used to suppress the undesirable vibration of long stay cables.

We have been particularly interested in receiving manuscripts that encompass the development of efficient and convenient composite nonlinear dampers; experimental investigation, advanced modeling and systematical theoretical analysis of nonlinear dynamic systems; and optimization of creative nonlinear dampers and damping mechanisms. Therefore, the papers we have received are on a wide variety of topics, which can be mainly divided into three categories: academic fundamental phase, current cutting-edge research and industrial application phase. It is worth mentioning that to broaden the scope of nonlinearity, apart from the nonlinear dissipative devices, the specific structures that contain nonlinear connections or express nonlinear behaviors and the base-isolated structures whose isolators would yield under large displacements are also the targets of this special issue. Moreover, to reinforce the point that linear dampers are capable of producing desirable damping performance under certain circumstances, the recent research pertaining to the linear dampers, including the tuned mass damper and eddy current tuned mass damper (the damping force produced by the eddy currents is proportional to the relative velocity), are also included in this special issue.

This special issue has already published 13 very high-quality papers. The author groups represent currently active researchers in the structural vibration control area. The topics are not only current (cutting-edge research) but also of great academic (fundamental phase) and industrial (applied phase) interest. The readers will observe that compared to linear dissipative devices, the application of nonlinear dissipative devices in civil engineering is just in its preliminary stages, with most of the research concentrating on the theoretical study, numerical simulation and experimental study. Hence, more further efforts should be made on the application phase of nonlinear dampers. The papers are cited below, with brief comments for each paper concerning the main topic and contributions of the paper.

Academic fundamental phase. Due to the fact that the traditional linear/nonlinear dampers cannot meet the demands of vibration attenuation in severe conditions, such as a power transmission tower undergoing multi-component seismic excitations, a submerged pipeline being subjected to seawater environments and a building structure withstanding debris flow, some authors have subsequently proposed new linear/nonlinear dissipative devices, which has enriched the academic fundamental research of linear/nonlinear dissipative devices. The papers in this category and corresponding comments are listed below:

(1)　Tian, L.; et al. Vibration Control of a Power Transmission Tower with Pounding Tuned Mass Damper under Multi-Component Seismic Excitations [1].

The very first submitted and accepted paper of this Special Issue proposes a new nonlinear dissipative device that can be applied to increase the seismic resistance of a power transmission tower. This device is namely the pounding tuned mass damper (Pounding TMD), which combines the impact damper and the tuned mass damper (TMD). The main contributions of this paper are as follows: (a) a three-dimensional finite element modal of a practical power transmission tower attached with TMD/Pounding TMD is established to verify the superior effectiveness of Pounding TMD over TMD; and (b) parametric analysis was carried out through this model, including mass ratio, ground motion intensity, gap and incident angle.

(2)　Chen, J.; et al. Experimental Study on Robustness of an Eddy Current-Tuned Mass Damper [2].

In this paper, the robustness of an eddy current tuned mass damper (ECTMD) is investigated experimentally through the vibration control of a cantilever beam, with comparison of its results to the robustness of a tuned mass damper. The experimental results indicate that the damping performance of the ECTMD is superior to that of the TMD, which is mainly due to its higher robustness under both free vibration and forced vibration.

(3)　Wang, W.; et al. Experimental Study on Vibration Control of a Submerged Pipeline Model by Eddy Current Tuned Mass Damper [3].

This paper utilizes an eddy current tuned mass damper to suppress the excessive vibration of submerged pipelines and validates the feasibility of eddy current damping in a seawater environment through an experimental study. The test results show that the damping provided by the eddy current in a seawater environment is only slightly varied compared to that in an air environment. Furthermore, with the optimal ECTMD control, the vibration response of the submerged pipeline is significantly decreased.

(4)　Li, P.; et al. Experimental Study on the Performance of Polyurethane-Steel Sandwich Structure under Debris Flow [4].

To strengthen the impact resistance of buildings subjected to debris flow, this paper proposes the use of a special material, which is namely polyurethane-steel sandwich composite, as the structural material, which generates the polyurethane-steel sandwich structure. The impact resistance of polyurethane-steel sandwich structure under debris flow is investigated by a series of impact loading tests, which allows for comparison with the test results of traditional steel frame structures. The test results demonstrate that: (a) the steel frame structure mainly depends on the impacted column to resist the impact loading; and (b) when subjected to debris flow, the polyurethane-steel sandwich structure exhibits superior performance in resisting the impact loading.

(5)   Wang, Z.; et al. Development of a Self-Powered Magnetorheological Damper System for Cable Vibration Control [5].

    In this paper, a new nonlinear dissipative device, which is the self-powered magnetorheological (MR) damper control system, is applied to attenuate the undesirable vibration of long stay cables. The vibration mitigation performance of the presented self-powered MR damper system is evaluated by model tests with a 21.6-m long cable. The experimental results show that: (a) the supplemental modal damping ratios of the cable in the first four modes can be significantly enhanced by the self-powered MR damper system, demonstrating the feasibility and effectiveness of the new smart passive system; and (b) both the self-powered MR damper and the generator are quite similar to a combination of a traditional linear viscous damper and a negative stiffness device, with the negative stiffness being able to enhance the mitigation efficiency against cable vibration.

    Current cutting-edge research. Since it is really common that the vibrating structures would present nonlinear behaviors when being subjected to strong excitations, we occasionally use the nonlinear deformation in the main structure to dissipate vibration energy. Undoubtedly, the nonlinear properties of vibrating structures should be considered when estimating the seismic performance of structures and evaluating the damping performance of dampers. In this sense, some scholars proposed new simulation tools for vibrating structures currently in a nonlinear stage, which complements the current cutting-edge research of the analysis methods for nonlinear vibrating structures. The papers in this category and corresponding comments are listed below:

(6)   Chikhaoui, K.; et al. Robustness Analysis of the Collective Nonlinear Dynamics of a Periodic Coupled Pendulums Chain [6].

The paper conducts the robustness analysis of a special nonlinear system, which is namely the periodic coupled pendulums chain, by a generic discrete analytical model. The main contribution of this paper is that the robustness analysis results demonstrate the benefits of the presence of imperfections in such periodic structures. To be more specific, imperfections can be utilized to generate energy localization that is suitable for several engineering applications, such as vibration energy harvesting.

(7)   Ye, J.; et al. Member Discrete Element Method for Static and Dynamic Responses Analysis of Steel Frames with Semi-Rigid Joints [7].

    This paper's objective is to investigate the complex behaviors of steel frames with nonlinear semi-rigid connections, including both static and dynamic responses, by a simple and effective numerical method that is based on the Member Discrete Element Method (MDEM). The advantages of the proposed simulation approach are as follows: (a) the modified MDEM can accurately capture the linear and nonlinear behavior of semi-rigid connections; and (b) the modified MDEM can avoid the difficulties of finite element method (FEM) in dealing with strong nonlinearity and discontinuity.

(8)   Mansouri, I.; et al. Prediction of Ultimate Strain and Strength of FRP-Confined Concrete Cylinders Using Soft Computing Methods [8].

    In this paper, the effectiveness of four different soft computing methods for predicting the ultimate strength and strain of concrete cylinders confined with fiber-reinforced polymer (FRP) sheets is evaluated, including radial basis neural network (RBNN), adaptive neuro fuzzy inference system (ANFIS) with subtractive clustering (ANFIS-SC), ANFIS with fuzzy c-means clustering (ANFIS-FCM) and M5 model tree (M5Tree). The comparison results show that the ANFIS-SC, performed slightly better than the RBNN and ANFIS-FCM in estimating the ultimate strain of confined concrete. On the other hand, M5Tree provided the most inaccurate strength and strain estimates.

(9) Wen, B.; et al. Soil-Structure-Equipment Interaction and Influence Factors in an Underground Electrical Substation under Seismic Loads [9].

This paper proposes a seismic response analysis method for underground electrical substations considering the soil–structure–equipment interactions, which is performed by changing the earthquake input motions, soil characteristics, electrical equipment type and structure depths. The numerical results indicate that: (a) as a boundary condition of soil–structure, the coupling boundary is feasible in the seismic response of an underground substation; (b) the seismic response of an underground substation is sensitive to burial depth and elastic modulus; (c) the oblique incidence of input motion has a slight influence on the horizontal seismic response, but has a significant impact on the vertical seismic response; and (d) the bottom of the side wall is the seismic weak part of an underground substation, so it is necessary to increase the stiffness of this area.

(10) Liu, C.; et al. Base Pounding Model and Response Analysis of Base-Isolated Structures under Earthquake Excitation [10].

To study the base pounding effects of the base-isolated structure under earthquake excitations, this paper proposes a base pounding theoretical model with a linear spring-gap element. The numerical analysis conducted through this model suggests that: (a) the model offers much flexibility in analyzing base pounding effects; (b) there is a most unfavorable clearance width between adjacent structures; and (c) the structural response increases with pounding and consequently, it is necessary to consider base pounding in the seismic design of base-isolated structures.

(11) Chen, Z.; et al. Application of the Hybrid Simulation Method for the Full-Scale Precast Reinforced Concrete Shear Wall Structure [11].

This paper proposes a new nonlinear seismic performance analysis method for the full-scale precast reinforced concrete shear wall structure based on hybrid simulation (HS). To be more specific, an equivalent force control (EFC) method with an implicit integration algorithm is employed to deal with the numerical integration of the equation of motion (EOM) and the control of the loading device. The accuracy and feasibility of the EFC-based HS method is verified experimentally through the substructure hybrid simulation tests of the pre-cast reinforced concrete shear-wall structure model. Because of the arrangement of the test model, an elastic non-linear numerical model is used to simulate the numerical substructure. The experimental results of the descending stage can be conveniently obtained from the EFC-based HS method.

Industrial application phase. Finally, based on both the theoretical and experimental academic research, the practical designs or optimum methods that are a valuable reference for actual engineering applications can be obtained. In this special issue, one paper proposes a design method for seismically isolated reinforced concrete frame-core tube tall building, while another paper puts forward an optimum method of tuned mass dampers for the pedestrian bridge, both of which are of great industrial interests. The papers in this category and corresponding comments are listed below:

(12) Li, A.; et al. Research on the Rational Yield Ratio of Isolation System and Its Application to the Design of Seismically Isolated Reinforced Concrete Frame-Core Tube Tall Buildings [12].

This paper proposes a high-efficiency design method based on the rational yield ratio of the isolation system and applies it to the design of the seismically isolated reinforced concrete (RC) frame-core tube tall buildings. The main contributions of this paper are as follows. (a) Through 28 carefully designed cases of seismically isolated RC frame-core tube tall buildings, the rational yield ratio of the isolation system for such buildings is recommended to be 2–3%. (b) Based on the recommended rational yield ratio, a high-efficiency design method is proposed for seismically isolated RC frame-core tube tall buildings. (c) The rationality, reliability and efficiency of the proposed method are validated by a case stay of a seismically isolated RC frame-core tube tall building with a height of 84.1 m, which is designed by the proposed design method.

(13) Shi, W.; et al. Application of an Artificial Fish Swarm Algorithm in an Optimum Tuned Mass Damper Design for a Pedestrian Bridge [13].

This paper proposes a new optimization method for the tuned mass damper (TMD), which can be applied to alleviate the vibration of pedestrian bridges based on the artificial fish swarm algorithm (AFSA). The optimization goal of this design method is to minimize the maximum dynamic amplification factor of the primary structure under external harmonic excitations. Through a case study of an optimized TMD based on AFSA for a pedestrian bridge, it was shown that the TMD designed based on AFSA has a smaller maximum dynamic amplification factor than the TMD designed based on other classical optimization methods, while the optimized TMD has a good effect in controlling the human-induced vibrations at different frequencies.

Conflicts of Interest: The authors declare no conflict of interest.

References

1. Tian, L.; Rong, K.; Zhang, P.; Liu, Y. Vibration control of a power transmission tower with pounding tuned mass damper under multi-component seismic excitations. *Appl. Sci.* **2017**, *7*, 477. [CrossRef]
2. Chen, J.; Lu, G.; Li, Y.; Wang, T.; Wang, W.; Song, G. Experimental study on robustness of an eddy current-tuned mass damper. *Appl. Sci.* **2017**, *7*, 895. [CrossRef]
3. Wang, W.; Dalton, D.; Hua, X.; Wang, X.; Chen, Z.; Song, G. Experimental study on vibration control of a submerged pipeline model by eddy current tuned mass damper. *Appl. Sci.* **2017**, *7*, 987. [CrossRef]
4. Li, P.; Liu, S.; Lu, Z. Experimental study on the performance of polyurethane-steel sandwich structure under debris flow. *Appl. Sci.* **2017**, *7*, 1018. [CrossRef]
5. Wang, Z.; Chen, Z.; Gao, H.; Wang, H. Development of a self-powered magnetorheological damper system for cable vibration control. *Appl. Sci.* **2018**, *8*, 118. [CrossRef]
6. Chikhaoui, K.; Bitar, D.; Kacem, N.; Bouhaddi, N. Robustness analysis of the collective nonlinear dynamics of a periodic coupled pendulums chain. *Appl. Sci.* **2017**, *7*, 684. [CrossRef]
7. Ye, J.; Xu, L. Member discrete element method for static and dynamic responses analysis of steel frames with semi-rigid joints. *Appl. Sci.* **2017**, *7*, 714. [CrossRef]
8. Mansouri, I.; Kisi, O.; Sadeghian, P.; Lee, C.-H.; Hu, J. Prediction of ultimate strain and strength of FRP-confined concrete cylinders using soft computing methods. *Appl. Sci.* **2017**, *7*, 751. [CrossRef]
9. Wen, B.; Zhang, L.; Niu, D.; Zhang, M. Soil–structure–equipment interaction and influence factors in an underground electrical substation under seismic loads. *Appl. Sci.* **2017**, *7*, 1044. [CrossRef]
10. Liu, C.; Yang, W.; Yan, Z.; Lu, Z.; Luo, N. Base pounding model and response analysis of base-isolated structures under earthquake excitation. *Appl. Sci.* **2017**, *7*, 1238. [CrossRef]
11. Chen, Z.; Wang, H.; Wang, H.; Jiang, H.; Zhu, X.; Wang, K. Application of the hybrid simulation method for the full-scale precast reinforced concrete shear wall structure. *Appl. Sci.* **2018**, *8*, 252. [CrossRef]
12. Li, A.; Yang, C.; Xie, L.; Liu, L.; Zeng, D. Research on the rational yield ratio of isolation system and its application to the design of seismically isolated reinforced concrete frame-core tube tall buildings. *Appl. Sci.* **2017**, *7*, 1191. [CrossRef]
13. Shi, W.; Wang, L.; Lu, Z.; Zhang, Q. Application of an artificial fish swarm algorithm in an optimum tuned mass damper design for a pedestrian bridge. *Appl. Sci.* **2018**, *8*, 175. [CrossRef]

applied sciences

MDPI

Article

Vibration Control of a Power Transmission Tower with Pounding Tuned Mass Damper under Multi-Component Seismic Excitations

Li Tian [1], Kunjie Rong [1], Peng Zhang [2] and Yuping Liu [1,*]

[1]   School of Civil Engineering, Shandong University, Jinan 250061, Shandong Province, China; tianli@sdu.edu.cn (L.T.); kunjierong@163.com (K.R.)
[2]   Transportation Equipment and Ocean Engineering College, Dalian Maritime University, Dalian 116026, Liaoning Province, China; peng1618@163.com
*   Correspondence: civil_sdu@163.com; Tel.:+86-178-6513-1119

Academic Editor: César M. A. Vasques
Received: 7 March 2017; Accepted: 2 May 2017; Published: 5 May 2017

Abstract: In this paper, the two-dimensional vibration controls of a power transmission tower with a pounding tuned mass damper (PTMD) under multi-component seismic excitations are analyzed. A three-dimensional finite element model of a practical power transmission tower is established in ABAQUS (Dassasult Simulia Company, Providence, RI, USA). The TMD (tuned mass damper) and PTMD are simulated by the finite element method. The response of the transmission tower with TMD and PTMD are analyzed, respectively. To achieve optimal design, the influence of the mass ratio, ground motion intensity, gap, and incident angle of seismic ground motion are investigated, respectively. The results show that the PTMD is very effective in reducing the vibration of the transmission tower in the longitudinal and transverse directions. The reduction ratio increases with the increase of the mass ratio. The ground motion intensity and gap have no obvious influence on the reduction ratio. However, the incident angle has a significant influence on the reduction ratio.

Keywords: power transmission tower; pounding tuned mass damper; multi-component seismic excitations; mass ratio; gap; incident angle

1. Introduction

The transmission tower is an important component of the transmission line, and the power transmission tower-line system is an important lifeline facility. The damage of a power transmission tower-line system may lead to the paralysis of the power grid. With the increasing height of transmission towers and the span of the transmission line, the seismic risk has increased and several failures have been reported during the past decades. During the 1992 Landers earthquake, about 100 transmission lines, and several transmission towers, failed in the city of Los Angeles [1]. In the 1994 Northridge earthquake, a number of transmission towers were destroyed, and the power system was greatly damaged [1]. During the 1995 Kobe earthquake, more than 20 transmission towers were damaged [2]. In the 2008 Wenchuan earthquake, more than 20 towers collapsed and a 220 kV transmission line in Mao County was destroyed [3–5]. As shown in Figure 1, the 2010 Haiti earthquake caused damage to transmission towers. During the 2013 Lushan earthquake, more than 39 transmission lines were destroyed [6]. Therefore, studies on the vibration control of power transmission towers needs to be conducted to improve and guarantee the safety of transmission lines.

Figure 1. The collapse of transmission towers during the Haiti earthquake.

Some research about the vibration control of a transmission tower under wind loading has been conducted at home and abroad [7–12]. However, there are few studies about the vibration control of transmission towers under earthquake excitation. In recent years, researchers have conducted studies regarding impact dampers. Ema et al. [13] investigated the performance of impact dampers from free damped vibration generated when a step function input was supplied to a leaf spring with a free mass. Collete [14] studied the vibration control capability of a combined tuned absorber and impact damper under a random excitation using numerical and experimental methods. Cheng et al. [15] researched the free vibration of a vibratory system equipped with a resilient impact damper. The results presented above show that the impact damper can reduce the response of structures. Due to space limitations, vibration control devices are not suitable for transmission towers. Therefore, a new type of vibration control device has been developed which combines the impact damper and tuned mass damper (TMD). Zhang et al. [16] proposed a new type of TMD, the pounding tuned mass damper (PTMD), to upgrade the seismic resistance performance of a transmission tower. Compared with TMD, the bandwidth vibration suppression of PTMD is larger, so the vibration reduction effect of PTMD is better than that of TMD. The PTMD has also been applied for vibration control of subsea pipeline structures [17–19] and traffic poles [20], and both simulation results and experimental results have demonstrated the effectiveness of the PTMD. However, in the previous studies, the PTMD has been simulated by a modified Hertz-contact model. Since the Hertz-contact model cannot be established in finite element modelling (FEM) software, such as ABAQUS, the primary structures were all simulated by simplified multi-mass models.

Based on the above research, two-dimensional vibration controls of a power transmission tower with a PTMD under multi-component seismic excitations are performed. A three-dimensional finite element model is created in ABAQUS according to practical engineering. The vibration reduction mechanism of the PTMD is introduced, and the PTMD is simulated using finite element software. To compare with the vibration reduction effect of the PTMD, the vibration control of the TMD is also conducted. A parametric study of the PTMD is carried out to provide a reference for the optimal design of a transmission tower with a PTMD.

2. Vibration Reduction Mechanism of PTMD

The equations of motion of structures with a PTMD can be expressed as [16]:

$$M\ddot{U}(t) + C\dot{U}(t) + KU(t) = -M\ddot{U}_g(t) + F_P\Delta P(t) \tag{1}$$

where, M, C, and K are the mass, damping, and stiffness of the structure, respectively; $\ddot{U}(t)$, $\dot{U}(t)$, and $U(t)$ are the vectors of the acceleration, velocity, and displacement of the structure, respectively;

$\ddot{U}_g(t)$ is the input ground motion acceleration in two horizontal directions; and $P(t)$ is the pounding force, which can be calculated as follows:

$$P = \begin{cases} \beta(u_1 - u_2 - g_p)^{3/2} + c_k(\dot{u}_1 - \dot{u}_2) & u_1 - u_2 - g_p > 0 (\dot{u}_1 - \dot{u}_2 > 0) \\ \beta(u_1 - u_2 - g_p)^{3/2} & u_1 - u_2 - g_p > 0 (\dot{u}_1 - \dot{u}_2 < 0) \\ 0 & u_1 - u_2 - g_p < 0 \end{cases} \tag{2}$$

where, β is the pounding stiffness coefficient that is obtained by the least squares optimization algorithm; u_1 and u_2 are the displacements of the pounding motion limiting collar and the mass block, respectively; $\dot{u}_1 - \dot{u}_2$ is the relative velocity; g_p is the impact gap; and c_k is the nonlinear impact damping coefficient, which can be expressed as follows:

$$c_k = 2\gamma\sqrt{\beta\sqrt{u_1 - u_2 - g_p}\frac{m_1 m_2}{m_1 + m_2}} \tag{3}$$

where, m_1 and m_2 are the mass of the two impact bodies, respectively; γ is the hysteretic damping ratio, which can be defined as:

$$\gamma = \frac{10.0623 - 10.0623e^2}{12.2743e^2 + 16e} \tag{4}$$

where, e is the Newtonian velocity recovery coefficient and is obtained by the falling ball test.

As can be seen from Equation (1), Δ is the location vector of the pounding force, and F_P is the direction of the pounding force:

$$F_P = \begin{cases} 1 & u_p - u_n - g_p > 0 \\ -1 & u_p - u_n - g_p < 0 \\ 0 & otherwise \end{cases} \tag{5}$$

where, u_p and u_n are the displacement of the PTMD and top node of the structure, respectively.

3. Modeling of a Transmission Tower with a PTMD

3.1. Structural Model

A SZ21-type transmission tower practical engineering example in Northeast China was selected as the research object. Figure 2 shows the practical graph of the transmission tower. The height of the tower is 53.9 m, and its weight is 20.23 tons. The tower size is shown in Figure 3. The main member and diagonal members of the transmission tower are made of Q235 and Q345 angle steels with elastic moduli of 206 GPa. A three-dimensional finite element model of the power transmission tower was established by using ABAQUS (Dassasult Simulia Company, Providence, RI, USA), as shown in Figure 4. The *X*, *Y*, and *Z* directions of the model are expressed as the longitudinal, transverse, and vertical directions of the structure, respectively. The members of the transmission tower are simulated by B31 elements, and the base nodes of the transmission tower are fixed at the ground. Based on the analysis of the dynamic characteristics, the frequencies in the *Y* and *X* directions of the transmission tower are analyzed. The first three natural frequencies in the *Y* direction are 1.768, 4.870, and 8.909 Hz, while the first natural frequencies in the *X* direction are 1.797, 4.954, and 9.774 Hz. The vibration modes that shape the transmission tower are shown in Figure 5.

Figure 2. Practical graph of the transmission tower.

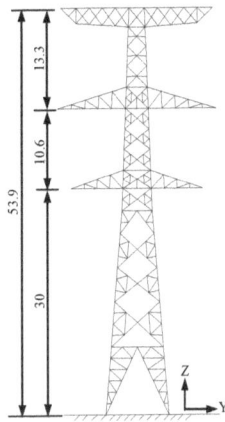

Figure 3. Tower size (m).

Figure 4. Three-dimensional finite element model of the transmission tower.

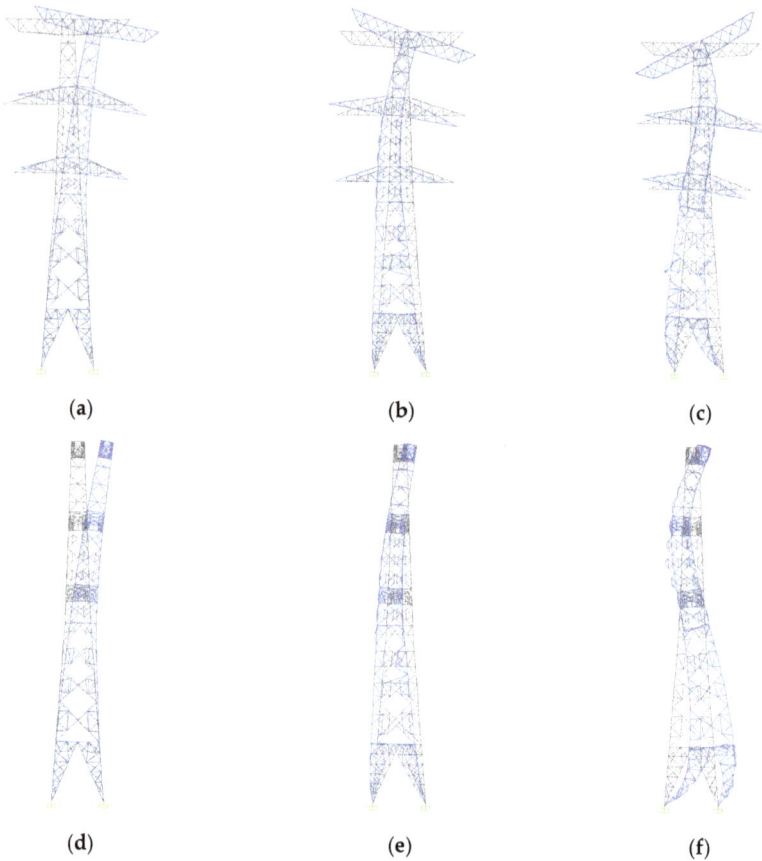

Figure 5. Vibration mode shapes of the transmission tower. (**a**) The first modal shape in the Y direction; (**b**) The second modal shape in the Y direction; (**c**) The third modal shape in the Y direction; (**d**) The first modal shape in the X direction; (**e**) The second modal shape in the X direction; (**f**) The third modal shape in the X direction.

3.2. Simulation of the PTMD

A PTMD can be obtained by the combination of a TMD and an impact damper, which has double the vibration reduction characteristics. The proposed PTMD is shown in Figure 6. The PTMD includes a cable, a mass block, a limiting device, and viscoelastic material, and the mass block is covered with viscoelastic material. The PTMD device is installed at the top of the tower by using the connecting plate, and the connecting plate is fixed on the angle steel of the tower by bolts. When the earthquake loads are small, the PTMD can be regarded as TMD. When the earthquake loads are large enough, the mass block will impact on the limiting device. Due to pounding energy dissipation, the PTMD has double the reduction characteristics, and the vibration reduction effect depends on the mass block, collision, and viscoelastic material.

The PTMD is simulated in ABAQUS. The mass ratio is 2%, and the mass of mass block is 404.7 kg. The mass block and limiting device are simulated by S3R elements. The spring element is adopted to simulate the cable, and the axial stiffness of the spring is 1900 kN/m. The axial stiffness of the spring is large enough so that the axial deformation can be ignored. The gap between the mass block and the limiting device is 0.02 m. The Mooney-Rivlin model is used for the viscoelastic material, and the mechanical constant C1 and C2 are 3.2×10^6 Pa and 8.0×10^5 Pa, respectively. The contact is defined as the surface-to-surface contact, and the penalty contact method is used as the contact algorithm. The length of the cable is determined by the natural period of the structure, which can be obtained from $l = T^2 g / 4\pi^2$. Design guidelines of the optimal parameter of the PTMD are described in Figure 7.

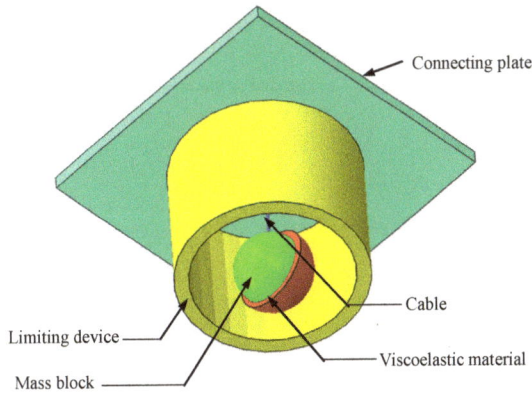

Figure 6. Schematic diagram of the PTMD.

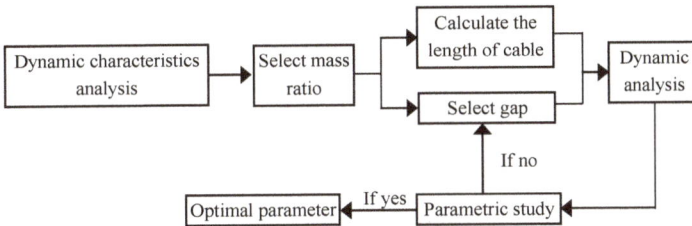

Figure 7. Design guidelines of the optimal parameters of the PTMD.

To verify the accuracy of the finite element simulation of the PTMD, the finite element model of the transmission tower with the PTMD is compared to Zhang's simplified model [16]. Figure 8 shows the time history curve of the top displacement of the transmission tower with the PTMD under the conditions of the El Centro earthquake. It can be seen that the two time history curves are slightly different, and the trend and maximum displacement are the same. Therefore, the finite element model of the PTMD is more accurate and can be used for further analysis.

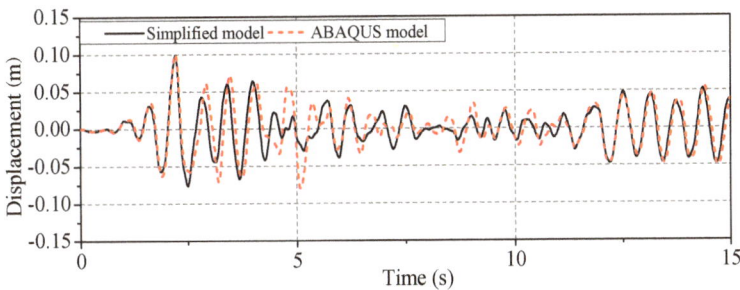

Figure 8. Displacement response of the transmission tower with the PTMD under the conditions of the El Centro earthquake.

4. Numerical Analysis and Discussion

4.1. Selection of Seismic Waves

Based on the Code for Seismic Design of Buildings [21], three typical natural seismic acceleration waves are selected, as listed in Table 1. The seismic category of the transmission tower is referred to as an eight-degree seismic design zone by the Code for Seismic Design of Buildings, so the peak ground acceleration is adjusted to 400 gal. Two horizontal components of seismic waves are applied along the longitudinal and transverse directions of the transmission tower simultaneously, and the maximum peak ground motion component of the seismic waves are input along the longitudinal direction of the structure.

Table 1. Seismic records.

ID	Earthquake	Event Date	Magnitude	Station
EQ1	Imperial Valley	18 May 1940	6.9	El Centro
EQ2	Northridge	17 January 1994	6.6	La-Baldwin Hills
EQ3	Kobe	16 January 1995	6.9	Oka

4.2. Vibration Control of the PTMD

The response of the transmission tower is shown in Figure 5 and the PTMD under multi-component seismic excitations is analyzed. To compare with the vibration reduction effect of the PTMD, the response of the transmission tower with the TMD is also carried out. The mass ratio between the PTMD and the transmission tower is 2%. The length of the cable is 0.08 m. The gap between the mass block and limiting device is 0.02 m. The vibration reduction ratios of the TMD and PTMD can be expressed as follows:

$$\eta_D = \frac{D_0 - D_c}{D_0} \times 100\% \tag{6}$$

$$\eta_A = \frac{A_0 - A_c}{A_0} \times 100\% \tag{7}$$

$$\eta_F = \frac{F_0 - F_c}{F_0} \times 100\% \tag{8}$$

where, η_D, η_A, and η_F are the vibration reduction ratios of displacement, acceleration, and axial forces, respectively; D_0, A_0, and F_0 are the maximum response of the displacement, acceleration and axial force of the transmission tower without control, respectively. D_c, A_c, and F_c are the maximum response of the displacement, acceleration, and axial force of the transmission tower with control, respectively.

The responses of the transmission tower with the TMD, PTMD, and without control were subjected to the conditions of the El Centro earthquake and are shown in Figure 9. It can be seen from the displacement and acceleration time history curves at the top of the transmission tower that the PTMD can effectively reduce the response of the displacement and acceleration. Due to double the vibration control characteristics of the PTMD, the vibration reduction effect of the PTMD is better than that of the TMD, and the response of the transmission tower with PTMD is always smaller than that of the TMD. Note that the vibration control of the PTMD is stable. The PTMD can reduce the maximum axial force of the transmission tower, and the vibration reduction effect is different along the height of the transmission tower.

Table 2 listed the vibration reduction ratio of the transmission tower under multi-component seismic excitations. The vibration reduction ratios of the transmission tower under different seismic excitations are different. Analyzing the vibration reduction ratio of the transmission tower under the El Centro earthquake conditions, the TMD can effectively reduce the peak value of displacements in the longitudinal and transverse directions by 21% and 26%, but the vibration reduction ratios of 29%

and 44% of the transmission tower with the PTMD are larger than those of with the TMD. The RMS (root mean square) reduction ratios of the displacements of the PTMD in the longitudinal and transverse directions are 54% and 54%, greater than those of the TMD which are 36% and 12%. The vibration reduction ratios of the acceleration peak values of the PTMD in the longitudinal and transverse directions are 37% and 26%, and the RMS reduction ratios in the longitudinal and transverse directions are 52% and 36%, which are larger than those of the TMD. In terms of axial force, the maximum axial force of the transmission tower with the PTMD is reduced by 28%, but the vibration reduction ratio of the TMD is only 10%. The results are similar to the response of the transmission tower under the Northridge and Kobe earthquake conditions shown in Table 2. It can be seen from the table that the reduction ratio of the PTMD is significantly larger than that of the TMD owing to the double reduction characteristics.

Table 2. Vibration reduction ratio of the transmission tower under multi-component seismic excitations.

Seismic Records	Direction	Damper	Displacement		Acceleration		Damper	Axial Internal Force	
			Peak (%)	RMS (%)	Peak (%)	RMS (%)		Peak (%)	RMS (%)
El Centro	X	TMD	21	36	33	38	TMD	10	15
		PTMD	29	54	37	52			
	Y	TMD	26	12	15	16	PTMD	28	37
		PTMD	44	54	26	36			
Northridge	X	TMD	57	66	44	54	TMD	26	25
		PTMD	63	71	51	58			
	Y	TMD	39	33	23	13	PTMD	31	30
		PTMD	52	47	28	18			
Kobe	X	TMD	33	65	54	53	TMD	8.0	15
		PTMD	47	74	70	54			
	Y	TMD	30	26	36	28	PTMD	27	29
		PTMD	54	70	65	55			

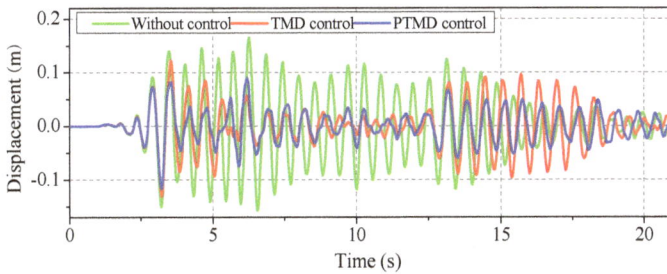

(a)

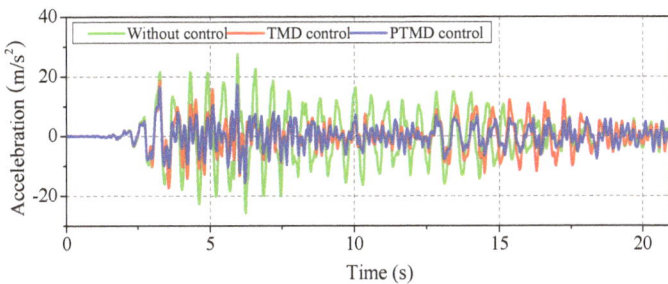

(b)

Figure 9. *Cont.*

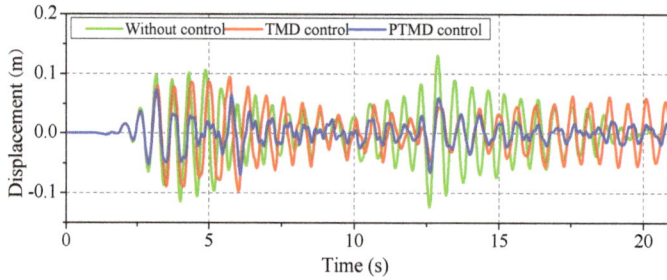

(c)

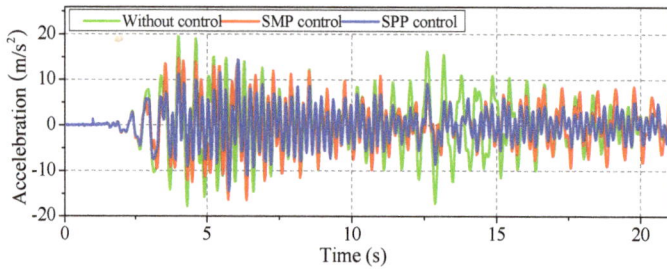

(d)

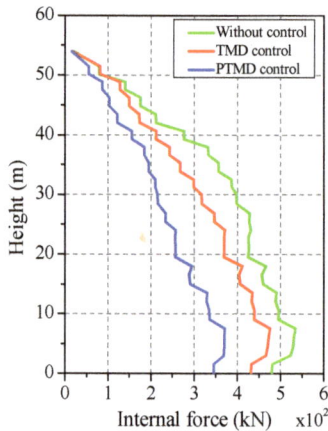

(e)

Figure 9. Dynamic response under the El Centro earthquake conditions. (**a**) Longitudinal displacement; (**b**) Longitudinal acceleration; (**c**) Transverse displacement; (**d**) Transverse acceleration; (**e**) Axial internal force.

4.3. Parametric Study

To obtain an optimal design of the PTMD, the effect of the mass ratio between the PTMD and the transmission tower, the effect of the ground motion intensity, the effect of the gap between the mass block and the limiting device, and the effect of the incident angle of the seismic ground motion are investigated, respectively. The El Centro earthquake is selected in this section. Unless mentioned otherwise, the peak ground acceleration of the El Centro earthquake is adjusted to 400 gal, and the mass ratio and gap are 2% and 0.02 m, respectively.

4.3.1. Effect of Mass Ratio

To investigate the effect of the mass ratio, ten different mass ratios are considered in the analysis, and the mass ratios are selected as 0.5%–5%, in increments of 0.5%, to cover the range of the change of the mass ratio. Figure 10 shows the vibration reduction ratios of the maximum displacement with different ratios. The reduction ratio of the PTMD increases gradually with the increase of the mass ratio until 2%. However, the increase of the reduction ratio of the PTMD is slow when the mass ratio is larger than 2%. The reduction ratios in the longitudinal and transverse directions with the change of the mass ratio have the same trend. Therefore, considering the effect of the reduction ratio and economics, 2% is selected as the optimal result.

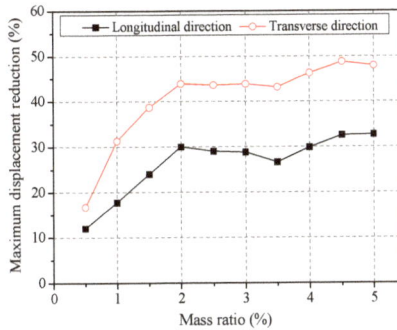

Figure 10. Vibration reduction ratios of the maximum displacement with the different mass ratios.

4.3.2. Effect of Seismic Intensity

To study the effect of the ground motion intensity, $125 \, \text{cm/s}^2$, $220 \, \text{cm/s}^2$, $400 \, \text{cm/s}^2$, and $620 \, \text{cm/s}^2$ peak ground acceleration are considered, respectively. The vibration reduction ratios of the maximum displacement with different intensities are shown in Figure 11. The reduction ratio decreases with the increase of the ground motion intensity. The reduction effect of the transmission tower with PTMD in the longitudinal direction is greater than that of in the transverse direction. The reduction ratio of the transmission tower with the PTMD is affected insignificantly by the change of the ground motion intensity.

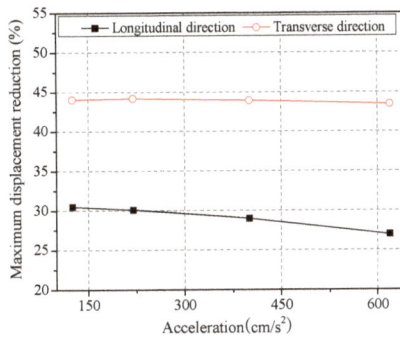

Figure 11. Vibration reduction ratios of the maximum displacement with different intensities.

4.3.3. Effect of the Gap

To obtain the effect of the gap between the mass block and the limiting device, nine different gaps are considered in this analysis, and the gaps are selected as 0.02–0.18 m, in increments of 0.02 m.

Figure 12 shows variation in the reduction ratios of the maximum displacement with different gaps. It can be seen that the reduction ratio increases first, and then decreases with the increase of the size of the gap, but the change of the reduction ratio is not obvious. The reason for this phenomenon is that the pounding energy dissipation is limited with few collisions when the gap is large. Therefore, the influence of the gap on the control effect of the PTMD is not significant.

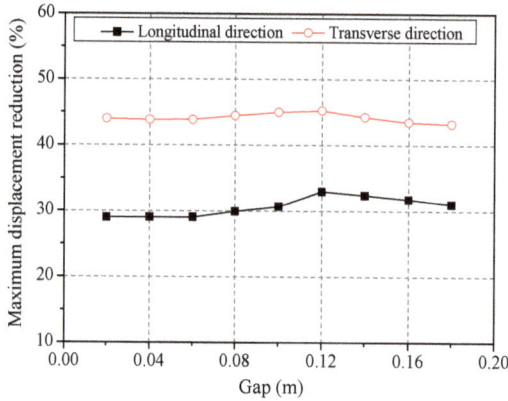

Figure 12. Variation reduction ratios of the maximum displacement with different gap sizes.

4.3.4. Effect of the Incident Angle

To investigate the effect of the incident angles, seven different incident angles are considered in the study, with the incident angle being from 0–90°, in increments of 15°. As can be seen in Figure 13, the vibration reduction ratios of the maximum displacement with different incident angles are given. With the increase of the incident angle, the reduction ratio in the longitudinal direction increases gradually owing to the decreasing ground motion intensity, and the maximum reduction ratio is 1.5 times that of the minimum reduction ratio. On the contrary, the reduction ratio in the transverse direction decreases with the increase of the incident angle, and the maximum reduction ratio is 3.0 times that of the minimum reduction ratio. Based on the above analysis, the incident angle has a significant influence on the reduction ratio. Therefore, the incident angle cannot be ignored for the analysis of the transmission tower with the PTMD.

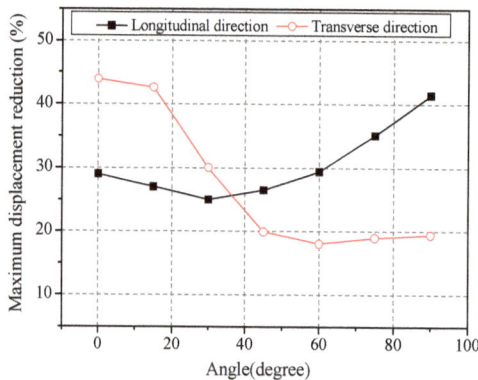

Figure 13. Vibration reduction ratios of the maximum displacement with different incident angles.

Appl. Sci. **2017**, *7*, 477

5. Conclusions

According to a 500 kV transmission line practical engineering example, a three-dimensional finite element model of the power transmission tower is established. The PTMD is simulated in ABAQUS. The vibration reduction mechanism of PTMD is introduced. The response of the transmission tower with a TMD and PTMD are performed, respectively. Based on the above analysis results, the following conclusions are drawn:

(1) Compared with the TMD, the PTMD is more effective in reducing the vibration of a transmission tower under multi-component seismic excitations. The vibration reduction ratios of the transmission tower with PTMD are varied with different seismic waves.

(2) The reduction ratios of the transmission tower with PTMD in the longitudinal and transverse directions have the same trend with the increase of mass ratio until 2%. The mass ratio of 2% is the optimal result.

(3) The reduction ratios of the transmission tower with the PTMD in the longitudinal and transverse directions decrease with the increase of the ground motion intensity, but the ground motion intensity has an insignificant influence on the reduction ratio.

(4) The reduction ratio of the transmission tower with the PTMD in the longitudinal and transverse directions increases first, and then decreases with the increase of the gap. The influence of the gap on the control effect of the PTMD is not significant.

(5) The reduction ratio in the longitudinal direction increases gradually with the increase of the incident angle. Compared with the reduction ratio in the longitudinal direction, the trend of the reduction ratio in the transverse direction is just the opposite. The incident angle has a significant influence on the reduction ratio.

Acknowledgments: This study were financially supported by the National Natural Science Foundation of China (No. 51578325 and 51208285), and Key Research and Development Plan of Shandong Province (No. 2016GGX104008).

Author Contributions: Li Tian and Kunjie Rong did the modelling work, analyzed the simulation date and wrote the paper. Peng Zhang and Yuping Liu revised and checked the paper.

Conflicts of Interest: The authors declare no conflict of interest.

References

1. Hall, F.J. *Northridge Earthquake of January 17, 1994: Reconnaissance Report*; Earthquake Engineering Research Institute: Oakland, CA, USA, 1995; Volume 11, pp. 212–215.
2. Luo, Q.F. Damages to life-line systems caused by Hyogoken Nanbu, Japan, earthquake and their recovery. *J. Catastrophol.* **1997**, *12*, 43–48.
3. Zhang, Z.Y.; Zhao, B.; Cao, W.W. Investigation and Preliminary Analysis of Damages on the Power Grid in the Wenchuan Earthquake of M8.0. *Electr. Power Technol. Econ.* **2008**, *20*, 1–4.
4. Yu, Y.Q. Investigation and Analysis of Electric Equipment Damage in Sichuan Power Grid Caused by Wenchuan Earthquake. *Power Syst. Technol.* **2008**, *32*, 1–6.
5. Zhang, D.C.; Zhao, W.B.; Liu, M.Y. Analysis on seismic disaster damage cases and their causes of electric power equipment in 5.12 Wenchuan earthquake. *J. Nanjing Univ. Technol.* **2009**, *31*, 44–48.
6. Liu, R.S.; Liu, J.L.; Yan, D.Q. Seismic damage investigation and analysis of electric power system in Lushan M_s 7.0 earthquake. *J. Nat. Disasters* **2013**, *22*, 83–90.
7. Kilroe, N. Aerial method to mitigate vibration on transmission towers. In Proceedings of the 2000 IEEE ESMO—2000 IEEE 9th International Conference on Transmission and Distribution Construction, Operation and Live-Line Maintenance Proceedings, Montreal, QC, Canada, 8–12 October 2000; pp. 187–194.
8. Battista, R.C.; Rodrigues, R.S.; Pfeil, M.S. Dynamic behavior and stability of transmission line towers under wind forces. *J. Wind Eng. Ind. Aerodyn.* **2003**, *91*, 1051–1067.
9. Park, J.H.; Moon, B.W.; Min, K.W. Cyclic loading test of friction-type reinforcing members upgrading wind-resistant performance of transmission towers. *Eng. Struct.* **2007**, *29*, 3185–3196.

10. Chen, B.; Zheng, J.; Qu, W. Control of wind-induced response of transmission tower-line system by using magnetorheological dampers. *Int. J. Struct. Stab. Dyn.* **2009**, *9*, 661–685.

11. Li, L.; Cao, H.; Ye, K. Simulation of galloping and wind-induced vibration control. *Noise Vib. Worldw.* **2010**, *41*, 15–21.

12. Tian, L.; Yu, Q.; Ma, R. Study on seismic control of power transmission tower-line coupled system under multicomponent excitations. *Math. Probl. Eng.* **2013**, *2013*, 1–12.

13. Ema, S.; Marui, E. A fundamental study on impact dampers. *Int. J. Mach. Tools Manuf.* **1994**, *34*, 407–421.

14. Collette, F.S. A combined tuned absorber and pendulum impact damper under random excitation. *J. Sound Vib.* **1998**, *216*, 199–213.

15. Cheng, C.C.; Wang, J.Y. Free vibration analysis of a resilient impact damper. *Int. J. Mech. Sci.* **2003**, *45*, 589–604.

16. Zhang, P.; Song, G.B.; Li, H.N. Seismic control of power transmission tower using pounding TMD. *J. Eng. Mech.* **2013**, *139*, 1395–1406.

17. Zhang, P.; Li, L.; Patil, D.; Singla, M.; Li, H.N.; Mo, Y.L.; Song, G. Parametric study of pounding tuned mass damper for subsea jumpers. *Smart Mater. Struct.* **2016**, *25*, 15–28.

18. Song, G.B.; Zhang, P.; Li, L.Y.; Singla, M.; Patil, D.; Li, H.N.; Mo, Y.L. Vibration control of a pipeline structure using pounding tuned mass damper. *J. Eng. Mech.* **2016**, *142*, 04016031.

19. Li, H.N.; Zhang, P.; Song, G.; Patil, D.; Mo, Y. Robustness study of the pounding tuned mass damper for vibration control of subsea jumpers. *Smart Mater. Struct.* **2015**, *24*, 095001.

20. Li, L.; Song, G.; Singla, M.; Mo, Y.L. Vibration control of a traffic signal pole using a pounding tuned mass damper with viscoelastic materials (II): Experimental verification. *J. Vib. Control* **2013**, *21*, 670–675.

21. GB 50011-2010. In *Code for Seismic Design of Buildings*; Ministry of Construction of the People's Republic of China and the State Quality Supervision and Quarantine Bureau: Beijing, China, 2010.

applied
sciences

MDPI

Article

Robustness Analysis of the Collective Nonlinear Dynamics of a Periodic Coupled Pendulums Chain

Khaoula Chikhaoui , Diala Bitar, Najib Kacem *and Noureddine Bouhaddi

Department of Applied Mechanics, FEMTO-ST Institute, CNRS/UFC/ENSMM/UTBM, Univ. Bourgogne
Franche-Comté, 25000 Besançon, France; khaoula.chikhaoui@femto-st.fr (K.C.); diala.bitar@femto-st.fr (D.B.);
noureddine.bouhaddi@univ-fcomte.fr (N.B.)
* Correspondence: najib.kacem@femto-st.fr; Tel.: +33-3-81-66-67-02

Received: 2 June 2017; Accepted: 28 June 2017; Published: 3 July 2017

Abstract: Perfect structural periodicity is disturbed in presence of imperfections. The present
paper is based on a realistic modeling of imperfections, using uncertainties, to investigate the
robustness of the collective nonlinear dynamics of a periodic coupled pendulums chain. A generic
discrete analytical model combining multiple scales method and standing-wave decomposition is
proposed. To propagate uncertainties through the established model, the generalized Polynomial
Chaos Expansion is used and compared to the Latin Hypercube Sampling method. Effects of
uncertainties are investigated on the stability and nonlinearity of two and three coupled pendulums
chains. Results prove the satisfying approximation given by the generalized Polynomial Chaos
Expansion for a significantly reduced computational time, with respect to the Latin Hypercube
Sampling method. Dispersion analysis of the frequency responses show that the nonlinear aspect of
the structure is strengthened, the multistability domain is wider, more stable branches are obtained
and thus multimode solutions are enhanced. More fine analysis is allowed by the quantification
of the variability of the attractors' contributions in the basins of attraction. Results demonstrate
benefits of presence of imperfections in such periodic structure. In practice, imperfections can be
functionalized to generate energy localization suitable for several engineering applications such as
vibration energy harvesting.

Keywords: nonlinear coupled pendulums; collective dynamics; robustness analysis; polynomial
chaos expansion

---

1. Introduction

In structural mechanics as well as in practically all fields of engineering, the periodicity
characterizes the structuring of many systems such as layered composites, crystal lattices, bladed
disks, turbines, multi-cylinder engines, ship hulls, aircraft fuselages, micro and nanoelectromechanical
systems, etc. Periodicity implies an infinite or finite geometrical repetition of a unit cell in one, two or
three dimensions and requires appropriate approaches to investigate it. Under the hypothesis of
perfect periodicity, many works provided interesting insights in the behavior of these structures. In
the context of wave propagation in periodic structures, the basic works performed in linear case
by Brillouin [1] and Mead [2] are based on the Floquet's principle or the transfer matrix in order to
compute propagation constants. Based on the transfer matrix theory, a combination of wave and
finite element approaches was proposed by Duhamed et al. [3] and used later by Goldstein et al. [4]
to calculate forced responses of waveguide structures. Casadei et al. [5] developed analytical and
numerical models based on the transfer matrix approach to investigate the dispersion properties
and bandgaps of a beam with a periodic array of airfoil-shaped resonating units bonded along
its length. Using the Floquet-Bloch' theorem, Gosse et al. [6] completely described the behavior
of a heat exchanger periodic structure only from the vibroacoustic knowledge of the basic unit.

Collet et al. [7] extended the analysis to two-dimensional periodic structures with complex damping configurations and underlined the reduced computational costs allowed by the Floquet-Bloch' theorem when representing whole structures by unit cell modeling. Recently, Droz et al. [8] combined the Wave Finite Element Method (WFEM) with Component Mode Synthesis (CMS) to evaluate the dispersion characteristics of two-dimensional periodic waveguides.

On the other hand, the wave propagation becomes considerably complicated when the governing wave equation contains nonlinear terms (i.e., contact, material or geometric nonlinearity). In this case, complex phenomena such as localization, solitons and breathers arise and traditional Floquet-Bloch and transfer matrix wave analyses are no longer applicable. In literature, other methodologies are developed to deal with nonlinear periodic structures such as perturbation approaches. For instance, Chakraborty and Mallik [9] investigated the harmonic wave propagation in one-dimensional periodic chain consisting of identical masses and weakly non-linear springs through single-frequency harmonic balance. They used a perturbation approach to calculate the propagation and attenuation constants. A straightforward perturbation analysis is applied by Boechler et al. [10] to investigate amplitude-dependent dispersion of a discrete one-dimensional nonlinear periodic chain with Hertzian contact. Otherwise, as an alternative to perturbation approach for strongly nonlinear systems, Georgiades et al. [11] proposed a combination of shooting and pseudo-arc-length continuation to examine nonlinear normal modes and their bifurcations in cyclic periodic structures. Moreover, Lifshitz et al. used a secular perturbation theory to calculate the response of a coupled array of nonlinear oscillators under parametric excitation in [12] and of N nonlinearly coupled micro-beams in [13] using discrete models. The method of multiple scales is used by Nayfeh [14] to construct a first-order uniform expansion in the presence of internal resonance for the governing equations of parametrically excited multi-degree-of-freedom systems with quadratic nonlinearities. Using the same methodology, Bitar et al. [15] investigated the collective dynamics of a periodic structure of coupled nonlinear Duffing-Van Der Pol oscillators under simultaneous parametric and external excitations. An analytico computational model was used to compute the frequency responses and the basins of attraction of two and three coupled oscillators. The authors demonstrated the importance of the multimode solutions and the robustness of their attractors. The multiple scales method was also used by Gutschmidt and Gottlied [16] in a continuum-based model to investigate the dynamic behavior of an array of N nonlinearly coupled micro-beams. Furthermore, Manktelow et al. [17] used the multiple scales method to investigate wave interactions in monoatomic mass-spring chain with a cubic nonlinearity. In [18], the method was combined with a finite-element discretization of a single unit cell, to study the wave propagation in continuous periodic structures subject to weak nonlinearities. The authors proposed later robust tools for wave interactions analysis in diatomic chain with two degrees of freedom per unit cell [19]. Recently, Andreassen et al. [20] studied the wave interactions in a periodically perforated plate through the two-dimensional dispersion characteristics, group velocities and internal resonances investigation. Romeo and Rega [21] identified the regions of existence of discrete breathers and guided their analysis using the nonlinear propagation region of chain of oscillators with cubic nonlinearity exhibiting periodic solutions. Furthermore, using the idea of harmonic balance in the periodic structures inspired from [9] the Harmonic Balance Method (HBM) was combined with multiple scales method in [22] in order to study the attenuation caused by weak damping of harmonic waves through a discrete periodic structure. The HBM was later used by Narisetti et al. [23] to analyze the influence of nonlinearity and wave amplitude on the dispersion properties of plane waves in strongly nonlinear periodic uniform granular media. Particularly, periodic coupled pendulum structure has been the purpose of several researches in literature. Marlin [24], for instance, proved several theorems on the existence of oscillatory, rotary, and mixed periodic motions of N coupled simple pendulums. Khomeriki and Leon [25] demonstrated numerically and experimentally the existence of three tristable stationary states. Jallouli et al. [26] investigated the nonlinear dynamics of a two-dimensional array of coupled pendulums under parametric excitation and, recently [27], the energy localization phenomenon in an array of coupled pendulums under simultaneous external and parametric excitations by means of a nonlinear Schrodinger equation. The authors show that adding an external excitation increases the existence region of solitons. Bitar et al. [28,29] investigated the collective nonlinear dynamics of perfectly periodic coupled pendulum structure under primary resonance using multiple scales and standing-wave decomposition. The authors studied the effects of modal interactions on the nonlinear dynamics. They highlighted the large number of multimode solutions and the bifurcation topology transfer between the modal intensities, in frequency domain. The analysis of the Basins of attraction illustrated the distribution of the multimodal solutions which increases by increasing the number of coupled pendulums. A detailed review was presented by Nayfeh et al. [30] dealing with the influence of modal interactions on the nonlinear dynamics of harmonically excited coupled systems. Besides, the study of collective nonlinear dynamics of coupled oscillators may serve to identify the Intrinsic Localized Modes (ILMs). ILMs are defined as localizations due to strong intrinsic nonlinearity within an array of perfectly periodic oscillators. Such localization phenomenon was studied by Dick et al. [31] in the context of microcantilever and microresonator arrays. Authors used the multiple scales method and other methods to construct nonlinear normal modes and suggested to realize an ILM as a forced nonlinear vibration mode.

It is important to note that the dynamic analysis of periodic structures is greatly simplified by assuming perfect periodicity. However, far from this mathematical idealization, imperfections, which can be due to material defects, manufacturing defaults, structural damage, ageing, fatigue, etc., and which reflect the reality of systems, can perturb the perfect arrangement of cells in a structure and change significantly the dynamic behavior from the predictions done under perfect periodicity hypothesis. In literature, primary works dealing with the issue of presence of imperfections in periodic structures treat it under the framework of disorder. Kissel [32], for instance, investigated the effects of disorder in one-dimensional periodic structure using Monte Carlo (MC) simulations. He used a transfer matrix modeling and the limit theorem of Furstenberg to compute products of random matrices for structures carrying a single pair of waves and the theorem of Oseledets for those carrying multiplicity of wave types. The results show that disorder causes wave attenuation and pronounced spatial localization of normal modes at frequencies near the bandgaps of the perfectly periodic associated structure. Statistical investigation of the effect of disorder on the dynamics of one-dimensional weakly/strongly coupled periodic structures, using the MC method, was carried out by Pierre et al. [33]. The effect of disorder is evaluated through the statistics of the localization factor reflecting the exponential decay of the vibration amplitude. An extension of the analysis from single degree of freedom bays to multimode bays which are more representative of periodic engineering structures was then presented in [34]. Impact of disorder on the vibration localization in randomly mistuned bladed disks was also discussed by Castanier et al. in the review paper [35]. Statistical investigations were made using both classical and accelerated MC methods. With the aim of computational cost saving of numerical analysis, CMS-based ROMs could then be used to calculate the mistuned forced response for each MC simulation, at relatively low cost. Moreover, to study the effects of the randomness of flexible joints on the free vibrations of simply-supported periodic large space trusses, Koch [36] combined an extended Timoshenko beam continuum model, MC simulations and first-order perturbation method. These works proved that the normal modes, which would be periodic along the length of a perfectly periodic structure, are localized in a small region when periodicity is perturbed. Moreover, Zhu et al. [37] studied the wave propagation and localization in periodic and randomly disordered periodic piezoelectric axial-bending coupled beams using a finite element model and the transfer matrix approach. The localization factor characterizing the average exponential rate of decay of the wave amplitude in the disordered periodic structure was computed using the Lyapunov exponent method. The authors proved that the wave propagation and localization can be altered by properly adjusting the structural parameters.

In the context of disorder in periodic coupled pendulums structures, Tjavaras and Triantafyllou [38] investigated numerically the effect of nonlinearities on the forced response of two disordered pendulums coupled through a weak linear spring. Disorder generates modal localization and reveals large sensitivity to small parametric variations. In [39], the authors demonstrated that an impurity introduced by longer pendulum in the chain of coupled parametrically driven damped pendulums supporting soliton-like clusters expands its stability region. Whereas impurity introduced by shorter pendulum defects simply repel solitons producing effective partition of the chain. Hai-Quing and Yi [40] developed a discrete theoretical model based on the envelope function approach to study analytically and numerically the effect of mass impurity on nonlinear localized modes in a chain of parametrically driven and damped nonlinear coupled pendulums. The influence of impurities on the envelope waves in a driven nonlinear pendulums chain has been investigated numerically under a continuum-limit approximation in [41] and then experimentally in [42].

Design of engineering structures with periodicity, nonlinearity and uncertainty is a complex challenge and the main aim of this work is to deal with. Under the hypothesis of small imperfections, the collective dynamics and the localization phenomenon due to the weak coupling of components is preserved. To investigate the collective dynamics of perfectly periodic nonlinear N degrees of freedom systems and control modal interactions between coupled components, previous works [12,15,28,29] proposed discrete analytical models combining the multiple scales method and standing-wave decomposition. The main objective of the present work is to extend these methodologies to the presence of imperfections by proposing a more generic discrete model. If, in particular, imperfections are taken into account in a probabilistic framework as parametric uncertainties modeled by random variables, uncertainty propagation methods must be applied. Uncertainties are thus propagated through the proposed generic model to evaluate the robustness of the collective dynamics against the randomness of the uncertain input parameters. The established generic discrete analytical model leads to a set of coupled complex algebraic equations. These equations are written according to the number and positions of the imperfections in the structure and then numerically solved using the Runge-Kutta time integration method. To propagate uncertainties through the established model, the statistical Latin Hypercube Sampling (LHS) method [43] is used as a reference with respect to which the efficiency of the generalized Polynomial Chaos Expansion (gPCE) [44,45] is evaluated.

Uncertainty effects on the nonlinear dynamics of two and three coupled pendulums chains are investigated in this paper. Dispersion analyses of the frequency responses, in modal and physical coordinates, and the basins of attraction are carried out. Moreover, in order to highlight the complexity of the multimode solutions in terms of attractors and bifurcation topologies, a thorough analysis through the basins of attractions is performed. The robustness of the multimode branches against uncertainties around a chosen frequency in the multistability domain is investigated.

2. Mechanical Model

Figure 1 illustrates a generic structure for N coupled pendulums of identical length l, mass m. and viscous damping coefficient c generated by the dissipative force acting on the supporting point of each one. The pendulums are coupled by linear springs of stiffness k and are subject to an external excitation $f \cos(\Omega t)$ each one. The inclination angle ϕ_n. from the equilibrium position quantifies the rotational displacement of the nth pendulum. Applied boundary conditions are such as the pendulum labeled 0. and $N + 1$ are fixed so that $\phi_0 = \phi_{N+1} = 0$. The periodicity of the structure is broken by presence of p pendulums containing parametric uncertainties which can, for instance, be the pendulum's length l_s as illustrated in Figure 1.

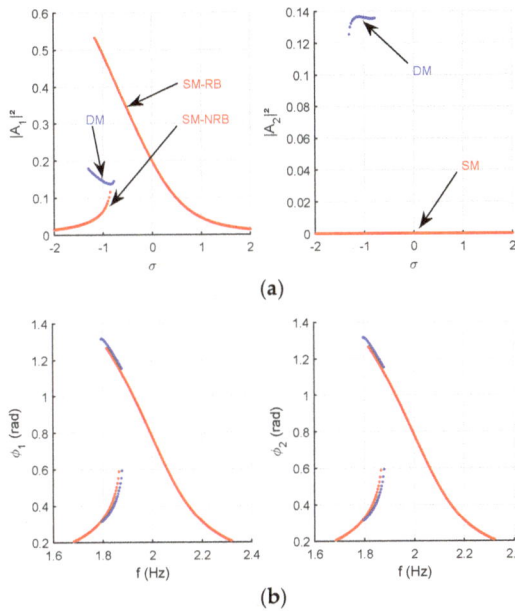

Figure 1. Periodic nonlinear coupled-pendulums chain with imperfection.

In perfect periodicity case, one can refer to works performed in [12,15,28,29] to investigate the collective dynamics of the periodic nonlinear coupled-pendulums chain. Nevertheless, such analyzes are no longer suitable if periodicity is disturbed. The main objective of the present work is to propose a generic model which is adapted to the presence of uncertainties.

Uncertainties are supposed to affect structural input parameters, here some pendulums' lengths, and to vary randomly. A probabilistic modeling of uncertainties, by random variables, is used and implies applying stochastic uncertainty propagation methods to evaluate the effect of the randomness in structural input parameters on the collective dynamics of the nonlinear coupled-pendulums chain.

Developing the generic model, through which uncertainties will be propagated, is based on the fact that the pendulums behave in different ways, depending on the position of each one with respect to uncertainties localization. Indeed, the equations of motion of the system are written according to the number and positions of uncertainties in the structure.

2.1. Equations of Motion

Applying the Lagrange approach leads to the equation of motion of the nth pendulum:

$$\ddot{\phi}_n + c_n\dot{\phi}_n + \omega_n^2\phi_n + k_n\mathcal{L}_c(\phi,\widetilde{\phi}) + \alpha_n\phi_n^3 = f_n\cos(\Omega t), \tag{1}$$

where $c_n = \frac{c}{m}$, $k_n = \frac{k}{ml^2}$, $\alpha_n = -\frac{g}{6l}$, $f_n = \frac{f}{ml^2}$ if the nth pendulum is deterministic and $c_n = \frac{c}{m}$, $k_n = \frac{k}{ml_s^2}$, $\alpha_n = -\frac{g}{6l_s}$, $f_n = \frac{f}{ml_s^2}$ if the length of the nth pendulum, of stochastic displacement $\widetilde{\phi}_n$, is uncertain.

Since linear coupling between pendulums is very weak and small imperfections are considered, each angular frequency ω_n is supposed to be equal to the eigenfrequency ω_0 ($\omega_n = \omega_0 = \sqrt{g/l}$).

The linear coupling term $\mathcal{L}_c(\phi,\widetilde{\phi})$ depends on the positions of the stochastic pendulums in the chain. If stochastic pendulums are not adjacent, four different configurations are distinguished:

a.   If the concerned pendulum is deterministic as well as its neighbors, $\mathcal{L}_c(\phi,\widetilde{\phi}) = 2\phi_n - \phi_{n-1} - \phi_{n+1}$;

b. If the concerned pendulum is deterministic but the previous one is stochastic, $\mathcal{L}_c(\phi, \tilde{\phi}) = 2\phi_n - \tilde{\phi}_{n-1} - \phi_{n+1}$;

c. If the concerned pendulum is deterministic but the following is stochastic, $\mathcal{L}_c(\phi, \tilde{\phi}) = 2\phi_n - \phi_{n-1} - \tilde{\phi}_{n+1}$;

d. If the stochastic pendulum is concerned, the deterministic displacement ϕ_n is replaced by the stochastic one, $\tilde{\phi}_n$, in Equation (1) such as

$$\ddot{\tilde{\phi}}_n + c_n \dot{\tilde{\phi}}_n + \omega_0^2 \tilde{\phi}_n + k_n \mathcal{L}_c(\phi, \tilde{\phi}) + \alpha_n \tilde{\phi}_n^3 = f_n \cos(\Omega t), \tag{2}$$

and $\mathcal{L}_c(\phi, \tilde{\phi}) = 2\tilde{\phi}_n - \phi_{n-1} - \phi_{n+1}$, in this case.

The displacement ϕ_n can be expressed as a sum of standing wave modes with slowly varying amplitudes [12,15,28,29]. Taking into account the boundary conditions $\phi_0 = \phi_{N+1} = 0$, the standing wave modes are:

$$u_n = \sin(nq_m) \text{ with } q_m = \frac{m\pi}{N+1}, \quad m = 1 \dots N. \tag{3}$$

The displacement ϕ_n of the nth pendulum is thus expressed as

$$\phi_n = \underbrace{\sum_{m=1}^{N} A_m \sin(nq_m) \exp(i\omega_0 t) + c.c.}_{\phi_{n0}} + \varepsilon\, \phi_{n1}, \tag{4}$$

if it is deterministic, and as

$$\tilde{\phi}_n = \underbrace{\sum_{m=1}^{N} \tilde{A}_m \sin(nq_m) \exp(i\omega_0 t) + c.c.}_{\tilde{\phi}_{0n}} + \varepsilon\, \tilde{\phi}_{1n}, \tag{5}$$

if it is affected by uncertainties.

2.2. Multiple Scales Method Applied to Stochastic Model

The multiple scales method [46,47] consists on replacing the single time variable by an infinite sequence of independent time scales ($T_i = \varepsilon^i t$), where ε is a dimensionless parameter assumed to be small, and eliminating secular terms in the fast time variable $T_0 = t$.

Limiting the study to a first order perturbation, ($T = T_1 = \varepsilon^1 t$), Equation (1) takes the form

$$\ddot{\phi}_n + \varepsilon\, c_n \dot{\phi}_n + \omega_0^2 \phi_n + \varepsilon\, \mathcal{L}_c(\phi, \tilde{\phi}) + \varepsilon\, \alpha_n \phi_n^3 = \varepsilon\, f_n \cos(\Omega t), \tag{6}$$

where the excitation frequency Ω is expressed as: $\Omega = \omega_0 + \varepsilon\, \sigma$, σ being the detuning parameter.

The solution of Equation (6) can generally be given by a formal power series expansion: $\phi_n = \sum_i \varepsilon^i \phi_{ni}$. Up to the order ε^1, the solution is of the form

$$\phi_n = \phi_{n0} + \varepsilon\, \phi_{n1}. \tag{7}$$

Its derivatives are given by

$$\dot{\phi}_n = \frac{d\phi_n}{dt} = \phi_{n0}^{(0,1)} + \varepsilon \left[\phi_{n0}^{(1,0)} + \phi_{n1}^{(0,1)} \right], \tag{8}$$

$$\ddot{\phi}_n = \frac{d^2\phi_n}{dt^2} = \phi_{n0}^{(0,2)} + 2\,\varepsilon\, \phi_{n0}^{(1,1)} + \varepsilon\, \phi_{n1}^{(0,2)}, \tag{9}$$

with $\phi_{n0}^{(0,1)} = \frac{\partial\phi_{n0}}{\partial t}$, $\phi_{n0}^{(0,2)} = \frac{\partial^2\phi_{n0}}{\partial t^2}$, $\phi_{n0}^{(1,0)} = \frac{\partial\phi_{n0}}{\partial T}$ et $\phi_{n0}^{(1,1)} = \frac{\partial^2\phi_{n0}}{\partial t\partial T}$.

Appl. Sci. **2017**, 7, 684

Substituting Equations (7)–(9) into Equation (6) and separating the terms with different orders of ε, one obtains a hierarchical set of equations. For the order ε^0, an unperturbed equation is obtained

$$\ddot{\phi}_{n0} + \omega_0^2 \phi_{n0} = 0. \tag{10}$$

The solution of Equation (10), which appears in every order in the expansion of the approximate solution, is expressed as

$$\phi_{n0} = A_n \, exp(i\omega_0 t) + c.c. \tag{11}$$

for the order ε^1, one obtains an equation of the form

$$\phi_{n1}^{(0,2)} + \omega_0^2 \phi_{n1} + c_n \phi_{n0}^{(0,1)} + 2\phi_{n0}^{(1,1)} + k_n(2\phi_{n0} - \phi_{n0-1} - \phi_{n0+1}) + \alpha_n \phi_{n0}^3 = \frac{f_n}{2} exp[i(\omega_0 t + \sigma T)]. \tag{12}$$

Substituting Equations (4) and (5) into Equation (12) leads to N equations of the form

$$\phi_{n1}^{(0,2)} + \omega_0^2 \phi_{n1} = \sum_{m=1}^{N} \left(m^{th} secular \, terms \right) e^{i\omega_0 t} + other \, terms, \tag{13}$$

where the secular terms (coefficients of $e^{i\omega_0 t}$) should be equated to zero to satisfy the condition of solvability of the multiple scales method. Projecting the response on the standing-wave modes implies to multiply all terms by $sin(nq_m)$ and sum over n. Consequently, a generic complex equation of the mth amplitude A_m is obtained:

$$2i\omega_0 A_m^{(1,0)} + i\omega_0 c_n A_m + k_n(2A_m - cos[q_m] * G_m)$$
$$+S * \frac{2k_n}{N+1} \sum_{n=1}^{N} sin[nq_m] \sum_{x=1}^{N} cos[nq_x] sin[q_x](A_x - \tilde{A}_x) + \frac{3}{4}\alpha_n \sum_{j,k,l} A_j A_k A_l^* \Delta_{jkl,m}^{(1)} - \tag{14}$$
$$\frac{1}{(N+1)} f_n \, exp(i\sigma T) \sum_{n=1}^{N} sin[nq_m] = 0,$$

where $\Delta_{jkl,m}^{(1)}$ is the delta function [12,15] defined in terms of the Kronecker deltas as

$$\Delta_{jkl,m}^{(1)} = \delta_{-j+k+l,m} - \delta_{-j+k+l,-m} - \delta_{-j+k+l,2(N+1)-m}$$
$$+\delta_{j-k+l,m} - \delta_{j-k+l,-m} - \delta_{j-k+l,2(N+1)-m} \tag{15}$$
$$-\delta_{j+k+l,m} - \delta_{j+k+l,2(N+1)-m} - \delta_{j+k+l,2(N+1)-m'}$$

with $\delta_{v,w}$ is the Kronecker delta equal to 1 if $v = w$ and to 0 otherwise. The functions G_m and S are defined in the Appendix A.

2.3. Uncertainty Propagation

To propagate uncertainties through the established model, one can use, in a probabilistic framework, stochastic uncertainty propagation methods. Statistical methods, such as the MC method [48] and the LHS method [43], are the most frequently used in the literature and are considered as reference since they permit to achieve a reasonable accuracy. The LHS method consists on generating a succession of deterministic computations $\left\{ A_m\left(\xi^{(n)}\right), n = 1, \ldots, N_{LHS} \right\}$ according to a set of random variables $\left\{ \xi^{(n)} \right\}_{n=1}^{N_{LHS}}$ to approximate the mth amplitude A_m. The LHS method permits to reduce the computing time required by the very time-consuming MC method by partitioning the variability space into regions of equal probability and picking up one sampling point in each region. Nevertheless, it remains computationally unaffordable since the accuracy level is proportional to the number of generated simulations. To overcome this prohibitive computational cost without a significant loss of accuracy, the gPCE is used in this work [44,45]. The gPCE combines multivariate polynomials

25

and deterministic coefficients. Indeed, it approximates the mth amplitude A_m using a decomposition, practically truncated by retaining only polynomials terms with degree up to p:

$$A_m = \sum_{j=0}^{P} (\hat{A}_m)_j \psi_j(\xi) = \hat{A}_m^T \Psi(\xi); \; P + 1 = \frac{(d+p)!}{d!p!}, \tag{16}$$

where $(\hat{A}_m)_j$ are the unknown deterministic coefficients and $\psi_j(\xi)$ the multivariate polynomials of d independent random variables $\xi = \{\xi_i\}_{i=1}^{d}$.

Solving the gPCE consists on computing the deterministic coefficients $(\hat{A}_m)_j$. To do this, one can use intrusive or non-intrusive approaches. The former implies model modifications. However the latter considers the initial model as a black box. The regression approach is one of the most commonly used non-intrusive methods. In its standard form, it consists in minimizing the difference between the gPCE approximate solution and the exact solution. The latter is a set of deterministic solutions $\left\{ \overline{A}_m \left(\xi^{(n)} \right), n = 1, \ldots, M \right\}$ corresponding to M realizations of random variables $\Xi = \left\{ \xi^{(n)} \right\}_{n=1}^{M}$ forming an experimental design (ED). The approximate solution takes, consequently, the form

$$\hat{A}_m = \left(\Psi^T \Psi \right)^{-1} \Psi^T \overline{A}_m = \Psi^+ \overline{A}_m, \tag{17}$$

where $\Psi_{nj} \equiv \left(\psi_j \left(\xi^{(n)} \right) \right)_{\substack{n=1,\ldots,M \\ j=0,\ldots,P}}$ is called the data matrix and Ψ^+ is its Moore-Penrose pseudo-inverse.

In order to ensure the numerical stability of the regression approximation, the matrix $\left(\Psi^T \Psi \right)$ must be well-conditioned. Therefore, the ED should be well selected and have the size $M \geq P + 1$.

The ED selection technique used in this work is based on two conditions:

(i)   classification of all possible combinations of the roots of the Hermite polynomial of degree $p + 1$ so as to maximize the variable [49–51]:

$$\zeta_M \left(\xi^{(n)} \right) = 2\pi^{-d/2} exp \left(- \frac{\| \xi^{(n)} \|^2}{2} \right). \tag{18}$$

(ii) minimization of the number:

$$\kappa = \| \left(\Psi^T \Psi \right)^{-1} \| \cdot \| \Psi^T \Psi \|, \tag{19}$$

where $\| \cdot \|$ is the 1-norm of the matrix, in order to ensure that the invertible matrix $\Psi^T \Psi$ is well-conditioned [51,52].

A number M of roots' combinations, which verify the conditions in Equations (18) and (19), create then the ED.

Statistical quantities, such as the first and second moments (the mean and the variance, respectively), could then be calculated to quantify the randomness of the stochastic responses.

2.4. Solving Procedure

To solve Equation (14), a transformation of the complex amplitude to Cartesian form is needed:

$$A_m = (a_m + ib_m) exp(i\sigma T). \tag{20}$$

Substituting Equation (20) into Equation (14) and simplifying by $exp(i\sigma T)$, one can obtain two generic equations for the real and imaginary parts of each amplitude A_m:

$$
\begin{aligned}
a_m^{(1,0)} &= \frac{\sigma}{2\omega_0}b_m - \frac{c_n}{2}a_m - \frac{k_n}{2\omega_0}(2b_m - cos[q_m]Im(G_m)) \\
&- S * \frac{1}{N+1}\frac{k_n}{\omega_0}\sum_{n=1}^{N} sin[nq_m]\sum_{x=1}^{N} cos[nq_x]sin[q_x](b_x - \tilde{b}_x) \\
&- \frac{3}{8}\frac{\alpha_n}{\omega_0}\sum_{j,k,l}[a_ja_kb_l + b_jb_kb_l]\Delta_{jkl,m}^{(1)},
\end{aligned}
\tag{21}
$$

$$
\begin{aligned}
b_m^{(1,0)} &= -\frac{\sigma}{2\omega_0}a_m - \frac{c_n}{2}b_m + \frac{k_n}{2\omega_0}(2a_m - cos[q_m]Re(G_m)) \\
&+ S * \frac{1}{N+1}\frac{k_n}{\omega_0}\sum_{n=1}^{N} sin[nq_m]\sum_{x=1}^{N} cos[nq_x]sin[q_x](a_x - \tilde{a}_x) \\
&+ \frac{3}{8}\frac{\alpha_n}{\omega_0}\sum_{j,k,l}[a_ja_ka_l + a_jb_kb_l]\Delta_{jkl,m}^{(1)} - \frac{1}{2(N+1)}\frac{f_n}{\omega_0}\sum_{n=1}^{N} sin[nq_m].
\end{aligned}
\tag{22}
$$

consequently, $2N(p + q + d)$ coupled algebraic equations are obtained.

Solving analytically these equations is very difficult or even impossible, especially in presence of uncertainties. To overcome this issue, numerical solving processes must be used. Subsequently, two possible configurations occur regarding including or not the stability analysis. To solve similar problem, Bitar et al. [15,29] applied the Asymptotic Numerical Method (ANM) [53–56] in graphical interactive software named MANLAB [57] and included stability analysis. The complexity of the study was underlined in presence of multiplicity of stable and unstable solutions. In the present work, accounting for uncertainties increases the number of stable and unstable branches and thus makes the solving ANM-based process very difficult, prohibitive or even impossible. To simplify the study, we choose to limit the solving process to the computation of stable solutions and to apply the Runge-Kutta time integration method to solve Equations (21) and (22).

3. Numerical Examples

Two numerical examples are considered in this section: two and three coupled pendulums chains. To make clear presentation and discussion of each example, deterministic study is presented at first. Stochastic results are then discussed compared to deterministic ones to evaluate the robustness of the collective dynamics of the considered structures against uncertainties.

In deterministic case, the design parameters of the perfectly periodic structure are listed in Table 1 [29].

Table 1. Design parameters of the periodic coupled pendulums chain.

m (kg)	l (m)	k (N·m)	c (kg·s^{-1})	f (N·m)	ω_0 (rad·s^{-1})
0.25	0.062	9.10^{-4}	0.16	0.01	12.58

In stochastic case, the length of the first pendulum is supposed to be uncertain and varies such as

$$
l_1 = l(1 + \delta_l\xi_l),
\tag{23}
$$

where δ_l is a chosen dispersion level and ξ_l a lognormal random variable.

Uncertainty propagation through the generic discrete model is performed using the LHS method and the gPCE. Effects of uncertainties on the collective dynamics of the structures are investigated through statistical analyses of the dispersions of the frequency responses and the basins of attraction.

The problem is numerically time-consuming when using the Runge-Kutta time integration method. It is also necessary to sufficiently refine frequency steps and vary initial conditions to obtain more stable solutions, although it is difficult to cover all possible solutions. These constraints impose minimizing as possible the number of samples for the LHS method. Therefore, 200 samples are generated.

To apply the gPCE method, a fourth order expansion is used ($p = 4$). The length of the first pendulum is considered to be an uncertain parameter ($d = 1$). Hence, $(p+1)^d = 5$ Hermite polynomial roots are chosen, according to which 5 deterministic solutions are computed. Consequently, the 200 samples required for the LHS method are reduced by 97.5%.

3.1. Example 1: Two Coupled Pendulums

Two coupled pendulums structure ($N = 2$) is considered here. $2N(p+q+d) = 8$ algebraic equations are generated with $p = 1$ stochastic pendulum, $q = 1$ deterministic pendulum neighbor of the stochastic one and $d = 0$ since no deterministic pendulum in the structure has deterministic neighbors.

3.1.1. Deterministic Study

If uncertainty is not taken into account, $p = 0$, $q = 0$ and $d = 1$. Consequently, 4 algebraic equations are generated. Deterministic responses in both modal and physical coordinates are illustrated in Figure 2.

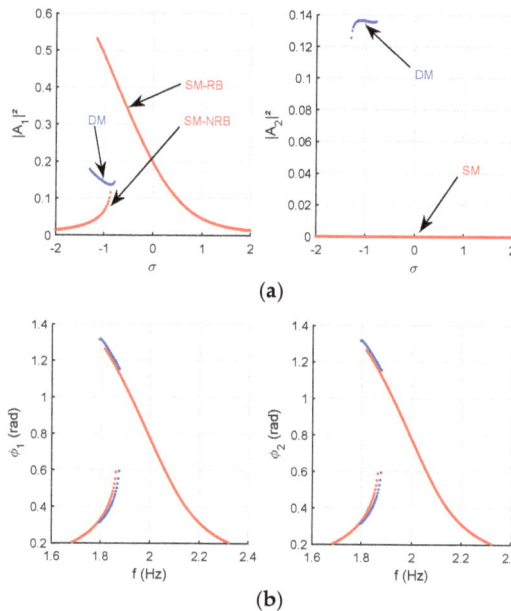

Figure 2. Deterministic (**a**) modal amplitudes $|A_1|^2$ and $|A_2|^2$ and (**b**) physical amplitudes ϕ_1 and ϕ_2 of the pendulums responses.

A multiplicity of stable solutions is generated, in the multistability domain, by modal interactions [58] between the responses which are driven by the collective dynamics. Three stable solutions could thus be obtained at several frequencies in the multistability domain [59–61]. They are classified as Single (SM) and Double mode (DM) solutions. The only SM branch, presented by red curve, corresponds to the null trivial solution of the second amplitude. Two SM branches are identified for the first amplitude: resonant branch (SM-RB) and non-resonant branch (SM-NRB) (Figure 2a). The DM branches, which are presented by blue curves, result from coupling between the pendulums. A correspondence in term of bifurcation points between the amplitudes with respect to each branch type is observed. It is generated by the bifurcation topology transfer. Identical physical responses reflect the symmetry between the pendulums.

More detailed illustration of the bifurcation topology transfer between the amplitudes $|A_1|^2$ and $|A_2|^2$ is allowed by the basins of attraction. The robustness of the attractors, corresponding to the multimode branches, and their contributions in the basins are investigated. In literature, Bartuccelli et al. [62] illustrated numerically the basins of attraction of a plane pendulum under parametric excitation. An experimental investigation of the basins of attraction of two fixed points of a nonlinear mechanical nano-resonator was illustrated by Kozinsky et al. [63]. Sliwa et al. [64] studied the basins of attraction of two coupled Kerr oscillators. Bitar et al. [15] studied the basins of attraction of two and three coupled oscillators under both external and parametric excitations. The basins of attraction are generally plotted in the phase plan $(\phi, \dot{\phi})$. In this work, the basins of attraction are plotted in the Nyquist plane (a, b) of real and imaginary parts of the responses amplitudes A_m [15,29] regarding the proposed solving methodology.

Fixing the detuning parameter on $\sigma = -1$, in the multistability domain, two and three stable solutions are obtained for $|A_2|^2$ and $|A_1|^2$, respectively. Two and three attractors correspond to these stable solutions, respectively, in the basins of attraction. Figure 3 illustrates the basins of attraction plotted in the Nyquist plane $((a_1)_0, (b_1)_0)$ for $(a_2)_0 = (b_2)_0 = 0.25$.

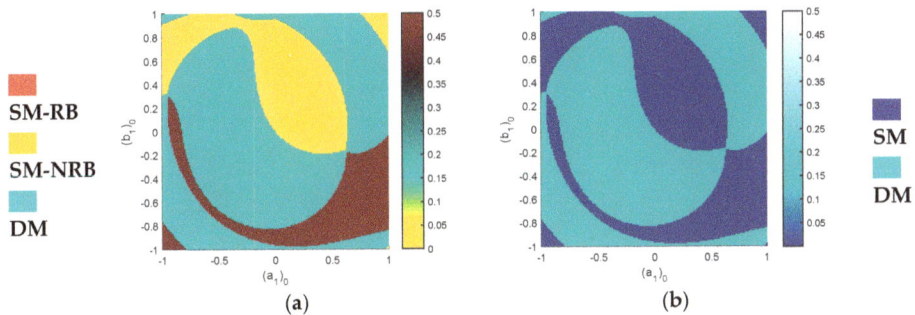

Figure 3. Basins of attraction of the deterministic amplitudes (a) $|A_1|^2$ and (b) $|A_2|^2$ in the Nyquist plane $((a_1)_0, (b_1)_0)$ for $\sigma = -1$ and $(a_2)_0 = (b_2)_0 = 0.25$.

When $|A_1|^2$ jumps between SM-RB and SM-NRB, $|A_2|^2$ is stabilized on SM. Similarly, a correspondence exists between the DM attractors. Subsequently, one can restrict the analysis to the basins of attraction of $|A_1|^2$.

Varying the detuning parameter ($\sigma = -1.2$, $\sigma = -1$ and $\sigma = -0.8$), the contribution of each attractor is illustrated quantitatively through the ratio of its size with respect to the global size of the basins of attraction. For $\sigma = -1.2$, the most robust attractors correspond to the SM-NRB; they dominate the basins with 52.5% of their total size, compared to 45.9% and only 1.6% for the SM-RB and DM attractors, respectively. Nevertheless, the SM-NRB attractors vanish for $\sigma = -0.8$ and the DM attractors become the most robust; their contribution is quantified by 64.5%, compared to a contribution of 35.5% of the SM-RB. For $\sigma = -1$, Figure 3, the DM attractors also dominate the basins of attraction; 60.9%. However, the contributions of the attractors of SM-RB and DM are quantified by 15.7% and 23.4%, respectively.

3.1.2. Stochastic Study

Impact of uncertainty is illustrated by the envelopes of the frequency responses amplitudes computed using the LHS method and the gPCE, in generalized and physical coordinates, for two dispersion levels: $\delta_l = 2\%$ and $\delta_l = 5\%$ (Figures 4–6).

In order to enable comparative study with deterministic responses, we use the initial conditions fixed in deterministic case, in the multistability domain. The discontinuity of the envelopes is due to the lack of initial conditions. Note that varying more the initial conditions makes computation prohibitive and it still difficult to cover all possible solutions and so obtain smooth curves.

Increasing the dispersion level affects more the responses variability. All multimode branches are larger in terms of amplitude and frequency. Consequently, the multimode solutions are enhanced and the multistability domain is wider. The uncertainty effect on each pendulum response depends on its position with respect to the stochastic one. The greatest impact is on the first pendulum responses (larger envelopes) since it contains the uncertain parameter. The second one is less affected by uncertainty. Modal localization is consequently generated on the first pendulum response.

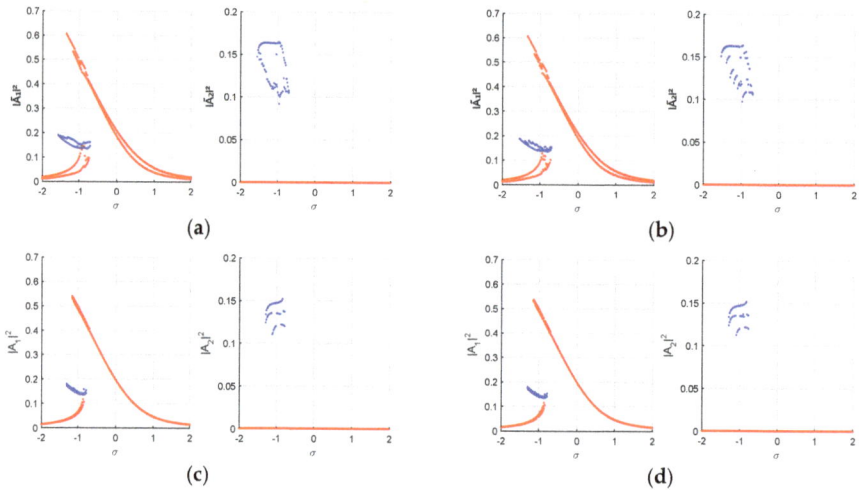

Figure 4. Envelopes of the amplitudes (**a,b**) $\left|\tilde{A}_1\right|^2$ and $\left|\tilde{A}_2\right|^2$ of the first pendulum modal response and (**c,d**) $|A_1|^2$ and $|A_2|^2$ of the second pendulum modal response computed using the LHS and gPCE methods, respectively, for $\delta_l = 2\%$.

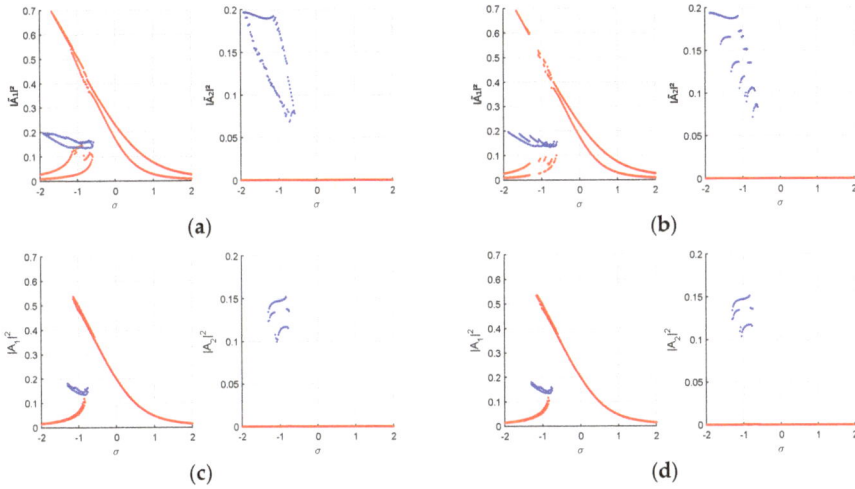

Figure 5. Envelopes of the amplitudes (**a**,**b**) $\left|\tilde{A}_1\right|^2$ and $\left|\tilde{A}_2\right|^2$ of the first pendulum modal response and (**c**,**d**) $|A_1|^2$ and $|A_2|^2$ of the second pendulum modal response computed using the LHS and gPCE methods, respectively, for $\delta_l = 5\%$.

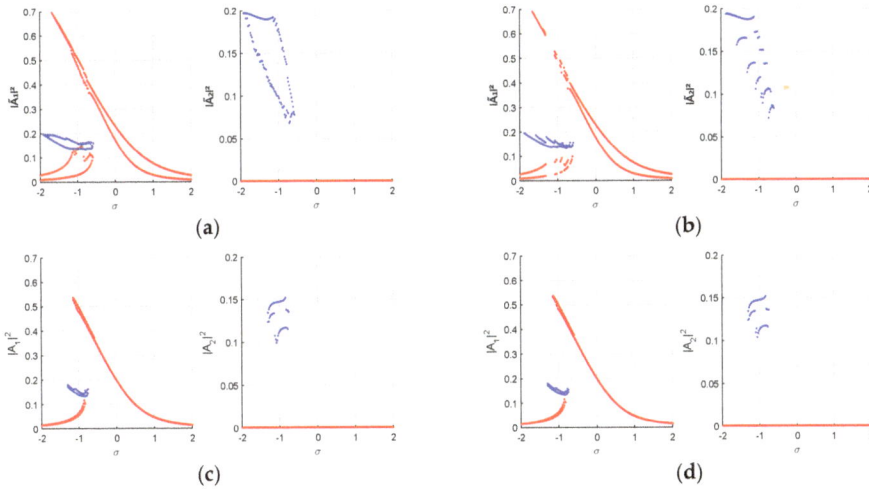

Figure 6. Envelopes of the amplitudes $\tilde{\phi}_1$ and ϕ_2, of the pendulums physical responses computed, for (**a**,**b**) $\delta_l = 5\%$ and (**c**,**d**) $\delta_l = 5\%$, using the LHS and gPCE methods, respectively.

Table 2 and Figures 7 and 8 illustrate the uncertainty effects through the intervals of variation of the amplitude ($\Delta\phi$) and frequency (Δf) ranges of the multistability domain. These intervals are computed between extreme bifurcation points, according to the example shown in Figure 7. $\tilde{\phi}_1$ is more affected by uncertainty than ϕ_2 since the first pendulum length is stochastic. More important variation is detected on Δf than on $\Delta\phi$. The ratio $\Delta f/\Delta\phi$ shows that the nonlinearity of the first pendulum response is strongly strengthened. Moreover, dispersion and nonlinearity are proportional.

Satisfying approximations are allowed by the gPCE with respect to the LHS method, considered as reference. Good agreement between modal and physical responses is proved by small accuracy errors. For $\delta_l = 5\%$, the gPCE errors increase, revealing the limits of the method against high level of uncertainty. It is important to note that some errors result from the luck of initial conditions making the detection of bifurcation points more difficult.

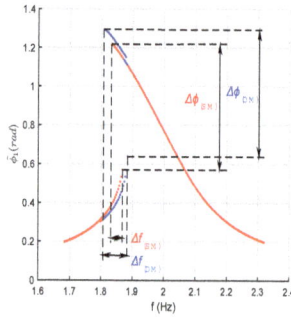

Figure 7. Example of measured amplitude and frequency ranges of the multistability domain according to the SM and DM.

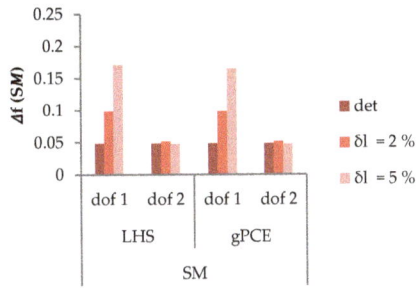

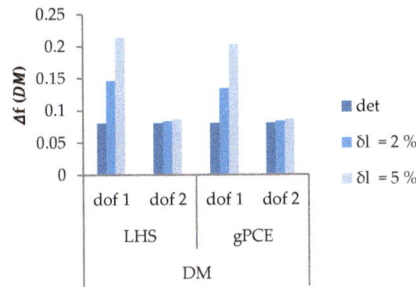

(a)

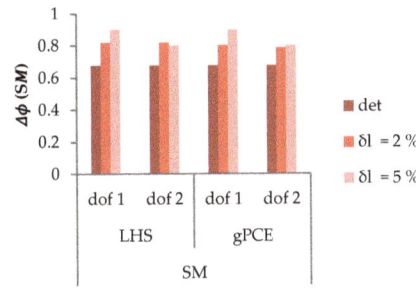

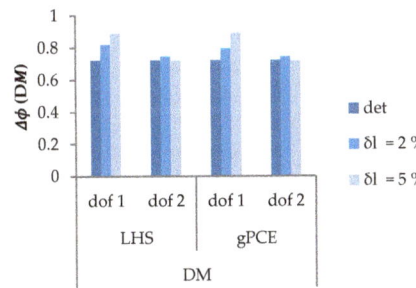

(b)

Figure 8. Evolution of (**a**) the frequency ranges and (**b**) the amplitude ranges of the multistability domain with respect to deterministic ones according to the SM and DM.

Table 2. Stochastic amplitude and frequency ranges of the multistability domain, according to the SM and DM, with respect to deterministic ranges.

Dof		SM				DM			
		LHS		gPCE		LHS		gPCE	
		2%	5%	2%	5%	2%	5%	2%	5%
Δf (%)	1	204.17	356.25	204.17	343.75	182.50	267.50	167.50	253.75
	2	106.25	97.92	106.25	97.92	103.75	107.50	103.75	107.75
$\Delta\phi$ (%)	1	120.86	133.14	118.49	132.69	113.57	123.13	110.11	123.55
	2	120.86	118.34	116.12	118.05	103.19	99.72	103.05	99.31
$\frac{\Delta f}{\Delta\phi}$ (%)	1	204.23	267.61	171.83	259.15	160.36	217.12	152.25	205.41
	2	87.32	83.10	91.55	830	100.90	107.21	100.90	108.11

In presence of uncertainty, a set of solutions is generated by the LHS method or the gPCE. Several basins of attraction are thus superposed. The envelopes of the attractors are quantified by the overlapped areas. For $\delta_l = 2\%$, the envelopes of the attractors of $\left|\tilde{A}_1\right|^2$ (first pendulum), computed using the LHS method, represent 64.57% of the basins. The gPCE approximation gives 62.34%. The envelopes of the attractors of $|A_1|^2$ (second pendulum) represent 9.67% of the basins (gPCE: 8.97%). For $\delta_l = 5\%$, the impact of uncertainty is more important. Indeed, the basins of attraction of $\left|\tilde{A}_1\right|^2$ are fully overlapped (100%). The size of the envelopes of the attractors of $|A_1|^2$, computed by the LHS method, increases to 19.04%. The gPCE approximate it by 17.55%. The gPCE gives a satisfying approximation of the basins' envelopes, although errors proportionally increase with dispersion level. The comparison of the envelopes of the basins of $\left|\tilde{A}_1\right|^2$ to those of $|A_1|^2$ highlights the modal localization generated on the first pendulum response.

Moreover, 96.4% of CPU time reduction is achieved when applying the gPCE, with respect to the LHS method. In fact, nearly 15.8 h are needed to compute the frequency responses using the LHS method. However, less than 0.6 hours is required for the gPCE implementation. More important time reduction ratio is achieved when computing the basins of attraction: 97.6%. Indeed, the 350.4 h required by the LHS method are reduced to only 8.4 h needed for the gPCE implementation. Note that all computations are made using Matlab on 8-Core Dual Processor with 64 GB of RAM (Random Access Memory).

For more detailed dispersion analysis, the attractors of SM-RB, SM-NRB and DM are plotted separately. The variability of the basins of attraction of $\left|\tilde{A}_1\right|^2$ is more important than the variability of the basins of $|A_1|^2$, which illustrates modal localization on first pendulum response. The dispersion of each attractor is illustrated by the variability of its contribution in the basins of attraction from a minimal size (color defined for the considered attractor in the deterministic case illustrated by Figure 3) to a maximal size bounded by the limit (white area) corresponding to other attractors' contributions. The envelopes are represented by black areas. Figure 9 shows the envelopes of the attractors of $\left|\tilde{A}_1\right|^2$, for $\delta_l = 2\%$, computed using the gPCE. Since, the basins of $\left|\tilde{A}_1\right|^2$ are fully overlapped for $\delta_l = 5\%$, Figure 10 illustrates the envelopes of the basins of $|A_1|^2$.

As shown in Figure 9, the DM attractors are the most robust. Table 3 compares the contributions of the attractors of SM-RB, SM-NRB and DM in the basins of attraction in deterministic and stochastic case (for $\delta_l = 2\%$ and $\delta_l = 5\%$). Table 3 lists both minimal sizes and envelopes of the attractors. Summing these two areas gives the global contribution of each attractor in the basins. Superposing the envelopes of the attractors of SM-RB, SM-NRB and DM gives the global envelopes.

In deterministic case, the DM attractors of the first pendulum response are the most robust. However, the contributions of the attractors of SM-RB and SM-NRB in the basins become more important for $\delta_l = 2\%$ and even more for $\delta_l = 5\%$. The SM-NRB attractors are in fact the most robust.

The DM attractors of the second pendulum response remain the most robust. This evolution of the attractors' contributions in the basins of attraction is due to modal localization generated by presence of uncertainty.

Table 3 proves that more accurate approximation of the variability of the basins of attraction is obtained using the gPCE for lower dispersion level ($\delta_l = 2\%$, here). The gPCE approximations of the basins of attraction are more affected than those of the frequency responses. The basins analysis is indeed more thorough.

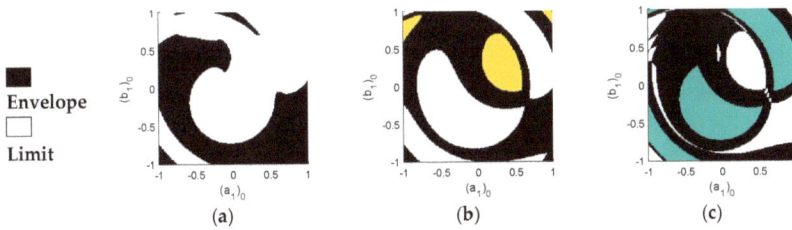

Figure 9. Envelopes of the attractors of (**a**) SM-RB, (**b**) SM-NRB and (**c**) DM of the modal intensity amplitude $\left|\tilde{A}_1\right|^2$ of the first pendulum response, in the Nyquist plane $((a_1)_0, (b_1)_0)$ for $\sigma = -1$ and $(a_2)_0 = (b_2)_0 = 0.25$, computed using the gPCE method for $\delta_l = 2\%$.

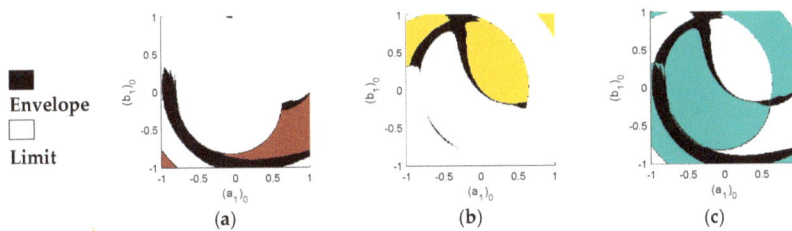

Figure 10. Envelopes of the attractors of (**a**) SM-RB, (**b**) SM-NRB and (**c**) DM of the modal intensity amplitude $|A_1|^2$ of the second pendulum response, in the Nyquist plane $((a_1)_0, (b_1)_0)$ for $\sigma = -1$ and $(a_2)_0 = (b_2)_0 = 0.25$, computed using the gPCE method for $\delta_l = 5\%$.

Table 3. Comparison of the contributions of the attractors of SM-RB, SM-NRB and DM in the basins of attraction computed using the LHS and gPCE methods, for deterministic case (det), $\delta_l = 2\%$ and $\delta_l = 5\%$.

				Basins of Attraction Size (%)			
				LHS		gPCE	
Modal Amplitude		**Attractor**	**Det**	ffi$_l$ = 2%	ffi$_l$ = 5%	ffi$_l$ = 2%	ffi$_l$ = 5%
$\left|\tilde{A}_1\right|^2$ (dof1)	Min	SM-RB	15.7	0	0	0	0
		SM-NRB	23.4	8.39	0	8.93	0
		DM	60.9	27.04	0	28.73	0
	Envelope	SM-RB	0	43.93	100	42.29	100
		SM-NRB	0	52.14	100	48.91	100
		DM	0	63.97	97.86	51.52	72.41
$|A_1|^2$ (dof2)	Min	SM-RB	15.7	12.73	9.91	12.89	10.03
		SM-NRB	23.4	22.13	21.75	22.20	22.16
		DM	60.9	55.47	49.30	55.94	50.26
	Envelope	SM-RB	0	6.81	11.55	6.32	11.04
		SM-NRB	0	2.86	7.49	2.65	6.50
		DM	0	9.67	19.04	8.97	17.55

3.2. Example 2: Three Coupled Pendulums

Three coupled pendulums structure ($N = 3$) is considered in this section. If uncertainty is not taken into account, 6 algebraic equations are generated ($p = 0$, $q = 0$ and $d = 1$). In stochastic case, $2N(p + q + d) = 18$ algebraic equations are generated with $p = 1$ stochastic pendulum, $q = 1$ deterministic pendulum neighbor of the stochastic one and $d = 1$ since one deterministic pendulum in the structure has deterministic neighbors.

3.2.1. Deterministic Study

Deterministic modal and physical responses are illustrated by Figure 11. The symmetry between the first and the third pendulum generates conformity between their physical responses.

Modal interactions and bifurcation topology transfer between the pendulums' responses, which are driven by the collective dynamics, generate up to six stable solutions at several frequencies in the multistability domain. The obtained stable branches are classified as Double (DM) and Triple mode (TM) solutions. The DM branches correspond to the solutions of the first and third amplitudes (red curves), whereas the TM solutions result from the coupling between the pendulums (blue curves). Three DM solutions (DM-RB, DM-NRB, DM-B1 and DM-B2) and two TM solutions (TM-B1 and TM-B2) are obtained, Figure 11a. A correspondence in term of bifurcation points between the amplitudes with respect to each branch type is observed. This is generated by the bifurcation topology transfer.

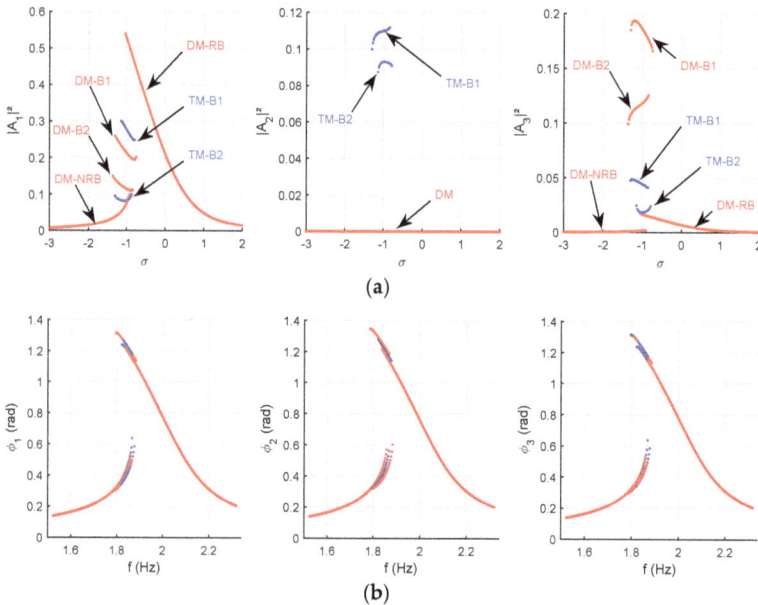

Figure 11. Deterministic (**a**) modal responses amplitudes $|A_1|^2$, $|A_2|^2$ and $|A_3|^2$ and (**b**) physical responses amplitudes ϕ_1, ϕ_2 and ϕ_3, of the pendulums.

Figure 12 illustrates the contributions of the attractors of the six stable branches in the basins of attraction. The basins are plotted for $\sigma = -1$, in the Nyquist plane $((a_1)_0, (b_1)_0)$ for arbitrary chosen initial conditions $(a_2)_0 = (b_2)_0 = 0.2$ and $(a_3)_0 = (b_3)_0 = 0.3$. Figure 12 shows also the responses amplitudes, which can be identified through the color bar.

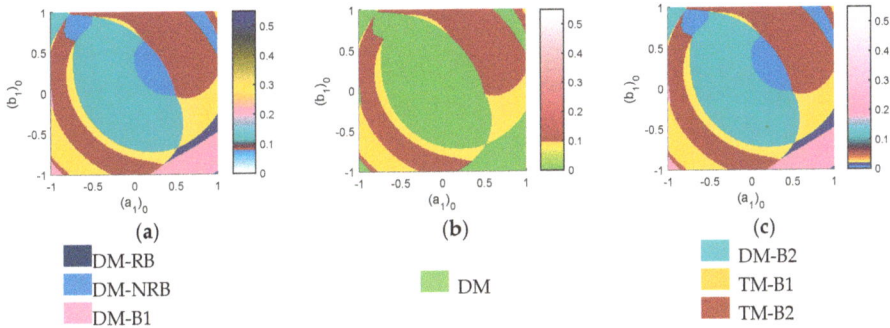

(a)   **(b)**   **(c)**

DM-RB   DM   DM-B2
DM-NRB   TM-B1
DM-B1   TM-B2

Figure 12. Basins of attraction of the deterministic amplitudes (**a**) $|A_1|^2$, (**b**) $|A_2|^2$ and (**c**) $|A_3|^2$ in the Nyquist plane $((a_1)_0, (b_1)_0)$ for $\sigma = -1$, $(a_2)_0 = (b_2)_0 = 0.2$ and $(a_3)_0 = (b_3)_0 = 0.3$.

The bifurcation topology transfer between the responses generates a correspondence between the attractors in the three basins of attraction. Indeed, when $|A_1|^2$ and $|A_3|^2$ jump between DM-RB, DM-NRB, DM-B1 and MD-B2, $|A_2|^2$ is stabilized on DM. Moreover, a correspondence exists between the TM attractors (TM-B1 and TM-B2) of the three pendulums responses.

The DM attractors are the most robust since they dominate the basins with 51% of their global size. The contribution of the TM attractors is nearly similar (49%). Focusing on DM, the DM-B2 attractors, occupying nearly 35% of the basins, are the most robust. Similarly, the TM-B2 attractor occupies 33%. The lowest contribution is of the DM-RB attractors representing less than 2% of the global basins size.

3.2.2. Stochastic Study

The greatest impact of uncertainty is obtained on the first pendulum responses since it contains the uncertain parameter. The second pendulum responses are less affected while the behavior of the third pendulum remains deterministic, because no coupling terms relate its equation with the one associated to the stochastic pendulum. Hence, the third pendulum responses are not presented hereafter. Figures 13–15 illustrate the envelopes of the responses amplitudes in generalized and physical coordinates, for $\delta_l = 5\%$ and $\delta_l = 2\%$.

For $\delta_l = 5\%$, all branches are larger in terms of amplitude and frequency ranges. Consequently, the multimode solutions are enhanced and the multistability domain is wider. Nevertheless, the overlap of envelopes makes detection of extreme statistics of adjacent branches more difficult. Modal localization is here generated on the response of the first pendulum containing uncertainty.

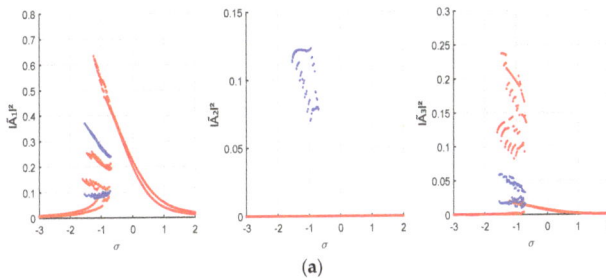

(a)

Figure 13. *Cont.*

(b)

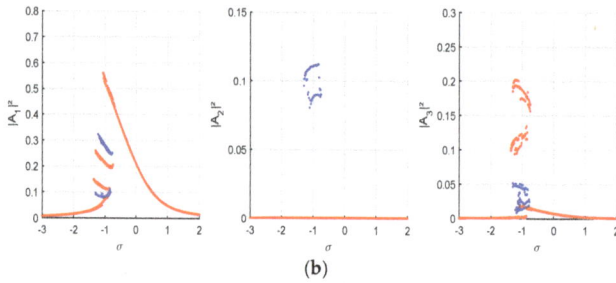

Figure 13. Envelopes of the modal intensity amplitude responses: (**a**) $\left|\tilde{A}_1\right|^2$, $\left|\tilde{A}_2\right|^2$ and $\left|\tilde{A}_3\right|^2$ of the first pendulum and (**b**) $|A_1|^2$, $|A_2|^2$ and $|A_3|^2$ of the second pendulum, computed using the gPCE method for $\delta_l = 2\%$.

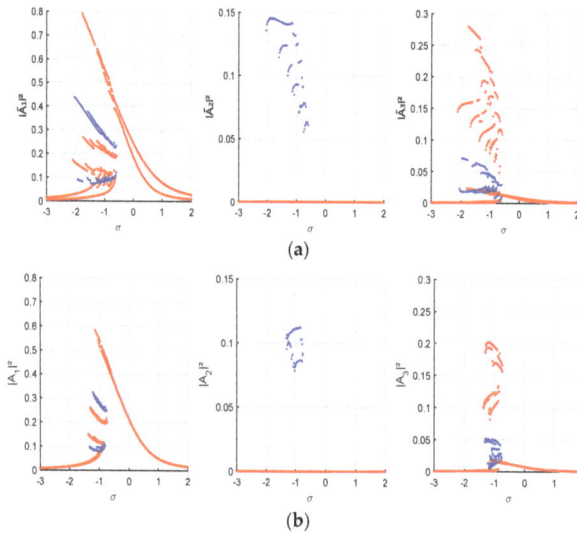

(a)

(b)

Figure 14. Envelopes of the modal intensity amplitude responses: (**a**) $\left|\tilde{A}_1\right|^2$, $\left|\tilde{A}_2\right|^2$ and $\left|\tilde{A}_3\right|^2$ of the first pendulum and (**b**) $|A_1|^2$, $|A_2|^2$ and $|A_3|^2$ of the second pendulum, computed using the gPCE method for $\delta_l = 5\%$.

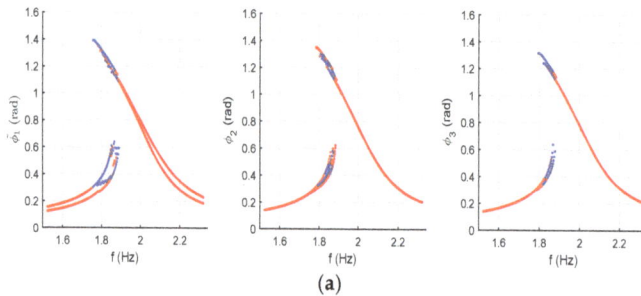

(a)

Figure 15. *Cont.*

38

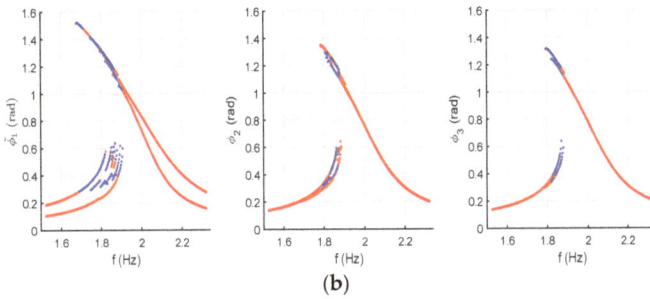

(b)

Figure 15. Envelopes of the physical amplitude responses $\widetilde{\phi}_1$, ϕ_2 and ϕ_3, computed using the gPCE method, for (a) $\delta_l = 2\%$ and (b) $\delta_l = 5\%$.

Figure 16 and Table 4 illustrate the uncertainty effects through the intervals $\Delta\phi$ and Δf of the multistability domain. More important variation is detected on Δf than on $\Delta\phi$. The ratios $\Delta f / \Delta\phi$ show that the nonlinear aspect of the structure is enhanced. Indeed, the nonlinearity of the first pendulum response is more strengthened than the one of second pendulum response.

Comparing uncertainty propagation methods, the gPCE allows generally an accurate approximation of the responses dispersion. Nevertheless, less accuracy is obtained for $\delta_l = 5\%$. Note that errors result also from lack of initial conditions and make detection of bifurcation points difficult. Furthermore, only 5 simulations are generated when applying the gPCE, compared to 200 samples used by the LHS method. A discontinuity of the envelopes' curves is thus detected, especially for $\delta_l = 5\%$.

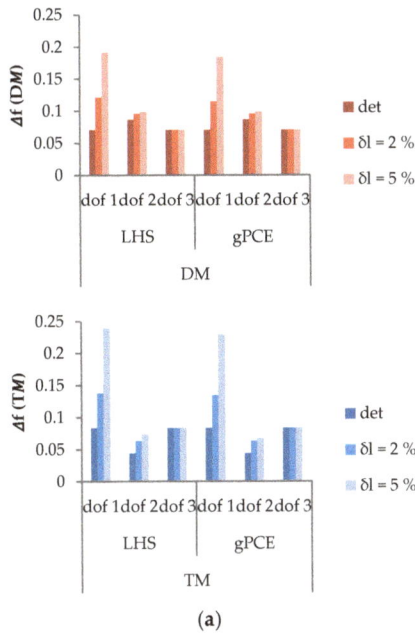

(a)

Figure 16. *Cont.*

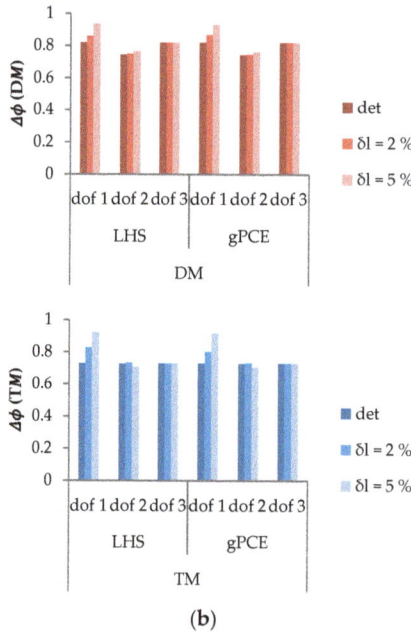

(b)

Figure 16. Evolution of (**a**) the frequency ranges and (**b**) the amplitude ranges of the multistability domain with respect to deterministic ones according to the DM and TM.

Table 4. Stochastic amplitude and frequency ranges of the multistability domain, according to the DM and TM, with respect to the deterministic ranges.

	Dof	DM				TM			
		LHS		gPCE		LHS		gPCE	
		2%	5%	2%	5%	2%	5%	2%	5%
Δf (%)	1	172.86	272.86	162.86	262.86	165.06	287.95	161.45	275.90
	2	110.47	113.95	110.47	1.95	143.18	165.91	143.18	152.27
$\Delta\phi$ (%)	1	104.76	114.15	105.73	113.66	113.29	126.44	109.73	125.75
	2	100.94	103.10	100.40	1.56	100.83	97.39	100.69	96.97
$\frac{\Delta f}{\Delta\phi}$ (%)	1	165.88	240.00	154.12	231.76	145.61	227.19	146.49	218.42
	2	109.48	110.34	109.48	111.21	0.98	168.85	140.98	155.74

Regarding the dispersion of the basins of attraction (Figure 17 and Table 5), the envelopes of the attractors of $\left|\tilde{A}_1\right|^2$, computed using the LHS method, cover 69.27% of the basins, for $\delta_l = 2\%$. The gPCE approximation gives 66.85%. For $\delta_l = 5\%$, the basins are fully overlapped. However, the envelopes cover only 7.58% of the basins of $|A_1|^2$, computed using the LHS method, for $\delta_l = 2\%$ and 16.40% for $\delta_l = 5\%$. The gPCE approximations are, respectively, 6.91% and 15.02%. Less accurate results are thus obtained by the gPCE for $\delta_l = 5\%$. To overcome this issue, one can elevate the polynomial order of the gPCE or use one of its recently developed variants [65–67].

As shown in Figure 17 and Table 5, a largest dispersion is detected on the DM-RB and DM-NRB attractors of $\left|\tilde{A}_1\right|^2$. Their contributions in the basins, approximated in deterministic case by 1.69% and 8.71% respectively, increase up to 14.5% and 47.5%, for $\delta_l = 2\%$. For $\delta_l = 5\%$, the attractors become totally overlapped (100%). The DM-B2 and TM-B2 attractors of $|A_1|^2$, remain the most robust against uncertainty. The robustness analysis of the basins of attraction illustrates the modal localization generated on one or more attractors, for different dispersion levels.

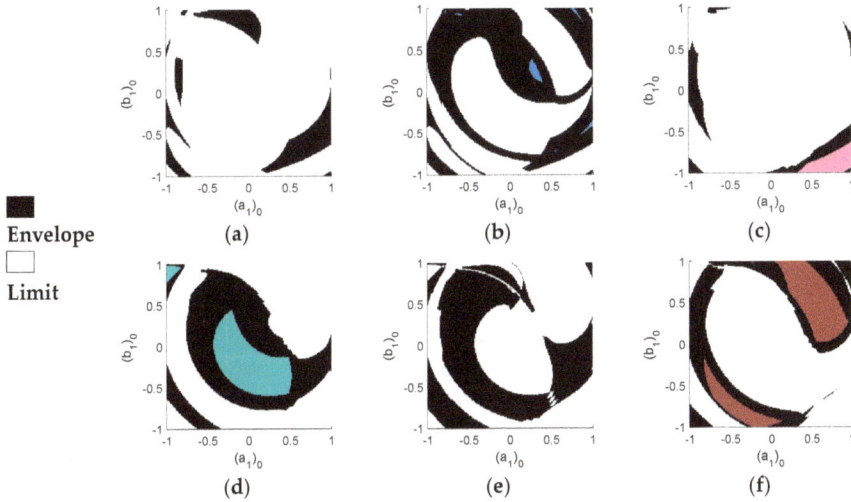

Figure 17. Envelopes of the attractors of (**a**) DM-RB, (**b**) DM-NRB, (**c**) DM-B1, (**d**) DM-B2, (**e**) TM-B1 and (**f**) TM-B2 of the amplitude $\left|\tilde{A}_1\right|^2$ of the first pendulum response, in the Nyquist plane $((a_1)_0, (b_1)_0)$ for $\sigma = -1$, $(a_2)_0 = (b_2)_0 = 0.2$ and $(a_3)_0 = (b_3)_0 = 0.3$, computed using the gPCE method for $\delta_l = 2\%$.

Table 5. Comparison of the contributions of different attractors in the basins of attraction computed using the LHS and gPCE methods, for deterministic case, $\delta_l = 2\%$ and $\delta_l = 5\%$.

Amplitude		Attractor	Det	LHS		gPCE	
				ffi$_l$ = 2%	ffi$_l$ = 5%	ffi$_l$ = 2%	ffi$_l$ = 5%
$\left|\tilde{A}_1\right|^2$ (dof1)	Min	DM-RB (%)	1.69	0	0	0	0
		DM-NRB (%)	8.71	0.43	0	0.82	0
		DM-B1 (%)	5.71	3.19	0	3.30	0
		DM-B2 (%)	34.87	12.72	0	13.77	0
		TM-B1 (%)	16.08	0	0	0	0
		TM-B2 (%)	32.	14.40	0	15.46	0
	Envelope	DM-RB (%)	0	14.53	100	12.21	99.93
		DM-NRB (%)	0	47.16	100	42.38	97.82
		DM-B1 (%)	0	8.12	25.50	7.01	10.30
		DM-B2 (%)	0	38.17	51.19	35.07	44.58
		TM-B1 (%)	0	47.31	69.55	43.44	44.27
		TM-B2 (%)	0	28.64	44.79	24.55	35.59
$\left|A_1\right|^2$ (dof2)	Min	DM-RB (%)	1.69	1.30	0.97	1.33	0.99
		DM-NRB (%)	8.71	9.59	9.08	9.65	9.30
		DM-B1 (%)	5.71	5.02	4.28	5.08	4.39
		DM-B2 (%)	34.87	32.91	30.72	33.12	30.96
		TM-B1 (%)	16.08	13.87	10.82	13.27	11.04
		TM-B2 (%)	32.94	31.05	28.72	31.27	29.29
	Envelope	DM-RB (%)	0	1.19	2.47	1.09	2.29
		DM-NRB (%)	0	2.63	6.97	2.37	6.29
		DM-B1 (%)	0	1.03	2.07	0.95	1.94
		DM-B2 (%)	0	3.27	6.48	3	5.98
		TM-B1 (%)	0	5.12	10.46	4.64	9.53
		TM-B2 (%)	0	3.22	7.61	2.91	6.73

On the other hand, the gPCE allows reducing the CPU time needed for the LHS method by nearly 95.2% (from 33.6 h down to 1.6 h) and 97.5% (from 489.6 h down to 12 h) when computing the frequency responses and the basins of attraction, respectively.

3.3. Overall Discussion

Three stable solutions are obtained at several frequencies in the multistability domain due to modal interactions between the responses of two coupled nonlinear pendulums, generated by the collective dynamics. However, up to six stable solutions are obtained if a three coupled pendulums chain is considered.

In presence of uncertainties, the stability of the system is enhanced. Indeed, a multiplicity of stable solutions is obtained and the multistability domain is wider in terms of frequency range and response amplitude. Moreover, the nonlinearity of the system is strengthened by increasing the ratios of the frequency intervals by the amplitude intervals computed between the extreme bifurcation points in the multistability domain.

The robustness of the attractors of different modes and modal branches is illustrated by the dispersion analysis of the basins of attraction and the quantitative study of the evolution of the distribution of these attractors from deterministic case to the case $\delta_l = 5\%$. Modal localization is generated by widening one or more special modal branches more than the others.

For a reduction of up to 97.5% of computing time compared to the LHS method, the gPCE method allows satisfying approximations of the frequency responses and the basins of attraction. These approximations are more accurate for the frequency responses since the dispersion analysis of the basins of attraction is more thorough. For high dispersion levels (>2%), some limits of the gPCE method are reached.

4. Conclusions

A generic discrete analytical model combining multiple scales method and standing-wave decomposition was proposed in this paper. The model accounts for parametric uncertainties and permits to investigate the robustness of the collective dynamics of nonlinear coupled pendulums chain. Uncertainties allow a realistic modeling of imperfections, which can disturb the perfect periodicity of real engineering structures. In probabilistic framework, dispersion analyses of the frequency responses and the basins of attraction of two and three coupled pendulums chains were performed. Statistical evaluations of the frequency responses, in both modal and physical coordinates, quantified the variability of the frequency and amplitude intervals of the multistability domain. The complexity of the frequency responses detected in terms of modal interactions and bifurcation topology transfer, in deterministic case, was amplified in presence of uncertainty. The envelopes of the multimode branches are enlarged, and even overlapped, making difficult the detection of the extreme statistics of adjacent branches. This complexity emphasizes the large number of multimode solutions, even for chains of only two and three coupled pendulums, and the modal localization generated by uncertainty. Moreover, the presence of imperfection in a periodic chain widens its multistability domain and strengthens its nonlinear aspect. Particularly, the stability and the nonlinearity of the responses of the pendulums containing uncertainty are strongly enhanced. Furthermore, the robustness analysis of the basins of attraction was investigated through the dispersion analysis of the attractors' contributions. Finer analysis was thus allowed highlighting the complexity of the problem and the modal localization generated on the response of the pendulum containing uncertainty. These advantages can serve as a hint of the important modal localization, expected, in presence of uncertainties, for large number of pendulums. Imperfections can be functionalized to generate energy localization suitable for several engineering applications such as vibration energy harvesting, mass sensing in micro and nanotechnology, energy scavenging or trapping applications.

The generalized Polynomial Chaos Expansion was compared to the Latin Hypercube Sampling method, considered as reference, for uncertainty propagation. Results proved the satisfying approximations given by the former method for up to 97% of computational time reduction with respect to the later.

The robustness analysis becomes, both analytically and numerically, more complex and prohibitive when large-size structures and high dispersion levels are considered. Future works will focus on these challenges in order to improve analytical and numerical analyses.

The robustness of the collective dynamics against the uncertainty of some chosen structural input parameters was discussed in this paper. Extending the study to parametric analysis, with the aim of investigating other possible influential factors with respect to system robustness, is an interesting perspective that will be addressed in future works.

Acknowledgments: This project has been performed in cooperation with the Labex ACTION program (contract ANR-11-LABX-01-01).

Author Contributions: This paper is developed by a continuous and synergic collaboration among all authors. All authors read, revised and approved the final manuscript.

Conflicts of Interest: The authors declare no conflict of interest.

Appendix A

The functions G_m and S are chosen according to the number and the positions of the uncertain pendulums in the chain, which is derived from the coupling. Four possible configurations (a–d) are defined, with respect to Section 2.1, as follows:

a. If the concerned pendulum is deterministic as well as its neighbors, $G_m = 2A_m$ and $S = 0$. In this case, the complex amplitude equation takes the form

$$2i\omega_0 A_m^{(1,0)} + i\omega_0 c_n A_m + 4k_n sin\left[\frac{q_m}{2}\right]^2 A_m + \frac{3}{4}\alpha_n \sum_{j,k,l} A_j A_k A_l^* \Delta_{jkl,m}^{(1)}$$
$$- \frac{1}{(N+1)} f_n \, exp(i\sigma T) \sum_{n=1}^{N} sin[nq_m] = 0,$$
(A1)

b. If the concerned pendulum is deterministic but the previous one is stochastic, $G_m = A_m + \tilde{A}_m$ et $S = -1$. Equation (14) is expressed as

$$2i\omega_0 A_m^{(1,0)} + i\omega_0 c_n A_m + k_n \left(2A_m - Cos[q_m]\left(A_m + \tilde{A}_m\right)\right)$$
$$+ \frac{2k_n}{N+1} \sum_{n=1}^{N} sin[nq_m] \sum_{x=1}^{N} cos[nq_x] sin[q_x]\left(\tilde{A}_x - A_x\right),$$
(A2)

c. If the concerned pendulum is deterministic but the following is stochastic, $G_m = A_m + \tilde{A}_m$ and $S = +1$. Equation (14) takes the form

$$2i\omega_0 A_m^{(1,0)} + i\omega_0 c_n A_m + k_n \left(2A_m - cos[q_m]\left(A_m + \tilde{A}_m\right)\right)$$
$$+ \frac{2k_n}{N+1} \sum_{n=1}^{N} sin[nq_m] \sum_{x=1}^{N} cos[nq_x] sin[q_x]\left(A_x - \tilde{A}_x\right)$$
$$+ \frac{3}{4}\alpha_n \sum_{j,k,l} A_j A_k A_l^* \Delta_{jkl,m}^{(1)} - \frac{1}{(N+1)} f_n \, exp(i\sigma T) \sum_{n=1}^{N} sin[nq_m] = 0,$$
(A3)

d. If the stochastic pendulum is concerned, $\tilde{A}_m$ replaces A_m in Equation (14), $G_m = 2\tilde{A}_m$ and $S = 0$. Equation (14) is thus written as

$$2i\omega_0 \tilde{A}_m^{(1,0)} + i\omega_0 c_n \tilde{A}_m + 2k_n \left(\tilde{A}_m - cos[q_m]A_m\right) + \frac{3}{4}\alpha_n \sum_{j,k,l} \tilde{A}_j \tilde{A}_k \tilde{A}_l^* \Delta_{jkl,m}^{(1)}$$
$$- \frac{1}{(N+1)} f_n \, exp[i\sigma T] \sum_{n=1}^{N} sin[nq_m] = 0$$
(A4)

The number of generated complex equations depends on the number and positions of stochastic pendulums. In fact, the N equations obtained in deterministic case become $N(p+q+d)$ in a stochastic one, where p is the number of stochastic pendulums, q the number of deterministic pendulums neighbors of the stochastic ones and $d = 1$ if the structure contains deterministic pendulums having deterministic neighbors and 0 otherwise.

References

1. Brillouin, L. Wave Propagation in Periodic Structures. In *Electric Filters and Crystal Lattices*; Dover Publications: Mineola, NY, USA, 1953.
2. Mead, D.J. Wave propagation in continuous periodic structures: Research contributions from Southampton, 1964–1995. *J. Sound Vib.* **1996**, *190*, 495–524. [CrossRef]
3. Duhamel, D.; Mace, B.R.; Brennan, M.J. Finite element analysis of the vibrations of waveguides and periodic structures. *J. Sound Vib.* **2006**, *294*, 205–220. [CrossRef]
4. Goldstein, A.L.; Arruda, J.R.F.; Silva, P.B.; Nascimento, R. Building Spectral Element Dynamic Matrices Using Finite Element Models of Waveguide Slices and Elastodynamic Equations. In Proceedings of the nternational Conference on Noise and Vibration Engineering (ISMA), Leuven, Belgium, 20–22 September 2010.
5. Casadei, F.; Bertoldi, K. Wave propagation in beams with periodic arrays of airfoil-shaped resonating units. *J. Sound Vib.* **2014**, *333*, 6532–6547. [CrossRef]
6. Gosse, G.; Pézerat, C.; Bessac, F. Periodic assembly of multi-coupled beams: Wave propagation and natural modes. *J. Acoust. Soc. Am.* **2008**, *123*. [CrossRef]
7. Collet, M.; Ouisse, M.; Ruzzene, M.; Ichchou, M.N. Floquet-Bloch decomposition for the computation of dispersion of two-dimensional periodic, damped mechanical systems. *Int. J. Solids Struct.* **2011**, *48*, 2837–2848. [CrossRef]
8. Droz, C.; Zhou, C.; Ichchou, M.N.; Lainé, J.-P. A hybrid wave-mode formulation for the vibro-acoustic analysis of 2D periodic structures. *J. Sound Vib.* **2016**, *363*, 285–302. [CrossRef]
9. Chakraborty, G.; Mallik, A.K. Dynamics of a weakly non-linear periodic chain. *Int. J. Nonlinear Mech.* **2001**, *36*, 375–389. [CrossRef]
10. Boechler, N.; Daraio, C.; Narisetti, R.K.; Ruzzene, M.; Leamy, M.J. Analytical and experimental analysis of bandgaps in nonlinear one dimensional periodic structures. In *IUTAM Symposium on Recent Advances of Acoustic Waves in Solids*; Wu, T.-T., Ma, C.-C., Eds.; Volume 26 of IUTAM Book Series; Springer: Dordrecht, The Netherlands, 2010; pp. 209–219.
11. Georgiades, F.; Peeters, M.; Kerschen, G.; Golinval, J.C.; Ruzzene, M. Modal Analysis of a Nonlinear Periodic Structure with Cyclic Symmetry. *AIAA J.* **2009**, *47*, 1014–1025. [CrossRef]
12. Lifshitz, R.; Cross, M.C. Response of parametrically driven nonlinear coupled oscillators with application to micromechanical and nanomechanical resonator arrays. *Phys. Rev. Lett. B* **2003**, *67*, 134302. [CrossRef]
13. Lifshitz, R.; Cross, M.C. *Nonlinear Dynamics of Nanomechanical and Micromechanical Resonators*; John Wiley & Sons: Hoboken, NJ, USA, 2008.
14. Nayfeh, A.H. The response of multidegree-of-freedom systems with quadratic nonlinearities to a harmonic parametric resonance. *J. Sound Vib.* **1983**, *90*, 237–244. [CrossRef]
15. Bitar, D.; Kacem, N.; Bouhaddi, N.; Collet, M. Collective dynamics of periodic nonlinear oscillators under simultaneous parametric and external excitations. *Nonlinear Dyn.* **2015**, *82*, 749–766. [CrossRef]
16. Gutschmidt, S.; Gottlieb, O. Nonlinear dynamic behavior of a microbeam array subject to parametric actuation at low, medium and large dc-voltages. *Nonlinear Dyn.* **2012**, *67*, 1–36. [CrossRef]
17. Manktelow, K.; Leamy, M.J.; Ruzzene, M. Multiple scales analysis of wavewave interactions in a cubically nonlinear monoatomic chain. *Nonlinear Dyn.* **2011**, *63*, 193–203. [CrossRef]
18. Manktelow, K.; Narisetti, R.K.; Leamy, M.J.; Ruzzene, M. Finite element based perturbation analysis of wave propagation in nonlinear periodic structures. *Mech. Syst. Signal Process.* **2013**, *39*, 32–46. [CrossRef]
19. Manktelow, K.; Leamy, M.J.; Ruzzene, M. Weakly nonlinear wave interactions in multi-degree of freedom periodic structures. *Wave Motion* **2014**, *51*, 886–904. [CrossRef]
20. Andreassen, E.; Manktelow, K.; Ruzzene, M. Directional bending wave propagation in periodically perforated plates. *J. Sound Vib.* **2015**, *335*, 187–203. [CrossRef]
21. Romeo, F.; Rega, G. Periodic and localized solutions in chains of oscillators with softening or hardening cubic nonlinearity. *Meccanica* **2015**, *50*, 721–730. [CrossRef]
22. Marathe, A.; Chatterjee, A. Wave attenuation in nonlinear periodic structures using harmonic balance and multiple scales. *J. Sound Vib.* **2006**, *289*, 871–888. [CrossRef]

23. Narisetti, R.K.; Ruzzene, M.; Leamy, M.J. Study of wave propagation in strongly nonlinear periodic lattices using a harmonic balance approach. *Wave Motion* **2012**, *49*, 394–410. [CrossRef]
24. Marlin, J.A. Periodic motions of coupled simple pendulums with periodic disturbances. *Int. J. Nonlinear Mech.* **1968**, *3*, 439–447. [CrossRef]
25. Khomeriki, R.; Leon, J. Tristability in the pendula chain. *Phys. Rev. E* **2008**, *78*, 057202. [CrossRef] [PubMed]
26. Jallouli, A.; Kacem, N.; Bouhaddi, N. Nonlinear dynamics of a 2D array of coupled pendulums under parametric excitation. In Proceedings of the 5th ECCOMAS Thematic Conference on Computational Methods in Structural Dynamics and Earthquake Engineering, Crete Island, Greece, 25–27 May 2015; pp. 1–8.
27. Jallouli, A.; Kacem, N.; Bouhaddi, N. Stabilization of solitons in coupled nonlinear pendulums with simultaneous external and parametric excitations. *Commun. Nonlinear Sci. Numer. Simul.* **2017**, *42*, 1–11. [CrossRef]
28. Bitar, D.; Kacem, N.; Bouhaddi, N. Multi-mode solutions in a periodic array of coupled nonlinear pendulums under primary resonance. In Proceedings of the 11th International Conference on Engineering Vibration, Ljubljana, Slovenia, 7–10 September 2015; pp. 1–9.
29. Bitar, D.; Kacem, N.; Bouhaddi, N. Investigation of modal interactions and their effects on the nonlinear dynamics of a periodic coupled pendulums chain. *Int. J. Mech. Sci.* **2017**, *127*, 130–141. [CrossRef]
30. Nayfeh, A.H.; Balachandran, B. Modal interactions in dynamical and structural systems. *Appl. Mech. Rev.* **1982**, *42*, 175–202. [CrossRef]
31. Dick, A.J.; Balachandran, B.; Mote, C.D., Jr. Intrinsic Localized Modes in Microresonator Arrays and Their Relationship to Nonlinear Vibration Modes. *Nonlinear Dyn.* **2008**, *54*, 13–29. [CrossRef]
32. Kissel, G.J. Localization in Disordered Periodic Structures. Ph.D. Thesis, Massachusetts Institute of Technology, Cambridge, MA, USA, 1988.
33. Pierre, C. Weak and strong vibration localization in disordered structures: A statistical investigation. *J. Sound Vib.* **1990**, *139*, 111–132. [CrossRef]
34. Cha, P.D.; Pierre, C. A statistical investigation of the forced response of finite, nearly periodic assemblies. *J. Sound Vib.* **1997**, *203*, 158–168. [CrossRef]
35. Castanier, M.P.; Pierre, C. Modeling and Analysis of Mistuned Bladed Disk Vibration: Status and Emerging Directions. *J. Propuls. Power* **2006**, *22*, 384–396. [CrossRef]
36. Koch, R.M. Structural Dynamics of Large Space Structures Having Random Parametric Uncertainties. *Int. J. Acoust. Vib.* **2003**, *8*, 95–103. [CrossRef]
37. Zhu, H.; Ding, L.; Yin, T. Wave Propagation and Localization in a Randomly Disordered Periodic Piezoelectric Axial-Bending Coupled Beam. *Adv. Struct. Eng.* **2013**, *16*, 1513–1522. [CrossRef]
38. Tjavaras, A.A.; Triantafyllou, M.S. Non-linear response of two disordered pendula. *J. Sound Vib.* **1996**, *190*, 65–76. [CrossRef]
39. Alexeeva, N.V.; Barashenkov, I.V.; Tsironis, G.P. Impurity-Induced Stabilization of Solitons in Arrays of Parametrically Driven Nonlinear Oscillators. *Phys. Rev. Lett.* **2000**, *84*, 3053–3056. [CrossRef] [PubMed]
40. Hai-Quing, X.; Yi, T. Parametrically driven solitons in a chain of nonlinear coupled pendula with an impurity. *Chin. Phys. Lett.* **2006**, *23*, 15–44. [CrossRef]
41. Chen, W.; Hu, B.; Zhang, H. Interactions between impurities and nonlinear waves in a driven nonlinear pendulum chain. *Phys. Rev. Lett.* **2002**, *64*, 134302. [CrossRef]
42. Zhu, Y.; Chen, W.; Lu, L. Experiments on the interactions between impurities and solitary waves in lattice model. *Sci. China Ser. G* **2003**, *46*, 460–467. [CrossRef]
43. Helton, J.C.; Davis, F.J. Latin hypercube sampling and the propagation of uncertainty in analyses of complex systems. *Reliab. Eng. Syst. Saf.* **2003**, *81*, 23–69. [CrossRef]
44. Xiu, D.; Karniadakis, G.E. The Wiener-Askey polynomial chaos for stochastic differential equations. *SIAM J. Sci. Comput.* **2002**, *24*, 619–644. [CrossRef]
45. Soize, C.; Ghanem, R. Physical systems with random uncertainties: Chaos representations with arbitrary probability measure. *SIAM J. Sci. Comput.* **2004**, *26*, 395–410. [CrossRef]
46. Kevorkian, J.; Cole, J.D. *Multiple Scale and Singular Perturbation Methods*; Springer: New York, NY, USA, 1996; Volume 114, ISBN 0-387-94202-5.
47. Nayfeh, A.H. *Perturbation Methods*; Wiley–VCH: Weinheim, Germany, 2004; ISBN 0-471-39917-5.
48. Rubinstein, R.-Y. *Simulation and the Monte Carlo Methods*; John Wiley & Sons: Hoboken, NJ, USA, 1981.

49. Berveiller, M.; Sudret, B.; Lemaire, M. Stochastic finite element: A non-intrusive approach by regression. *Eur. J. Comput. Mech.* **2006**, *15*, 81–92. [CrossRef]
50. Sudret, B. Global sensitivity analysis using polynomial chaos expansions. *Reliab. Eng. Syst. Saf.* **2008**, *93*, 964–979. [CrossRef]
51. Chikhaoui, K.; Bouhaddi, N.; Kacem, N.; Guedri, M.; Soula, M. Uncertainty quantification/propagation in nonlinear models: Robust reduction—Generalized polynomial chaos. *Eng. Comput.* **2017**, *34*. [CrossRef]
52. Blatman, G.; Sudret, B. An adaptive algorithm to build up sparse polynomial chaos expansions for stochastic finite element analysis. *Probabilist. Eng. Mech.* **2010**, *25*, 183–197. [CrossRef]
53. Cochelin, B. A path-following technique via an asymptotic-numerical method. *Comput. Struct.* **1994**, *53*, 1181–1192. [CrossRef]
54. Vannucci, P.; Cochelin, B.; Damil, N.; Potier-Ferry, M. An asymptotic-numerical method to compute bifurcating branches. *Int. J. Numer. Methods Eng.* **1998**, *41*, 1365–1389. [CrossRef]
55. Kacem, N.; Baguet, S.; Hentz, S.; Dufour, R. Computational and quasi-analytical models for non-linear vibrations of resonant MEMS and NEMS sensors. *Int. J. Nonlinear Mech.* **2011**, *46*, 532–542. [CrossRef]
56. Jallouli, A.; Kacem, N.; Bourbon, G.; Le Moal, P.; Walter, V.; Lardies, J. Pull-in instability tuning in imperfect nonlinear circular microplates under electrostatic actuation. *Phys. Lett. A* **2016**, *380*, 3886–3890. [CrossRef]
57. Karkar, S.; Arquier, R.; Cochelin, B.; Vergez, C.; Thomas, O.; Lazarus, A. Manlab 2.0, An Interactive Continuation Software. Available online: http://manlab.lma.cnrs-mrs.fr (accessed on 16 March 2011).
58. Abed, I.; Kacem, N.; Bouhaddi, N.; Bouazizi, M.L. Multi-modal vibration energy harvesting approach based on nonlinear oscillator arrays under magnetic levitation. *Smart Mater. Struct.* **2016**, *25*, e025018. [CrossRef]
59. Kacem, N.; Hentz, S.; Baguet, S.; Dufour, R. Forced large amplitude periodic vibrations of non-linear Mathieu resonators for microgyroscope applications. *Int. J. Nonlinear Mech.* **2011**, *46*, 1347–1355. [CrossRef]
60. Mahmoudi, S.; Kacem, N.; Bouhaddi, N. Enhancement of the performance of a hybrid nonlinear vibration energy harvester based on piezoelectric and electromagnetic transductions. *Smart Mater. Struct.* **2014**, *23*. [CrossRef]
61. Kacem, N.; Baguet, S.; Hentz, S.; Dufour, R. Nonlinear phenomena in nanomechanical resonators: Mechanical behaviors and physical limitations. *Méc. Ind.* **2010**, *11*, 521–529. [CrossRef]
62. Bartuccelli, M.V.; Gentile, G.; Georgiou, K.V. On the Dynamics of a Vertically Driven Damped Planar Pendulum. *Proc. R. Soc. A* **2001**, *457*, 3007–3022. [CrossRef]
63. Kozinsky, H.W.C.; Postma, I.; Kogan, A.; Husain, O.; Roukes, M.L. Basins of attraction of a nonlinear nanomechanical resonator. *Phys. Rev. Lett.* **2007**, *99*, 4. [CrossRef] [PubMed]
64. Sliwa, I.; Grygiel, K. Periodic orbits, basins of attraction and chaotic beats in two coupled Kerr oscillators. *Nonlinear Dyn.* **2012**, *67*, 755–765. [CrossRef]
65. Sinou, J.-J.; Didier, J.; Faverjon, B. Stochastic nonlinear response of a flexible rotor with local nonlinearities. *Int. J. Nonlinear Mech.* **2015**, *74*, 92–99. [CrossRef]
66. Jacquelin, E.; Adhikari, S.; Sinou, J.-J.; Friswell, M.I. Polynomial chaos expansion in structural dynamics: Accelerating the convergence of the first two statistical moment sequences. *J. Sound Vib.* **2015**, *356*, 144–154. [CrossRef]
67. Chouvion, B.; Sarrouy, E. Development of error criteria for adaptive multi-element polynomial chaos approaches. *Mech. Syst. Signal Process.* **2016**, *66–67*, 201–222. [CrossRef]

applied
sciences

MDPI

Article

Member Discrete Element Method for Static and Dynamic Responses Analysis of Steel Frames with Semi-Rigid Joints

Jihong Ye [1],* and Lingling Xu [2]

[1] State Key Laboratory for Geomechanics & Deep Underground Engineering,
    China University of Mining and Technology, Xuzhou 221116, China
[2] Key Laboratory of Concrete and Pre-Stressed Concrete Structures of the Ministry of Education,
    Southeast University, Nanjing 210018, China; xllfyc@163.com
*   Correspondence: jhye@cumt.edu.cn; Tel.: +86-516-8359-0609

Academic Editor: Zheng Lu
Received: 27 May 2017; Accepted: 7 July 2017; Published: 11 July 2017

Featured Application: The MDEM can naturally deal with strong nonlinearity and discontinuity. A unified computational framework is applied for static and dynamic analyses. The method is an effective tool to investigate complex behaviors of steel frames with semi-rigid connections.

Abstract: In this paper, a simple and effective numerical approach is presented on the basis of the Member Discrete Element Method (MDEM) to investigate static and dynamic responses of steel frames with semi-rigid joints. In the MDEM, structures are discretized into a set of finite rigid particles. The motion equation of each particle is solved by the central difference method and two adjacent arbitrarily particles are connected by the contact constitutive model. The above characteristics means that the MDEM is able to naturally handle structural geometric nonlinearity and fracture. Meanwhile, the computational framework of static analysis is consistent with that of dynamic analysis, except the determination of damping. A virtual spring element with two particles but without actual mass and length is used to simulate the mechanical behaviors of semi-rigid joints. The spring element is not directly involved in the calculation, but is employed only to modify the stiffness coefficients of contact elements at the semi-rigid connections. Based on the above-mentioned concept, the modified formula of the contact element stiffness with consideration of semi-rigid connections is deduced. The Richard-Abbort four-parameter model and independent hardening model are further introduced accordingly to accurately capture the nonlinearity and hysteresis performance of semi-rigid connections. Finally, the numerical approach proposed is verified by complex behaviors of steel frames with semi-rigid connections such as geometric nonlinearity, snap-through buckling, dynamic responses and fracture. The comparison of static and dynamic responses obtained using the modified MDEM and those of the published studies illustrates that the modified MDEM can simulate the mechanical behaviors of semi-rigid connections simply and directly, and can accurately effectively capture the linear and nonlinear behaviors of semi-rigid connections under static and dynamic loading. Some conclusions, as expected, are drawn that structural bearing capacity under static loading will be overestimated if semi-rigid connections are ignored; when the frequency of dynamic load applied is close to structural fundamental frequency, hysteresis damping of nonlinear semi-rigid connections can cause energy dissipation compared to rigid and linear semi-rigid connections, thus avoiding the occurrence of resonance. Additionally, fracture analysis also indicates that semi-rigid steel frames possess more anti-collapse capacity than that with rigid steel frames.

Keywords: steel frames with semi-rigid joints; Member Discrete Element Method; nonlinear semi-rigid connection; geometric nonlinearity; snap-through buckling; dynamic response; fracture

1. Introduction

Beam-to-column joints are frequently simplified to be fully rigid or pinned connections in structural analyses and designs, while real beam-to-column joints are semi-rigid connections between the two extreme cases. Compared with rigid connections, semi-rigid connections will reduce total stiffness of steel frames and make lateral displacement increase, thereby causing a second-order effect [1]. Therefore, the rigidity of a semi-rigid joint has a significant effect on the strength and displacement response of steel frames. At present, approaches for investigating mechanical behaviors of semi-rigid joints fall into three categories: experiment, mathematical model and numerical simulation. For the first two approaches, an amount of mature research has been carried out, and a variety of linear or nonlinear semi-rigid connection models were proposed [2–6]. Furthermore, Eurccode-3 (EC3) [7] provides the M-θ curves for some specified beam-to column joints of steel structures. Based on the M-θ curves of EC3, Keulen et al. [8] introduced the half initial secant stiffness approach to simplify the full characteristics of nonlinear connection stiffness curve into a bilinear curve in practical design, and analysis results of the two cases were in good agreement. Thus, numerical simulation methods of semi-rigid joints have increasingly attracted the attentions of numerous researchers in recent years.

The finite element method (FEM) is the most commonly used method in the numerical simulation of semi-rigid joints. Lui and Chen [9], Ho and Chen [10] considered structural geometric nonlinearity by stability functions, and analyzed nonlinearity and buckling behaviors of semi-rigid beam-to-column joints in conjunction with the geometric stiffness matrix and updated Lagrangian approach. Yau and Chan [11] presented a beam-column element with springs connected in series for geometric and material nonlinear analysis of steel frames with semi-rigid connections, as well as proposed an efficient approach to trace the equilibrium path of steel frames. Zhou and Chan [12] performed second-order analysis of steel frames with semi-rigid connections by a displacement-based finite-element approach. Chui and Chan [13,14], Sophianopoulos [15] investigated the dynamic responses of steel frames with semi-rigid joints, where a semi-rigid joint was simulated by a zero-length spring element. Li et al. [16] and Rodrigues et al. [17] presented a multi-degree of freedom spring system to represent physical models of semi-rigid connections and took semi-rigid connections as new elements. This system transferred axial forces, shearing forces as well as bending moments. All the above results are based on the conventional finite-element method. However, in the static or dynamic analyses especially for strong nonlinear and discontinuous problems, the method has some limits such as difficult modeling, plenty of unknowns, long consuming time and convergence difficulty. In order to shorten consuming time, Nguyen and Kim [18,19] proposed a numerical procedure based on the beam-column method, and performed nonlinearly elastic and inelastic time-history analysis of three-dimensional semi-rigid steel frames by using the based-displacement finite element method. Compared to SAP2000(Computers and Structures, Inc., California, CA, USA), the procedure captured the second-order effect by using only one element per member, and reduced computational time. Nevertheless, it failed to model fracture behaviors of steel frames with semi-rigid joints. Lien and Chiou [20], Yu and Zhu [21] carried out an investigation on dynamic collapse of semi-rigid steel frames using vector formed intrinsic finite-element (VFIFE) method. A zero-length spring element was employed to simulate the semi-rigid joints. However, the spring element is involved in the calculation, which makes that the method has a cumbersome calculation process and long consuming time. In addition, for discontinuous issues, this method needs to regenerate a new particle at the point where fracture occurs, causing the computation time to increase again.

Different from the aforementioned methods derived from continuum mechanics, the theoretical basis of Member Discrete Element Method (MDEM) presented by Ye team [22] according to conventional Particle Discrete Element [23] is non-continuum mechanics. In the MDEM, a member is represented by a set of finite particle rather than by continuum. The motion equation of each particle is solved by the central difference method and arbitrarily adjacent particles are connected by the contact constitutive model. The difference between the MDEM and conventional DEM lies in the establishment of the contact constitutive model. Therefore, the MDEM not only extends the application range of conventional DEM, but also inherits its advantages such as no assembly of stiffness matrix and iterative solution of motion equations, no special treatment for nonlinear issues, no re-meshing for fracture problems, and no convergence difficulty. At present, the method has been successfully applied to geometric nonlinearity [24] and nonlinear dynamic analysis [25] of member structures as well as progressive collapse simulation of reticulated shells [22]. These results show that the MDEM is a simple and effective tool to process geometrically large deformation, material nonlinearity and fracture issues. However, no report that the MDEM is adopted for static and dynamic response analysis of steel frames with semi-rigid joints exists.

A simple and effective numerical approach is proposed based on the MDEM for static and dynamic response analysis of steel frames with semi-rigid joints. Brief details of the basic theory for the MDEM are first presented including particle motion equations, particle internal forces, geometric nonlinearity modelling and fracture modeling. The semi-rigid connection is simulated by a virtual spring element, and then the modified formula of the contact element stiffness adjacent to the semi-rigid connection is derived. The nonlinear behaviors of semi-rigid connections are captured by the bilinear semi-rigid connection model and Richard-Abbort four-parameter model. Furthermore, its hysteresis behavior is traced by the independent hardening model. The correctness and accuracy of the numerical procedure proposed are verified by complex behaviors analyses of steel frames with semi-rigid connections, such as geometric nonlinearity, snap-through buckling, dynamic responses and fracture.

2. Member Discrete Element Method (MDEM)

2.1. Particle Motion Equations and Internal Forces

The Member Discrete Element Method (MDEM) presented by Ye team [22] discretizes a steel frame into a set of finite rigid particles, and adjacent particles (e.g., Particles A and B) are connected by the contact element (i.e., contact constitutive model), as shown in Figure 1. The criterion for the number of particles in the MDEM is consistent with that for beam elements in FEM, which is detailed in Reference [26]. The method not only overcomes the difficulties of conventional FEM in processing strong nonlinearity and discontinuity, but also eliminates the shortcomings with low precision, long consuming time and poor applicability of conventional DEM. The calculation framework of the MDEM consists of two parts: particle motion equations and particle internal forces, which is briefly elaborated below.

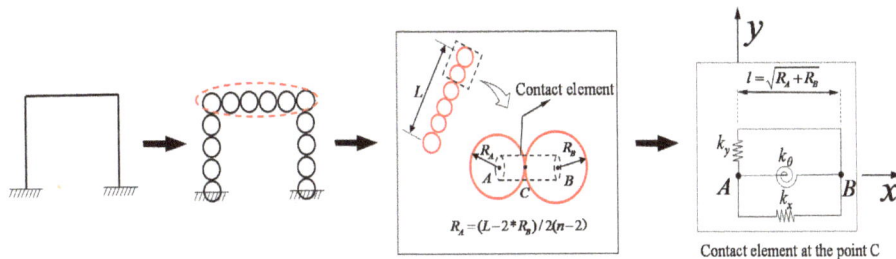

Figure 1. Discretization procedure of a plane steel frame.

2.1.1. Particle Motion Equations

The motion of each particle follows the Newton's second law at any time t. If damping term is supplemented to make structures static, the motion equations for an arbitrarily particle can be expressed as

$$M\ddot{U}(t) + C_M \dot{U}(t) = F^{ext}(t) + F^{int}(t), \tag{1}$$

$$I\ddot{\theta}(t) + C_I\dot{\theta}(t) = M^{ext}(t) + M^{int}(t), \tag{2}$$

where M and I are the equivalent mass and moment of inertia lumped at an arbitrarily particle, respectively; $\ddot{U}$ and $\ddot{\theta}$ denote translational and rotational acceleration vectors in the global coordinate system, respectively; $\dot{U}$ and $\dot{\theta}$ indicate translational and rotational velocity vectors in the global coordinate system, respectively; F^{int} and M^{int} are internal force and moment vectors acting at an arbitrarily particle, respectively; F^{ext} and M^{ext} external force and moment vectors acting at an arbitrarily particle, respectively; and C_M and C_I denote translational and rotational damping matrix, respectively. Additionally, C_M and C_I equal αM and αI in the dynamic analyses respectively while both are virtual terms under static loading. α is mass proportional damping coefficient.

If the central finite difference algorithm is employed to solve Equations (1) and (2), the translational velocity and acceleration of a particle can be approximated as

$$\dot{U}(n) = \frac{U(n+1) - U(n-1)}{2\Delta t}, \tag{3}$$

$$\ddot{U}(n) = \frac{U(n+1) - 2U(n) + U(n-1)}{\Delta t^2}, \tag{4}$$

where $U(n-1)$, $U(n)$ and $U(n+1)$ are the displacements of a particle at step $n-1$, n and $n+1$, respectively; and Δt is a constant time increment.

Substituting Equations (3) and (4) into Equation (1) yields

$$U(n+1) = \left(\frac{2\Delta t^2}{2 + \alpha\Delta t}\right)\frac{\left(F^{ext} + F^{int}\right)}{M} + \left(\frac{4}{2 + \alpha\Delta t}\right)U(n) + \left(\frac{-2 + \alpha\Delta t}{2 + \alpha\Delta t}\right)U(n-1), \tag{5}$$

Equation (5) is the formula of the translational displacements of a particle.

2.1.2. Particle Internal Forces

From the discretization procedure of a three-dimension steel frame in Figure 1, it can be found that the establishment of MDEM model is different from that of the conventional DEM model. In the MDEM, the radius of the particles at both ends (R_B) of all the members is first given, and then the radii of the particles in the members (R_A) are determined via equally dividing the remaining lengths of these members. The internal forces arising at the contact points (e.g., Point C) between connected particles (e.g., Particles A and B) need be first calculated, and then they are exerted reversely to these particles (e.g., Particles A and B), thereby acquiring internal forces of an arbitrary particle in Figure 1. The internal force and moment increment at a contact point can be obtained by contact constitutive model, which are

$$\begin{cases} \Delta f(n) = k\Delta u(n) \\ \Delta m(n) = k_\theta \Delta \theta'(n) \end{cases}, \tag{6}$$

where Δf and Δm are the incremental internal force and moment of inertia arising at a contact point, respectively; k and k_θ indicate translational and rotational contact stiffness coefficients, respectively; and Δu and $\Delta \theta'$ denote pure translational and rotational displacements of a contact point in a time step Δt respectively, which are in the local coordinate system.

The key point of the DEM is the determination of contact stiffness coefficients (i.e., k and k_θ) in Equation (6). The coefficients are obtained mainly from experiences or experiments for the conventional DEM, while the MDEM gives the formulas of contact stiffness coefficients. Prof. Ye team [22] supplemented rotational springs into contact constitutive model (Figure 1), which make that the force state of each contact element (i.e., equivalent beam AB in Figure 1) is the same as that of the spatial beam element with three translational and three rotational degrees of freedoms in the FEM.

According to the strain energy equivalent principle and simple beam theory, the formulas of contact stiffness coefficients for three-dimension member structures can be expressed as [24]

$$
k = \begin{bmatrix} EA/l & & \\ & 12EI_y/l^3 & \\ & & 12EI_z/l^3 \end{bmatrix} \qquad k_\theta = \begin{bmatrix} GI_x/l & & \\ & EI_y/l & \\ & & EI_z/l \end{bmatrix}, \qquad (7)
$$

where l is the distance between the two particle centers; A, E and G denote the area, elastic modulus and modulus of shearing of each member, respectively; and I_x, I_y and I_z are the cross sectional moments of inertia of a member about the x, y and z axes, respectively.

2.2. MDEM for Modelling Geometric Nonlinearity

Structural geometric nonlinearity comprises rigid body motion, large rotation and large deformation. For the FEM, special treatments and complex modifications are necessary to deal with problems of this kind, such as the arc-length method and co-rotational coordinate approach [27]. However, in the MFEM, particles are assumed to be rigid, and thus rigid body motion of particles can be removed naturally in the process of solving the pure displacement of contact points using the rigid body kinematics. Furthermore, the movement of each particle is represented numerically by a time-stepping algorithm. In the algorithm, the assumption of small deformation is satisfied within each time step, which makes that large deformation and large rotation issues are naturally handled [28]. Thus, the unified procedure is employed in the MDEM for geometric linear and nonlinear problems.

2.3. MDEM for Modeling Facture Behavior

Fracture behavior possesses strong discontinuity. In the FEM, fracture is modeled by defining birth-death elements or failure elements, which may cause the non-conservation of mass and convergence difficulty [26]. Nevertheless, in the MDEM, a structure is represented by a set of finite rigid particle rather than by continuum, and each particle is in a dynamic equilibrium. Given an appropriate fracture criterion only, Fracture behavior can be simulated [22]. The fracture is macroscopically represented by the separate of two adjacent particles in Figure 1. The Bending strain is take as the fracture criteria in this study, which is denoted as

$$
|\varepsilon_{\theta t}| \geq \varepsilon_{\theta p}, \qquad (8)
$$

where $\varepsilon_{\theta t}$ and $\varepsilon_{\theta p}$ are the bending strain at time t and ultimate bending strain, respectively. The ultimate bending strain is set to be 0.0004 according to relative experiments and studies [29].

3. Member Discrete Element Modeling for Semi-Rigid Connections

3.1. Virtual Zero-Length Spring Element

In this study, a semi-rigid joint is simulated by a multi-degree-of-freedom spring element with two particles but without actual mass and length. The basic concept of the spring element is similar to that of References [13,18]. The moment-rotation relationship of the zero-length spring element is expressed as a diagonal tangent stiffness matrix in other numerical approaches taking the FEM as a representation. In addition, the stiffness matrix needs to be integrated with structural total stiffness matrix. However, in the MDEM, the zero-length spring element is not directly involved in the calculation, but is employed only to modify the stiffness coefficient of each contact element adjacent to the semi-rigid connection. In addition, then the modified stiffness coefficient of the contact element is directly applied in the next time step. The whole procedure is simple and effective. Furthermore, computational time of the MDEM after considering semi-rigid connections has no increase. This is not only the nature difference between the zero-length spring element of the MDEM and that of other methods, but also the advantage of the MDEM processing semi-rigid connection problem.

The MDEM model of a simple frame is displayed in Figure 2. If the joint D (i.e., particle D) is a semi-rigid connection, the processing mode of the semi-rigid connection is given in Figure 3, that is, a zero-length spring element is supplemented at the particle. The spring element contains two particles E and F with same coordinates, and its translational and rotational stiffness matrix can be expressed as

$$k_s = \begin{bmatrix} R_x & & \\ & R_y & \\ & & R_z \end{bmatrix} \qquad k_{s\theta} = \begin{bmatrix} R_{\theta x} & & \\ & R_{\theta y} & \\ & & R_{\theta z} \end{bmatrix}, \tag{9}$$

where R_n and $R_{\theta n}$ are the translational and rotational stiffness vectors of a zero-length spring element with respect to n axis ($n = x, y, z$). In addition, for a two-dimensional problem, the zero-length spring element can degrade to three degrees of freedom which is two translational degrees of freedom and one rotational degree of freedom.

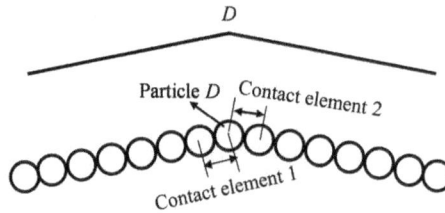

Figure 2. MDEM (Member Discrete Element Method) model of a simple steel frame.

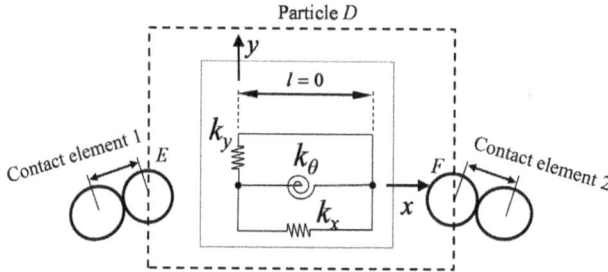

Figure 3. Zero-length spring element at a semi-rigid connection.

The comparison of Figures 1 and 3 shows that the force state of the zero-length spring element is identical with that of the contact element, except that the former is dummy while the length of the latter is related to connected particles. This characteristic makes that the MDEM is able to simulate a semi-rigid connection by directly modifying the contact element stiffness.

Only the bending stiffness of a semi-rigid joint is frequently taken into account in steel frames. The relationship between the moment and the relative rotation of a semi-rigid connection can be expressed as follow:

$$\Delta M = R_\theta \Delta \theta_r, \tag{10}$$

where R_θ is the linear or nonlinear stiffness of a semi-rigid connection (i.e., zero-length spring element stiffness), which is obtained by experiments or mathematic models; $\Delta \theta_r$ denotes the relative relation between particles E and F.

Taking the semi-rigid connection of this type in Figure 2 as an example, the derivation detail of the rotational stiffness coefficient of the contact element 1 is given. When the joint D is rigid, the rotational stiffness of the members 1 and 2 are $K_{\theta 1} = E_1 I_1 / L_1$ and $K_{\theta 2} = E_2 I_2 / L_2$, respectively, L_1 and L_2 are the lengths of the members 1 and 2. When the joint D is semi-rigid, the semi-rigid connection stiffness is assumed to be linearly distributed to the contact elements 1 and 2 in accordance with the stiffness of the members 1 and 2. Therefore, the contributions of the semi-rigid connection to the contact elements 1 and 2 are $K_{\theta 2} \cdot R_{\theta}/(K_{\theta 1} + K_{\theta 2})$ and $K_{\theta 1} \cdot R_{\theta}/(K_{\theta 1} + K_{\theta 2})$, respectively, denoted m_1 and m_2. Assuming that two particles of the contact element 1 have a relative rotation $\Delta\theta$, the deformation energy caused by the relative rotation is

$$J_c^{\theta} = \frac{1}{2}k_{\theta 1}'(\Delta\theta)^2 + \frac{1}{2}m_1(\Delta\theta)^2, \tag{11}$$

where $k_{\theta 1}'$ is the modified rotational stiffness coefficient of a contact element.

$\Delta\theta$ is so small in a time step that the length of the contact element 1can still be considered equal to the distance of two particle centers. The strain energy produced by pure bending deformation of the equivalent beam composed of two particles is expressed as

$$J_b^{\theta} = \int_0^L \frac{1}{2}E_1 I_1 \kappa^2 dx = \int_0^L \frac{1}{2}E_1 I_1 \left(\frac{\Delta\theta}{l_1}\right)^2 dx = \frac{1}{2}\frac{E_1 I_1}{l_1}(\Delta\theta)^2, \tag{12}$$

When the joint D is semi-rigid, the rotational stiffness coefficient of the contact element 1 is obtained according to the equivalent principle of strain energy (i.e., $J_c^{\theta} = J_b^{\theta}$), which is

$$k_{\theta 1}' = \frac{E_1 I_1}{l_1} - m_1 = k_{\theta 1} - m_1, \tag{13}$$

where $k_{\theta 1}$ is the rotational stiffness coefficient of the contact element 1 when the joint D is rigid.

The rotational stiffness coefficient of the contact element 2 can be obtained in the same way. In the MDEM, Equation (13) is the general formula for modifying rotational stiffness of the contact elements at the semi-rigid connections. For three most commonly used semi-rigid connection types for steel structures, specific formulas for modifying rotational stiffness are listed in Table 1.

Table 1. Specific formulas for modifying rotational stiffness for three semi-rigid connection types.

Semi-Rigid Connection Types	Configuration	Formulas
Elastic support		$k_{\theta}' = k_{\theta} - 1/2R_{\theta}$
Couple-bar joint		$k_{\theta}' = k_{\theta} - 1/2R_{\theta}$ (for two same bars)
Beam-column joint		$k_{\theta}' = R_{\theta}$ (for semi-rigid connections on the beam)

3.2. Semi-Rigid Connection Models

Equation (10) is the general formula of semi-rigid connection models. When $R_{s\theta}$ in the formula is a variable, the semi-rigid connection model is nonlinear. Two different models are employed to capture the nonlinear behaviors of semi-rigid connections in this study. The first one is the bilinear semi-rigid connection model presented by Keulen et al. [8]. The model is adopted in the classic examples to verify the correctness of the approach proposed. However, the accuracy of the model can only meet engineering practice. In order to more accurately capture the nonlinear behaviors and energy dissipation of semi-rigid connections, a number of nonlinear semi-rigid models have been developed. Of these, the most representatives are the Kishi–Chen three-parameter power model [4], the Richard–Abbott four-parameter model [5], the Chen–Lui exponential model [9], and the Ramberg–Osgood model [30]. The results of Chan and Chui [14] as well as Nguyen and Kim [19] indicate that the effect of different models on the behavior of steel frames is not significant. Therefore, only the Richard–Abbott four-parameter model [5] is applied in this study, the formula of the model can be expressed as

$$M = \frac{(R_{ki} - R_{kp})}{\left[1 + \left(\frac{(R_{ki} - R_{kp})|\theta_r|}{M_0}\right)^n\right]^{1/n}} + R_{kp}|\theta_r|, \tag{14}$$

where M and θ_r are the moment and rotation of the semi-rigid connection, respectively; R_{ki}, R_{kp} and M_0 are the initial stiffness, the strain-hardening stiffness and the reference moment of the semi-rigid connection, respectively, which are acquired by experiments [5]; n is the parameter defining the shape.

3.3. Cyclic Behavior Modelling of Semi-Rigid Connections

The independent hardening model shown in Figure 4 is commonly used to model the cyclic behavior of semi-rigid connections under cyclic loading because of its simple implementation [13,18,21], which is also employed in this study. The virgin moment-rotation curve is shaped by the connection model in Equation (14). The instantaneous tangent stiffness is defined by taking derivative of Equation (14), which is expressed as follow:

$$R_{kt} = (dM/d\theta_r)_t = (R_{ki} - R_{kp})/\left[1 + (R_{ki} - R_{kp}) \cdot |\theta_r|/M_0)^n\right]^{1/n} + R_{kp}, \tag{15}$$

The specific loading-unloading criteria of the model are detailed in [13,18,21].

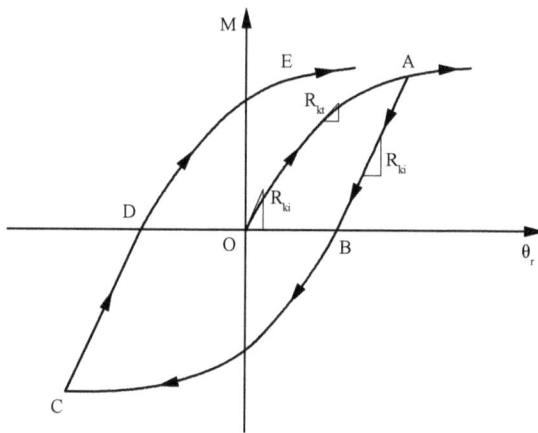

Figure 4. Independent hardening model.

3.4. Computational Procedures for Static and Dynamic Analysis of Steel Frames with Semi-Rigid Joints

The computational procedure of the MDEM is concise and straight-forward. From the particle motion equations (i.e., Equations (1) and (2)), it can be found that each particle is in dynamic equilibrium, and the difference between static analysis and dynamic analysis lies in the determination of damping. Therefore, the computational framework of static analysis is consistent with that of dynamic analysis, and the unified MDEM flowchart for static and dynamic analysis of steel frames with semi-rigid joints is displayed in Figure 5. Additionally, all examples in this study were computed by FORTRAN language. In addition, the computational hardware environment is 4.0 GHz intel8-core CPU, 16 G RAM and ST2000DM001-1ER164 hard disk.

Start

1. Input initial structural parameters, material properties and so on
2. Assemble element stiffness matrix and lumped mass matrix
3. Calculate contact stiffness coefficients between two adjacent particles according to Eq.(7)

Consider semi-rigid connections ?　　Yes

No

1. Solve motion equations of particles (i.e. Eqs.(1) and (2) at time t=0
2. Store particle displacement, velocity and acceleration, move to next time step $t+\Delta t$

Modify the stiffness coefficients of contact elements at the semi-rigid connections according Eq.(13)

Update particle coordinate, calculate the pure deformation (i.e. Δu and $\Delta \theta$) of contact points by using the rigid body kinematics

Are semi-rigid connections nonlinear ?　　Yes

No

Calculate the internal forces of contact element according to Eq.(6) (i.e. contact constitutive model)

1. Calculate the instantaneous tangent stiffness of semi-rigid connections by Eq.(15), and obtain the modified contact element stiffness at time $t+\Delta t$ by substituting it into Eq.(13)
2. Determine the force state of each semi-rigid connection according to independent hardening model in section 2.3

No

Reach the fracture criteria (Eq.(8) ?　　Yes　　The internal forces of the contact element at the fracture are set to be zero

No

1. Calculate particle internal forces by exerted reversely internal forces of the contact element to connected particles
2. Substitute particle internal forces into Eqs.(1) and (2), and store particle displacement, velocity and acceleration at the time step $t+\Delta t$

No

$t > T$

End

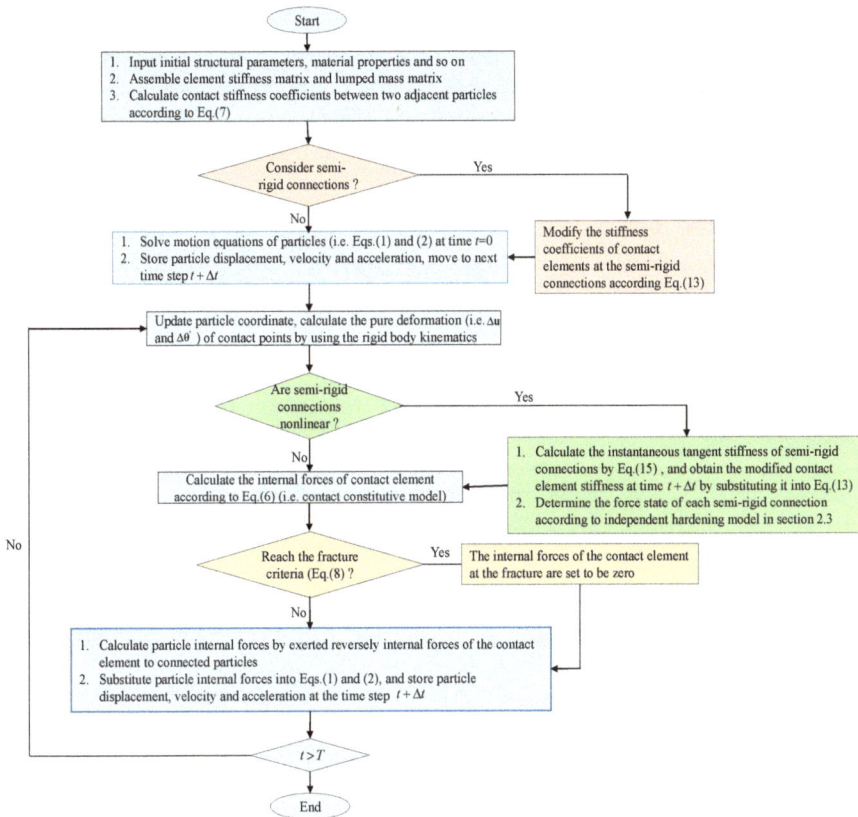

Figure 5. MDEM flowchart for static and dynamic analysis of steel frames with semi-rigid connections.

4. Examples

The applicability and accuracy of the procedure proposed are verified by various structural behaviors such as geometric nonlinearity, snap-through buckling, dynamic response and fracture. Of these, the beam-column joint is assumed to be linear (e.g., in Examples 3.1–3.4), bilinear (e.g., in Examples 3.5) or nonlinear semi-rigid connection (e.g., in Examples 3.6) accordingly. The geometrically nonlinear analysis of a column with elastic support under static loading is performed in Example 3.1 to preliminarily test the correctness of the proposed procedure. The static analysis of a beam with elastic supports is first carried out in Example 3.2, and then dynamic response of this structure is supplemented to emphasize the accuracy of the proposed procedure under dynamic loading. Example 3.3 further employed the proposed procedure into the static and dynamic response of steel frames with linear semi-rigid connections to validate the applicability of the proposed procedure in steel frames. In Example 3.4, the snap-through buckling behavior of steel frames is investigated in detail to reflect the advantages of the proposed procedure. Different from the above examples, Example 3.5 introduces the bilinear semi-rigid connection model proposed by Keulen et al. [8] into the MDEM for the static analysis of single-story and three-story frames with semi-rigid connections. In order to capture the nonlinear behavior and hysteresis performance of semi-rigid connections more accurately, the Example 3.6 adopts the Richard-Abbort four-parameter model [5] given in Section 2.2 for dynamic analysis of the double-span, six-story Vogel steel frame under cyclic loading. In addition, the effect of semi-rigid connections on the collapse behavior of the structure is studied.

4.1. Geometrically Nonlinear Analysis of a Column with Elastic Support

A column with elastic support is frequently used as a benchmark problem to verify the correctness of the approaches modelling semi-rigid connections. Geometrically nonlinear analysis of this structure has been performed by many researchers (Zhou and Chan [12]; Lien et al. [20]) using different methods. As shown in Figure 6, the properties of the column are listed below: cross section A = 0.1 m × 0.1 m, length L = 3.2 m, and elastic modulus E = 210 GPa. An axial concentrated load p was applied to the column top (i.e., the free end), meanwhile a lateral load $0.01p$ was added in the consideration of geometric imperfection. The material was assumed to be perfectly elastic. The support in the column bottom was set as rigid and semi-rigid in the rotational direction, respectively. The rotation-resistant stiffness k of the semi-rigid support is $10EI/L$.

In the MDEM, the column was discretized into 11 particles (i.e., 10 elements). When the support is semi-rigid, the stiffness of the bottom contact element in the rotational direction is determined to be $EI/l - 5EI/L$ according to Table 1, where l equals the distance between the centers of two particles of the contact element. Figure 6 indicates that the results obtained using the modified MDEM agree well with the existing results (Zhou and Chan [12]; Lien et al. [20]) in the two cases of fixed end and semi-rigid supports. That also illustrates that the modified MDEM can accurately and effectively deal with the geometrically nonlinearity of a column with elastic support.

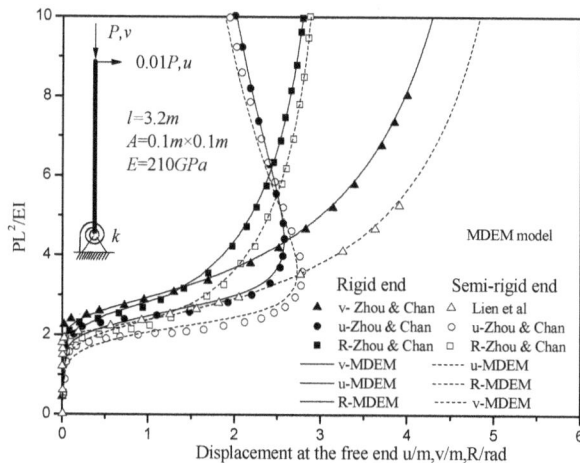

Figure 6. Load-displacement curves at the free end of the column.

4.2. Static and Dynamic Response of a Beam with Elastic Ends

The schematic configuration of a beam with elastic ends is displayed in Figure 7. The properties of the beam are listed as follow: cross section A = 25.4 mm × 3.175 mm, length L = 508 mm, and elastic modulus E = 207 GPa. A gradually increasing concentrated load P is applied at the mid-span point. Mondkar and Powell [31] as well as Yang and Saigal [32] analyzed the static and dynamic response of the beam with fixed ends. The static and dynamic response of the beam with semi-rigid ends was investigated by Chui and Chan [13] and Lien et al. [20] using the conventional FEM and the VFIFE, respectively. The rotation-resistant stiffness k of the semi-rigid ends is EI/L.

In this study, the beam was represented by 51 particles (i.e., 50 contact elements). The ends were modeled as rigid and semi-rigid in the rotational direction, respectively. The semi-rigid behavior was simulated by the contact elements at the ends with the stiffness in the rotational direction of $49EI/L$ which was obtained according to Table 1. Figure 7 shows that the load-displacement curves of the modified MDEM are consistent with the published results (Mondkar and Powell [29]; Chui and Chan [13]; Lien et al. [21]); the stiffness of the ends has a significant effect on structural stiffness, and the stiffness of the beam with semi-rigid ends is obviously reduced. In addition, the stiffness of the beam increases with an increase of the load, the reason is that the tensile reaction at the supports induces the axial tension arising in the beam.

The dynamic response analysis of the same beam was further performed. A sudden load was applied at the mid-span point, that is, the load p was linearly increased to p_{max} (640 lb) within a very short time (10 μs), and then the p_{max} is maintained until the calculation ended ($t = 5$s), as shown in Figure 8. During the analysis, the material was assumed to be perfectly elastic, and the viscous damping was ignored. It can be found from Figure 8 that the analysis results of the modified MDEM is identical with published results, and the minor error is due to the inevitable numerical method itself. Moreover, the stiffness of the ends affects significantly on the period and amplitude of displacement time history curves. The structural stiffness decreases with semi-rigid supports, whereas both the period and amplitude increase accordingly.

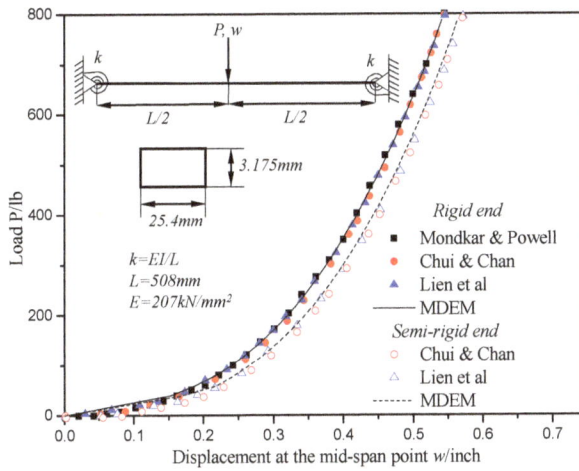

Figure 7. Displacement at the mid-span point of the beam under static loading.

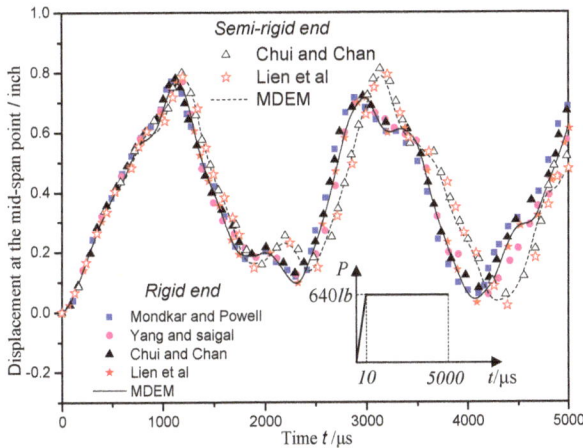

Figure 8. Displacement time history curves at the mid-span point of the beam under sudden loading.

4.3. Static and Dynamic Response Analysis of Steel Frames with Linear Semi-Rigid Connections

Figure 9 displays the configuration and loading of a single-bay two-story steel frame with W14 × 48 steel beams and W12 × 96 steel columns. The elastic modulus is 200 GPa. A vertical concentrated load P was applied to four joints, meanwhile a horizontal load $0.01p$ was added to consider the geometric imperfection or wind load. Several studies (Lui and Chen [9], Zhou and Chan [12]; Lien et al. [20]) carried out stability analysis of the structure in the two cases of rigid and semi-rigid connections using different simulation methods. In the above analysis, the bending-resistant stiffness of semi-rigid beam-column joints was taken to be 3.48×10^7 kN-m/rad.

The MDEM discretized the single-bay two-story steel frame into 28 particles (i.e., 28 contact elements). Of these, each column was represented by 4 contact elements while every beam by 6 contact elements. The bending-resistant stiffness of the contact elements at the beam ends was modified to be 3.48×10^7 kN-m/rad according to Table 1. The horizontal displacement of steel frames with linear semi-rigid joints is shown in Figure 9. The figure shows that the load-displacement relationship of the modified MDEM is in a good agree with the published results. Additionally, the effect of semi-rigid connections on the structural behavior is great, and structural loading capacity may be overestimated if semi-rigid connections are ignored.

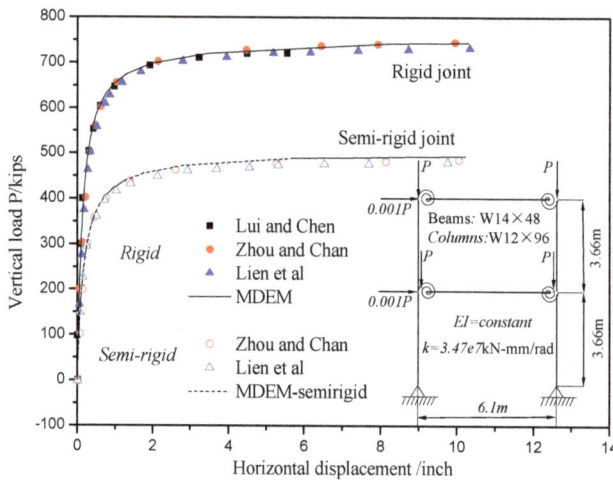

Figure 9. Horizontal displacement of steel frames with linear semi-rigid joints.

Another single-bay two-story steel frame with linear semi-rigid joints was adopted to analyze the dynamic response. The configuration and loading of this structure are shown in Figure 10. All the beams and columns are W8 × 48 and the Young's modulus is 205 GPa. The material of all the frame members was assumed to be perfectly elastic and the viscous damping was ignored. The initial geometric imperfection ψ was taken to be 1/438 rad. The lumped masses of 5.1 T and 10.2 T were set at the top of the columns and in the middle of the beams, respectively. The stiffness of semi-rigid connections was 23,000 kN-m/rad. In this study, each beam was simulated by 9 particles (i.e., 8 contact elements), while every column was discretized into 4 particles (3 contact elements). As the stiffness of each frame member is same, the bending-resistant stiffness of the contact elements at the semi-rigid joints can be determined to be 11,500 kN-m/rad by Equation (13) not Table 1. The vertical static loads of 50 KN and 100 KN were first applied on the beam-column joints and in the middle of beams to consider the second-order effects, and then the horizontal forces were added suddenly at each floor during 0.5 s until the calculation ended. The horizontal displacement on the top is shown in Figure 11. The comparison indicates the displacement responses of the MDEM and existing results (Chan and

Chui [13], Nguyen and Kim [18]) are consistent, and the influence of semi-rigid connections on the periods and the amplitudes of displacement responses of steel frames is more significant as well as the amplitude increase is up to 60%.

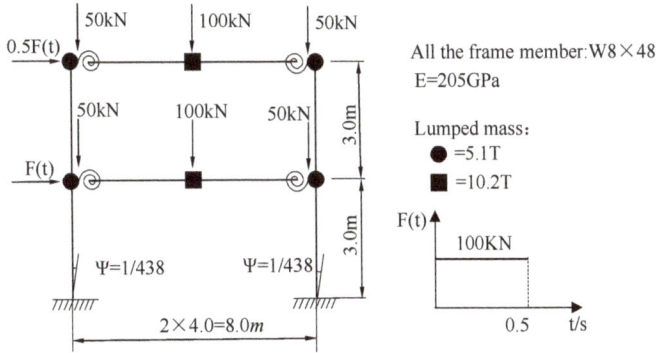

Figure 10. The configuration and loading of the structure adopted in dynamic analysis.

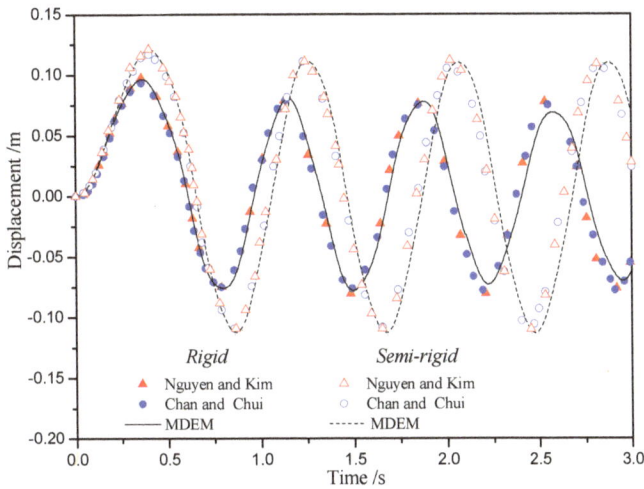

Figure 11. Displacement time history curves of a steel frame with linear semi-rigid connections.

4.4. Snap-Through Buckling Analysis of the Williams Toggle Frame with Linear Semi-Rigid Connections and Supports

The Williams Toggle frame is frequently taken as a typical example for snap-through buckling analysis. The structure has the obviously nonlinear bifurcation instability, that is, each bifurcation point may have one or more equilibrium paths and each path represents the extreme point instability or snap instability [12,33]. Many researchers (Zhou and Chan [12], Lien et al. [20]) performed static stability analysis of the steel frame with linear semi-rigid connections and supports, and the bending-resistant stiffness of each semi-rigid connection or support k is 10EI/L. The properties of the Williams Toggle frame is as follow: cross section $A = 19.13$ mm $\times$ 6.17 mm; the Young's modulus $E = 71$ GPa.

Negative or zero structural stiffness may occur at the instant of buckling, resulting that special treatments for conventional FEM is necessary to investigate snap-through buckling behavior. In the FEM, the most commonly used approach is the arc-length method [26]. However, in the MDEM, no negative or zero stiffness issues occur because of no assembly of stiffness matrix, and only displacement control method need be applied to trace the complete equilibrium path. In this study, each member was discretized into 13 particles (i.e., 12 contact elements), and the applied displacement increment was $\Delta u = 0.001$ mm. The bending-resistant stiffness of the contact elements at the semi-rigid joints was determined to be 7EI/L by Table 1. Snap-through buckling analyses of the MDEM and the existing results (Yang and Chiou [31]; Experimental result [34]; Zhou and Chan [12]; Lien et al. [20]) are displayed in Figure 12. The comparison shows that the modified MDEM can accurately carried out the snap-through buckling analysis of the Williams Toggle frame with semi-rigid connections and supports. Furthermore, semi-rigid connections and support have a significant effect on the critical load and post-buckling behavior.

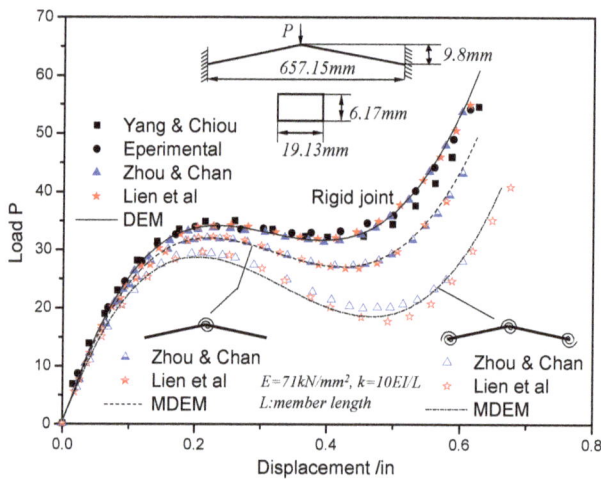

Figure 12. Displacement response of the Williams Toggle frame with linear semi-rigid connections and supports.

4.5. Static Analysis of a Steel Portal Frame with Bilinear Semi-Rigid Connections

The bilinear semi-rigid connection model presented by Keulen et al. [8] is adopted to simulate the nonlinear behavior of semi-rigid connections. Figure 13 shows the detail configuration of a steel portal frame. The beam-column joints are bolted by flush endplates, which are obviously nonlinear semi-rigid connections. Keulen et al. [8] carried out the second-order elastic-plastic analysis using the ANSYS software (ANSYS, Inc., Pittsburgh, PA, USA). Del Savio et al. [35] and Lien et al. [20] were further performed static analysis using different numerical methods.

The steel portal frame was made of S335 steel, IPE360 and HEA260 sections [8]. The span and height were 7.2 m and 3.6 m, respectively. A concentrated load F was applied at the joints and in the beam, as shown in Figure 13. The horizontal loads was set to be αF to consider the initial geometrical imperfection and wind load, where α was taken to be 0.1, 0.15, 0.2, 0.3 and 0.5 successively. The semi-rigid supports was assumed to be perfectly elastic-plastic, its initial stiffness is 23,000 kN-m/rad, and the moment capacity was 200 kN-m. The moment-rotation relation of beam-column joints, proposed by Keulen et al. [8], is shown in Figure 14. The joints were simulated using the full characteristic nonlinear model and a bilinear model. The connection model applied in this study is the bilinear model in Figure 14.

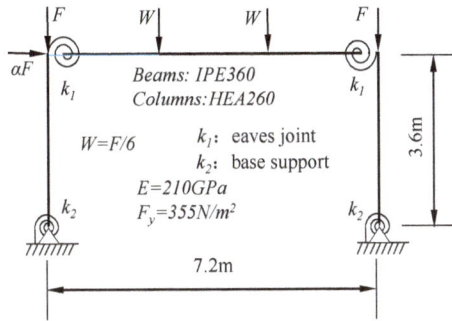

Figure 13. A steel portal frame with semi-rigid connections and supports.

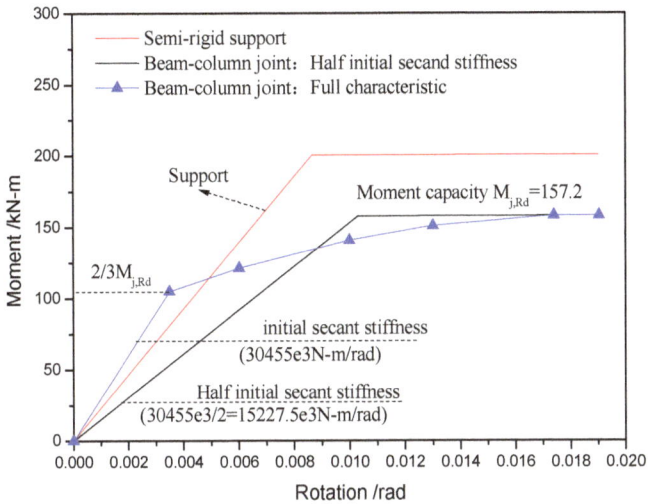

Figure 14. Load-moment curves.

In this study, there were 36 particles and 36 contact elements for this structure, and the radius of each particle was 0.2 m. The bending-resistant stiffness of the contact elements at the semi-rigid joints was obtained by Equation (13), while that of the contact elements at the supports could be determined according to Table 1. The comparison given in Figure 15 shows that the modified MDEM can accurately simulate the nonlinear behavior of semi-rigid connections and supports of a single-story steel portal frame. Moreover, the connection behavior obtained using simplified half initial secant stiffness approach has a good agreement with the full characteristic analysis. Taking $\alpha = 0.2$ as an example, the load-moment curves of joints and supports are shown in Figure 16. It can be found from the figure that the semi-rigid beam-column joint on the right top first reached the plastic state, and a plastic hinge formed.

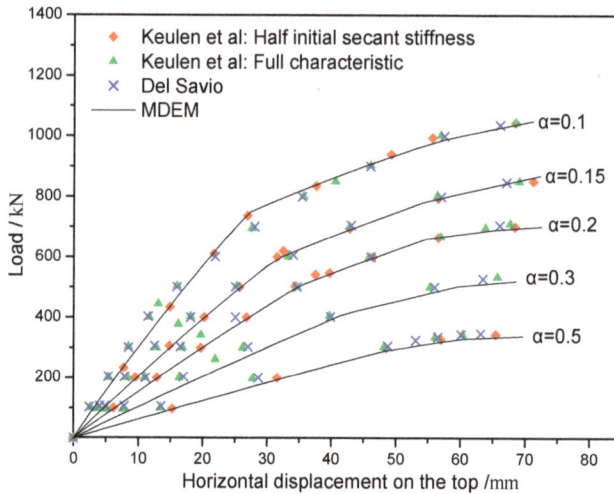

Figure 15. Displacement response on the top of the steel frame.

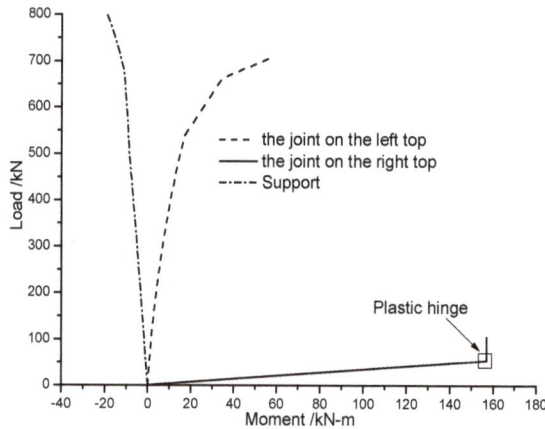

Figure 16. Load-moment curves of joints and supports.

The static analysis of a three-story steel portal frame was further performed, as shown in Figure 17. The nonlinear behavior of the beam-column joint was simulated using the bilinear model simplified by the half initial secant stiffness approach. The procedure of determining the bending-resistant stiffness of contact elements at the semi-rigid connection was the same as the single-story frame. The horizontal load was set to 0.15*F*. Figure 18 shows the moment-rotation relation of the beam-column joint. The analysis result of the MDEM and the published results (Keulen et al. [8] and Del Savio et al. [33]) are displayed in Figure 19. The figure indicates that the modified MDEM possesses considerable accuracy for the nonlinear behavior simulation of the semi-rigid connection of multi-story steel portal frames. Additionally, from the load-moment curves on the right joints and supports (Figure 20), it is found that for three-story steel portal frame with semi-rigid joints and supports, the beam-column joint on the right side of first floor reached the plastic state and a plastic hinge formed. When the load continues to increase to 336 kN and 403 kN accordingly, the plastic hinge in turn occurred at the beam-column joints on the right side of second floor and third floor.

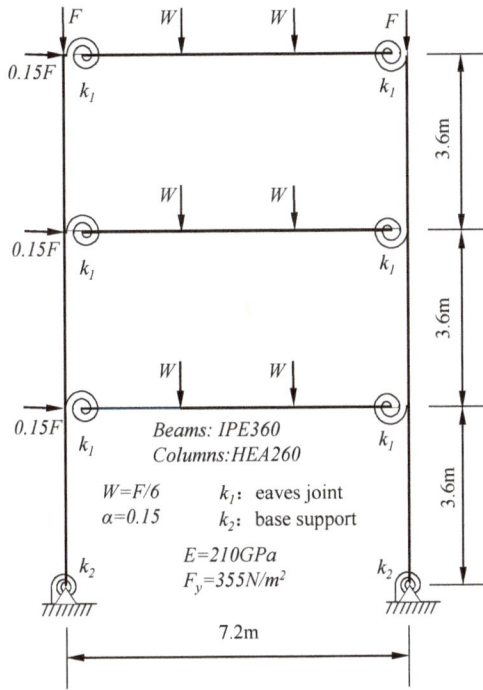

Figure 17. A single-span three-story steel portal frame with semi-rigid connections and supports.

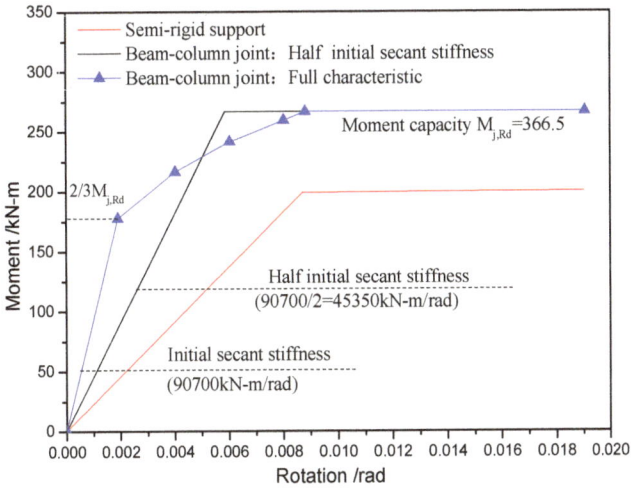

Figure 18. Load-moment curves.

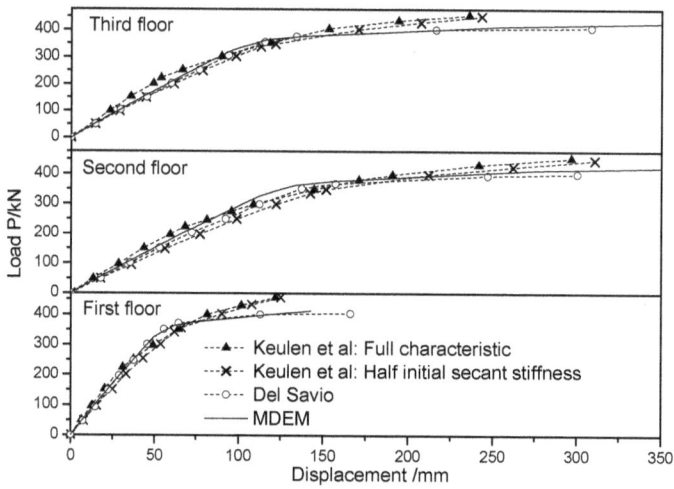

Figure 19. Load-Displacement curves at each floor of the steel frame.

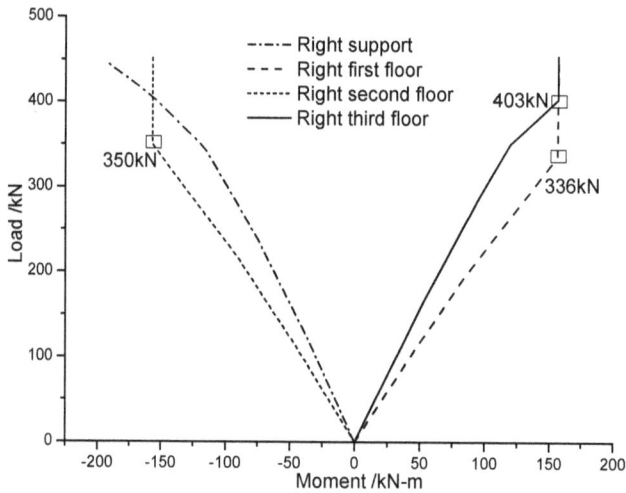

Figure 20. Load-Moment curves at the right joints and support.

4.6. Dynamic Analysis of the Two-Span, Six-Story Vogel Steel Frame with Nonlinear Semi-Rigid Connections

Figure 21 shows the configuration and loading of the two-span, six-story Vogel steel frame. Each span and each floor height were 6 m and 3.75 m, respectively. The section types of beams were IPE240, IPE300, IPE360 and IPE400, while the section types of columns were HEB160, HEB200, HEB220, HEB240 and HEB260. An initial geometric imperfection of 1/450 was taken into account. The distributed loads of 31.7 kN/m and 49.1 kN/m applied on the beams were converted into lumped masses at the joints. The elastic modulus and the Poisson ratio were 205 GPa and 0.3. In order to more intuitively study the energy consumption of semi-rigid connections, the material of all the frame members was assumed to be perfectly elastic, and the viscous damping was ignored.

In this study, there were 10 contact elements for each column while 16 contact elements for every beam. The nonlinear behavior of the semi-rigid connection was simulated by the Richard-Abbott model. The parameters of the model were as follow: $R_{ki} = 12336.86\,\text{kN} \cdot \text{m/rad}$; $R_{kp} = 112.97\,\text{kN} \cdot \text{m/rad}$; $M_0 = 180\,\text{kN} \cdot \text{m}$; $n = 1.6$. The instantaneous stiffness of the contact elements at semi-rigid connections was determined by Equations (13) and (15). The frequency of the dynamic force ω is 1.66 rad/s, which was close to the fundamental natural frequency of this structure. The dynamic response analysis of the frame was performed at three cases of rigid connection, linear semi-rigid connection and nonlinear semi-rigid connection, as shown in Figure 22. The comparison of the numerical results obtained using the modified MDEM and those presented by Chui and Chan [13] illustrates that the two match well, and the modified MDEM can accurately capture the nonlinear behavior of the semi-rigid connection. In addition, it is observed that resonance occurs in the frames with rigid or linear semi-rigid connections, but not in the frame with nonlinear semi-rigid connections, the reason is that the hysteresis damping of the nonlinear connection causes energy dissipation (Figure 23).

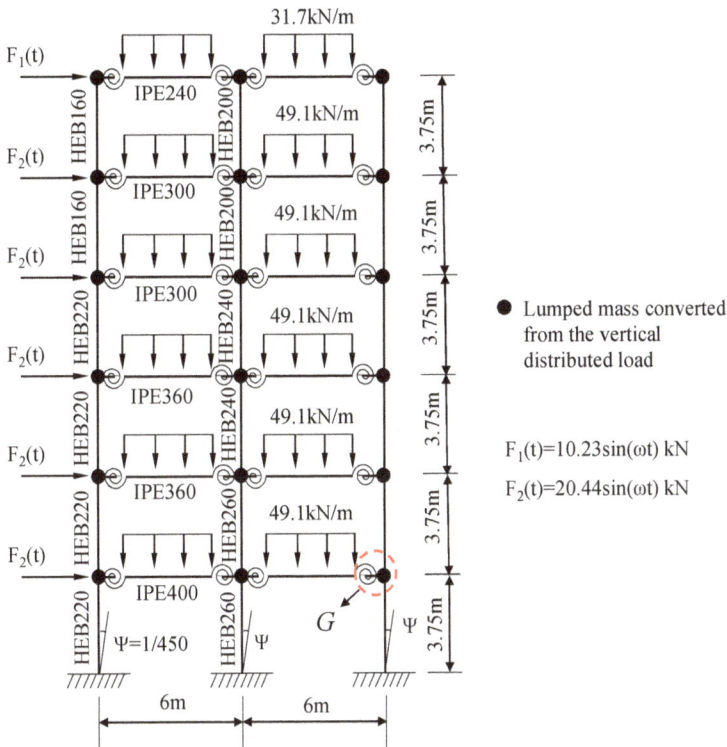

Figure 21. Configuration and loading of the two-span, six-story Vogel steel frame.

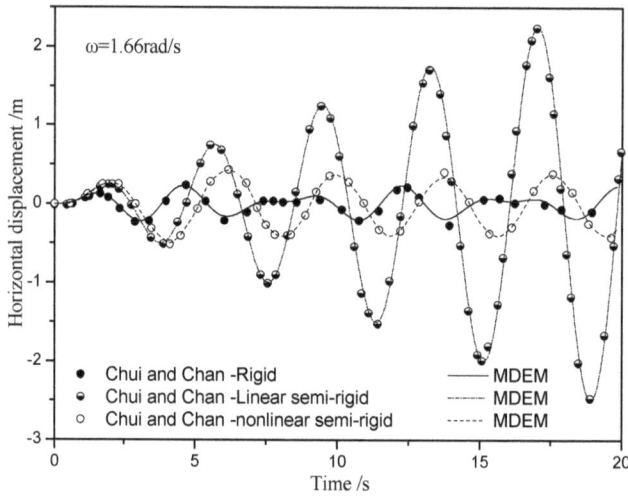

Figure 22. Horizontal displacement response of the top joint.

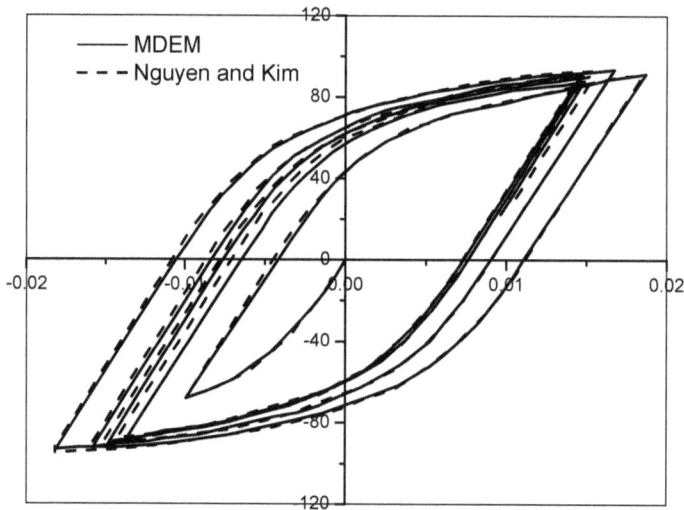

Figure 23. Hysteresis loops at the semi-rigid connection G.

In order to simulate the collapse behavior of this structure, the ultimate strain was taken as the fracture criterion. The material was elastic during the calculation, and the beam-column joints were assumed to be linear semi-rigid connections. The sine wave with the frequency of 1.66 rad/s was applied. The failure processes of steel frames with different connections are shown in Figures 24 and 25. It can be seen from the figures that the displacement at the top of the frame with semi-rigid connection is obviously larger than that of the frame with rigid connection, and the fracture time of the frame with rigid connection is earlier. It also indicates that the ductility of the semi-rigid connection is better than that of rigid connection. Steel frames with semi-rigid connections have more anti-collapse ability than those with rigid connection.

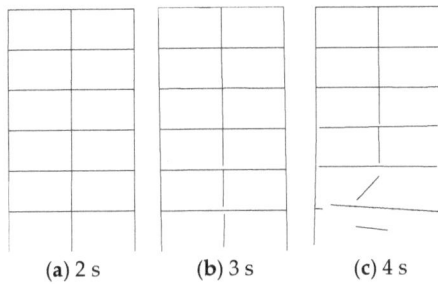

(a) 2 s   (b) 3 s   (c) 4 s

Figure 24. Failure processes of the steel frame with rigid connection.

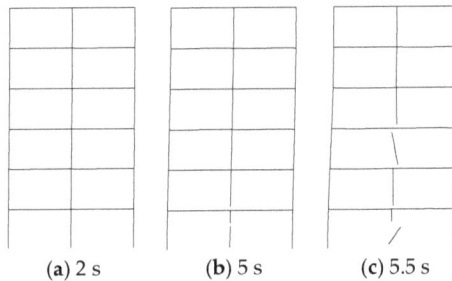

(a) 2 s   (b) 5 s   (c) 5.5 s

Figure 25. Failure processes of the steel frame with semi-rigid connection.

5. Conclusions

(1)   This paper presented an effective numerical approach for static and dynamic behavior simulation of steel frames with semi-rigid joints based on the Member Discrete Element Method (MDEM). In the MDEM, a structure is discretized into a set of finite rigid particles, as well as geometric nonlinearity and fracture behaviors can be naturally captured. A virtual spring element without actual length is applied to simulate the semi-rigid connection. On this basis, the modified formula of the contact element stiffness at the semi-rigid connection is derived. Finally, the numerical approach proposed is verified by complex behaviors of steel frames with semi-rigid connections such as geometric nonlinearity, snap-through buckling, dynamic responses and fracture. In addition, compared with other numerical approaches taking the FEM as a representation, the approach proposed is simple and feasible for simulating the semi-rigid connections because the zero-length spring element is not directly involved in the calculation.

(2)   The comparison of the analysis results of the proposed approach and the existing researches shows that the modified MDEM can accurately capture linear and nonlinear behaviors of semi-rigid connections. Some common conclusions can be drown as follow: the semi-rigid connections may significantly reduce structural stiffness, and structural bearing capacity under static loading will be overestimated if the semi-rigid connections are ignored; When the frequency of dynamic load applied is close to structural fundamental frequency, resonance occurs in the frames with rigid or linear semi-rigid connections, but not in the frame with nonlinear semi-rigid connections, the reason is that the hysteresis damping of the nonlinear connection causes energy dissipation. Fracture behavior analysis also indicates that frames with semi-rigid connection possess more anti-collapse capacity.

(3)   The MDEM can avoid the difficulties of finite element method (FEM) in dealing with strong nonlinearity and discontinuity. A unified computational framework is applied for static and dynamic analyses. The method is simple and generally good, which is an effective tool to

investigate complex behaviors of steel frames with semi-rigid connections. In the follow-up study, material nonlinearity will be taken into account to simulate the collapse process of steel frames with semi-rigid connections under strong earthquake or impact loading action.

Acknowledgments: This work is supported by the National Science Found for Distinguished Young Scholars of China (Grant No. 51125031) and A Project Funded by the Priority Academic Program Development of Jiangsu Higher Education Institutions (CE02-1-31).

Author Contributions: Jihong Ye and Lingling Xu modified the Member Discrete Element Method, analyzed the numerical results, and wrote the paper.

Conflicts of Interest: The authors declare no conflict of interest.

References

1. Nguyen, P.-C.; Kim, S.-E. Second-order spread-of-plasticity approach for nonlinear time-history of space semi-rigid steel frames. *Finite Elem. Anal. Des.* **2015**, *105*, 1–15. [CrossRef]
2. Abdalla, K.M.; Chen, W.F. Expanded database of semirigid steel connections. *Comput. Struct.* **1995**, *56*, 553–564. [CrossRef]
3. Brown, N.D.; Anderson, D. Structural properties of composite major axis end plate connections. *J. Constr. Steel Res.* **2001**, *57*, 327–349. [CrossRef]
4. Chen, W.F.; Kishi, N. Semirigid steel beam-to-column connections- data-base and modeling. *J. Struct. Eng. ASCE* **1989**, *115*, 105–119. [CrossRef]
5. Richard, R.M.; Abbott, B.J. Versatile elastic-plastic stress–strain formula. *J. Eng. Mech. Div. ASCE* **1975**, *101*, 511–515.
6. Wang, X.; Ye, J.; Yu, Q. Improved equivalent bracing model for seismic analysis of mid-rise CFS structures. *J. Constr. Steel Res.* **2017**, *136*, 256–264. [CrossRef]
7. Eurocode 3. *Design of Steel Structures-Joints Inbuilding Frames*; ENV-1993-1-1:1992/A2; Annex, J., Ed.; European Committee for Standardization (CEN): Brussels, Belgium, 1998.
8. Keulen, D.C.; Nethercot, D.A.; Snijder, H.H.; Bakker, M.C.M. Frame analysis incorporating semirigid joint action: Applicability of the half initial secant stiffnes approach. *J. Constr. Steel Res.* **2003**, *59*, 1083–1100. [CrossRef]
9. Lui, E.M.; Chen, W.F. Analysis and behavior of flexibly-jointed frames. *Eng. Struct.* **1986**, *8*, 107–118. [CrossRef]
10. Ho, W.M.G.; Chan, S.L. Semibifurcation and bifurcation analysis of flexibly connected steel frames. *J. Struct. Eng. ASCE* **1991**, *117*, 2298–2318. [CrossRef]
11. Yau, C.Y.; Chan, S.L. Inelastic and stability analysis of flexibly connected steel frames by springs-in-series model. *J. Struct. Eng.* **1994**, *120*, 2803–2819. [CrossRef]
12. Zhou, Z.H.; Chan, S.L. Self-equilibrating element for second-order analysis of semirigid jointed frames. *J. Eng. Mech.* **1995**, *121*, 896–902. [CrossRef]
13. Chui, P.P.T.; Chan, S.L. Transient response of moment-resistant steel frames with flexible and hysteretic joints. *J. Constr. Steel Res.* **1996**, *39*, 221–243. [CrossRef]
14. Chan, S.L.; Chui, P.T. *Non-Linear Static and Cyclic Analysis of Steel Frames with Semi-Rigid Connections*; Elsevier: Amsterdam, The Netherlands, 2002.
15. Sophianopoulos, D.S. The effect of joint flexibility on the free elastic vibration characteristics of steel plane frames. *J. Constr. Steel Res.* **2003**, *59*, 995–1008. [CrossRef]
16. Li, T.Q.; Choo, B.S.; Nethercot, D.A. Connection element method for the analysis of semirigid frames. *J. Constr. Steel Res.* **1995**, *32*, 143–171. [CrossRef]
17. Raftoyiannis, I.G. The effect of semi-rigid joints and an elastic bracing system on the buckling load of simple rectangular steel frames. *J. Constr. Steel Res.* **2005**, *61*, 1205–1225. [CrossRef]
18. Nguyen, P.-C.; Kim, S.-E. Nonlinear elastic dynamic analysis of space steel frames with semi-rigid connections. *J. Constr. Steel Res.* **2013**, *84*, 72–81. [CrossRef]
19. Nguyen, P.-C.; Kim, S.-E. Nonlinear inelastic time-history analysis of three-dimensional semi-rigid steel frames. *J. Constr. Steel Res.* **2014**, *101*, 192–206. [CrossRef]

20. Lien, K.H.; Chiou, Y.J.; Hsiao, P.A. Vector form intrinsic finite-element analysis of steel frames with semi-rigid joints. *J. Struct. Eng.* **2012**, *138*, 327–336. [CrossRef]
21. Ying, Y.; Xing, Y.Z. Nonlinear dynamic collapse analysis of semi-rigid steel frames based on the finite particle method. *Eng. Struct.* **2016**, *118*, 383–393.
22. Ye, J.; Qi, N. Progressive collapse simulation based on DEM for single-layer reticulated domes. *J. Constr. Steel Res.* **2017**, *128*, 721–731.
23. Cundall, P.A.; Strack, O.D.L. A discrete element model for granular assemblies. *Geotechnique* **1979**, *29*, 47–65. [CrossRef]
24. Ye, J.; Qi, N. Collapse process simulation of reticulated shells based on coupled DEM/FEM model. *J. Build. Struct.* **2017**, *38*, 52–61. (In Chinese)
25. Qi, N.; Ye, J. Nonlinear dynamic analysis of space frame structures by discrete element method. *Appl. Mech. Mater.* **2014**, *638–640*, 1716–1719. [CrossRef]
26. Xu, L.; Ye, J. DEM algorithm for progressive collapse simulation of single-layer reticulated domes under multi-support excitation. *J. Earthq. Eng.* **2017**. [CrossRef]
27. Wang, X.C. *Finite Element Method*; Tsinghua University Press: Beijing, China, 2003.
28. Qi, N.; Ye, J.H. Geometric nonlinear analysis with large deformation of member structures by discrete element method. *J. Southeast Univ. (Nat. Sci. Ed.)* **2013**, *43*, 917–922. (In Chinese)
29. Lynn, K.M.; Isobe, D. Finite element code for impact collapse problem. *Int. J. Numer. Method. Eng.* **2007**, *69*, 2538–2563. [CrossRef]
30. Ramberg, W.; Osgood, W.R. *Description of Stress-Strain Curves by Three Parameters*; National Advisory Committee for Aeronautics: Washington, DC, USA, 1943.
31. Mondkar, D.P.; Powell, G.H. Finite element analysis of nonlinear static and dynamic response. *Int. J. Numer. Method. Eng.* **1977**, *11*, 499–520. [CrossRef]
32. Yang, T.Y.; Saigal, S. A simple element for static and dynamic response of beams with material and geometric nonlinearities. *Int. J. Numer. Method. Eng.* **1984**, *20*, 851–867. [CrossRef]
33. Yang, Y.B.; Chiou, H.T. Analysis with beam elements. *J. Eng. Mech.* **1987**, *113*, 1404–1419. [CrossRef]
34. Williams, F.W. An approach to the nonlinear behaviour of the members of a rigid jointed plane framework with finite deflection. *Q. J. Mech. Appl. Maths* **1964**, *17*, 451–469. [CrossRef]
35. Del Savio, A.A.; Andrade, S.A.L.; Vellasco, P.C.G.S.; Martha, L.F. A nonlinear system for semirigid steel portal frame analysis. In Proceedings of the 7th International Conference on Computational Structures Technology, Tecnologiaem Computação Gráfica, Rio de Janeiro, Brazil, 10–11 January 2004; Volume 1, pp. 1–12.

applied
sciences

MDPI

Article

Prediction of Ultimate Strain and Strength of FRP-Confined Concrete Cylinders Using Soft Computing Methods

Iman Mansouri [1], Ozgur Kisi [2], Pedram Sadeghian [3], Chang-Hwan Lee [4] and Jong Wan Hu [5,6,*]

1   Department of Civil Engineering, Birjand University of Technology, Birjand 97175-569, Iran; mansouri@birjandut.ac.ir
2   School of Natural Sciences and Engineering, Ilia State University, Tbilisi 0162, Georgia; ozgur.kisi@iliauni.edu.ge
3   Department of Civil and Resource Engineering, Dalhousie University, 1360 Barrington Street, Halifax, NS B3H 4R2, Canada; Pedram.Sadeghian@dal.ca
4   Research Institute of Structural Engineering & System, DongYang Structural Engineers Co., Ltd., Seoul 05836, Korea; chlee@dysec.co.kr
5   Department of Civil and Environmental Engineering, Incheon National University, Incheon 22012, Korea
6   Incheon Disaster Prevention Research Center, Incheon National University, Incheon 22012, Korea
*   Correspondence: jongp24@incheon.ac.kr; Tel.: +82-32-835-8463

Received: 13 June 2017; Accepted: 18 July 2017; Published: 25 July 2017

Abstract: This paper investigates the effectiveness of four different soft computing methods, namely radial basis neural network (RBNN), adaptive neuro fuzzy inference system (ANFIS) with subtractive clustering (ANFIS-SC), ANFIS with fuzzy c-means clustering (ANFIS-FCM) and M5 model tree (M5Tree), for predicting the ultimate strength and strain of concrete cylinders confined with fiber-reinforced polymer (FRP) sheets. The models were compared according to the root mean square error (RMSE), mean absolute relative error (MARE) and determination coefficient (R^2) criteria. Similar accuracy was obtained by RBNN and ANFIS-FCM, and they provided better estimates in modeling ultimate strength of confined concrete. The ANFIS-SC, however, performed slightly better than the RBNN and ANFIS-FCM in estimating ultimate strain of confined concrete, and M5Tree provided the worst strength and strain estimates. Finally, the effects of strain ratio and the confinement stiffness ratio on strength and strain were investigated, and the confinement stiffness ratio was shown to be more effective.

Keywords: fiber reinforced polymer; concrete; column; confinement; stress; strain; model

1. Introduction

In recent years, the strengthening of existing concrete structures using externally bonded composite sheets of fiber reinforced polymer (FRP) has gained significant popularity. One common technique is wrapping unidirectional FRPs around the circumference of a concrete column to increase its axial strength and ductility. It is well-known that a concrete core expands laterally under uniaxial compression, but such expansion is confined by the FRP. Therefore, the core is subjected to a three-dimensional compressive state of stress in which the performance of the concrete core is significantly influenced by the confining pressure [1–5].

Many researchers studied the behavior of FRP-confined concrete and proposed a variety of confinement models for the ultimate condition of confined concrete under uniaxial compression loadings [6,7]. The majority of FRP confinement models are design-oriented and were developed using a regression analysis [8–12]. There have also been several analysis-oriented models developed based on the mechanics of confinement and strain compatibility between concrete and the FRP wrap [13–15]. Recently, a new category of models has been proposed based on soft computing methods, such as artificial neural networks, generic algorithms, and fuzzy logic. Models in this category can handle complex databases containing a large number of independent variables, identify the sensitivity of input parameters, and provide mathematical solutions between dependent and independent variables [16]. Pham and Hadi [16] proposed the utilization of neural networks to compute the strain and compressive strength of FRP-confined columns, and the results show agreement between proposed neural network models and experimental data. Also, there are several studies related to design-oriented and analysis-oriented models [9,17–29].

Lim et al. [30] proposed a new model for evaluating the ultimate condition of FRP-confined concrete using genetic programming (GP). The model was the first to establish the ultimate axial strain and hoop rupture strain expressions for FRP-confined concrete on the basis of evolutionary algorithms. The results showed that the predictions from the suggested model aligned with a database compiled by the authors. The proposed models provided improved predictions compared to the existing artificial intelligence models. The model proved that more accurate results can be achieved in explaining and formulating the ultimate condition of FRP-confined concrete. The model assessment presented in that study clearly illustrated the importance of the size of the test databases and the selected test parameters used in the development of artificial intelligence models on their overall performance.

This paper studies the capability of four soft computing techniques for predicting the ultimate strength and strain of FRP-confined concrete cylindrical specimens. The computing techniques include radial basis neural network (RBNN), adaptive neuro fuzzy inference system (ANFIS) with subtractive clustering (ANFIS-SC), ANFIS with fuzzy c-means clustering (ANFIS-FCM), and M5 model tree (M5Tree).

2. Overview of Soft Computing Approaches

2.1. Radial Basis Function Neural Network

Artificial neural networks (ANNs) are inspired by biological neural networks. ANNs include a set of processing components, called neurons, which operate in parallel processes and transmit information to other neurons, similar to the functioning of a biological brain. ANNs are an efficient method for modeling complex input-output relationships and can learn relationships directly from the data being modeled [31]. The nonlinearity within a radial basis function (RBF) network can be selected from a few classic nonlinear functions. The hidden layer carries out a fixed nonlinear transformation with no adjustable variables, and it maps the input onto a new layer. The output layer then performs a linear combination on this new layer, and the only adjustable variables are the weights of this linear combiner [32]. A general RBF network is schematically illustrated in Figure 1.

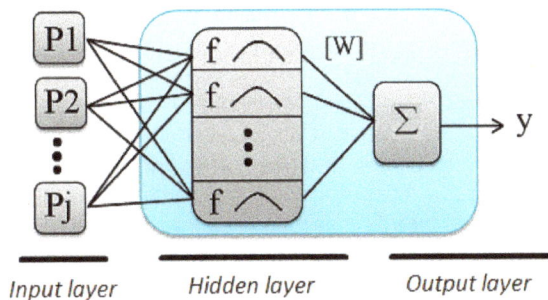

Figure 1. Schematic of a RBF network [31].

The radial basis function neural network (RBFN) model, which includes an input, an output and a single hidden layer, was developed by Powell [33] and Broomhead and Lowe [34]. In this model, the number of input and output nodes is similar to that of multi-layer perceptron (MLP) neural networks, and was selected using the nature of real input and output parameters. However, the rate of learning in RBFN is much faster than the MLP method. The output of RBFN can be calculated with the following equation:

$$Y = \sum_{p=1}^{p} W_p \, \theta \left(\|X - X_p\| \right) \tag{1}$$

In the equation, W_p is the weight connecting the output nodes and hidden nodes, θ represents the radial basis function, X and Y are the input and output variables, X_p indicates the center of each hidden node which is dependent on the input data, and $\|X - X_p\|$ represents the Euclidean metric between hidden and input nodes. Each group of input nodes which has identical information as the input is indicated by one hidden node, and the transformation related to any node within the hidden layer is named a Gaussian function [35]. More detailed information about RBNN theory can be obtained from Haykin [36].

2.2. Adaptive Neuro Fuzzy Inference System

ANFIS is a combined intelligent system including ideas from neural networks and fuzzy control, combining the advantages of both. Fuzzy logic is a superset of typical logic that has been improved to operate with uncertain data and the theory of partial truth [37,38]. The most important disadvantage of fuzzy logic is the lack of a systematic approach to choosing membership function variables and designing the fuzzy rules. On the other hand, ANN has the ability to learn its structure from the input-output sets.

Jang [39] introduced ANFIS as a universal approximation which can estimate any real continued function on a compact data set with the desired precision [40,41]. In terms of function, ANFIS can be considered as equivalent to fuzzy inference systems, and the ANFIS system used can be considered as comparable to the Sugeno first-order fuzzy model [42]. A simple example is presented below, in which a fuzzy inference system was assumed with two inputs of x and y and one output of z. For this example, the typical rule set of a first-order Sugeno fuzzy model, which possess two simple fuzzy If-Then rules, is as follows:

$$\text{Rule 1}: \text{ IF x is A1, y is B1 and z is C1 THEN } f_1 = p_1 x + q_1 y + r_1 z + t_1 \tag{2}$$

$$\text{Rule 2}: \text{ IF x is A2, y is B2 and z is C2 THEN } f_2 = p_2 x + q_2 y + r_2 z + t_2 \tag{3}$$

In the rule set, p_1, q_1, r_1 and p_2, q_2, r_2 are the variables of the THEN-part of the first-order Sugeno fuzzy model.

The ANFIS system uses a hybrid-learning algorithm to update parameters [43]. This algorithm is composed of two methods: the least squares approach and the gradient descent method. The function of the gradient descent approach is to adjust the variables of premise non-linear membership function, and the function of least squares method is to determine the resultant linear variables $\{p_i, q_i, r_i\}$. The learning process of this system has two steps. The first step includes the identification of consequent variables by the least squares method, while the prior variables are assumed to be fixed for the running cycle by the training set. After that, the error signals will spread backwards. In this part, the function of the gradient descent method includes updating the premise variables by minimizing the cost function, while the resultant variables stay fixed. Jang [39] presented the details of this algorithm and mathematical foundations of the hybrid learning algorithm.

In the present paper, two different ANFIS methods, including ANFIS with subtractive clustering (ANFIS-SC) and ANFIS with fuzzy c-means clustering (ANFIS-FCM),are utilized as modeling techniques. Subtractive clustering (SC) is an extension of the mountain clustering method suggested by Yager and Filev [44]. In this method, the data are clustered by evaluating the potential of data in the specification space. FCM is the modified K-means algorithm; it has some restrictions and may not operate properly with large data sets. FCM minimizes within cluster variance and the classification of data using the clustering algorithm [45]. FCM works by minimizing the squared error function.

2.3. M5 Model Tree

Quinlan [46] explained the M5Tree, which includes a regression function at the terminal nodes. In fact, the M5 algorithm employs the idea of splitting the parameter space into subspaces and building a local linear regression model in each. The splitting follows the concept used in building a decision tree (see Figure 2).

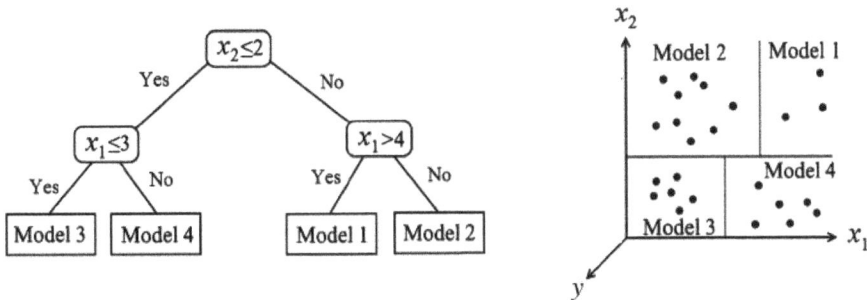

Figure 2. M5 tree; splitting the input space into subspaces and the resultant diagram [47].

The divide-and-conquer method is usually utilized to construct these types of tree-based models. Constructing a model tree entails two distinct steps. The first step includes a splitting criterion for the creation of a decision tree. In the M5Tree approach, this criterion is dependent on the standard deviation of the class values and then obtaining the decrement that can be expected in this error. The standard deviation reduction (SDR) can be calculated with the following equation:

$$SDR = sd(T) - \sum \frac{|T_i|}{T} sd(T_i) \tag{4}$$

In the equation, T indicates a set of instances that achieves the node, T_i is the subset of instances that have the ith result of the potential set, and sd is the standard deviation. The disarticulating proceeding causes the data in parent nodes to have more standard deviation compared to child nodes so that these data are purer. Once all the feasible splits are assessed, M5 selects the one that provides the maximum value for the expected error decrease. This grouping usually leads to a tree-like structure that should be trimmed by substituting a sub-tree, for example, with a leaf. In the next step, this tree growth must be trimmed and the sub-trees replaced with regression functions. In this method, the variable space is divided into several areas (subspaces), and a linear regression model is created for each of them. Quinlan [46] provides further details on M5Tree.

3. Results

In this study, four different soft computing techniques, i.e., RBNN, ANFIS-SC, ANFIS-FCM and M5 Tree, are employed for predicting strength (f'_{cc}/f'_{co}) and strain ($\varepsilon_{cc}/\varepsilon_{co}$). For the model simulations, a MATLAB neural network and fuzzy tool boxes are used, and 519 experimental data are adopted from Sadeghian and Fam [48,49]. Figure 3 shows the usual test setup for confined cylinders. Also, the statistical analyses of the input data employed in this study are summarized in Table 1. The data are used to develop models for ultimate strength and strain based on the strain ratio (ρ_ε) and the confinement stiffness ratio (ρ_K) inputs, which are defined as follows [48]:

$$f_l = \rho_K \rho_\varepsilon f'_{co} = \frac{2E_f \varepsilon_{h,rup} t}{D} \tag{5}$$

$$\rho_K = \frac{2E_f t}{(f'_{co}/\varepsilon_{co})D} \tag{6}$$

$$\rho_\varepsilon = \frac{\varepsilon_{h,rup}}{\varepsilon_{co}} \tag{7}$$

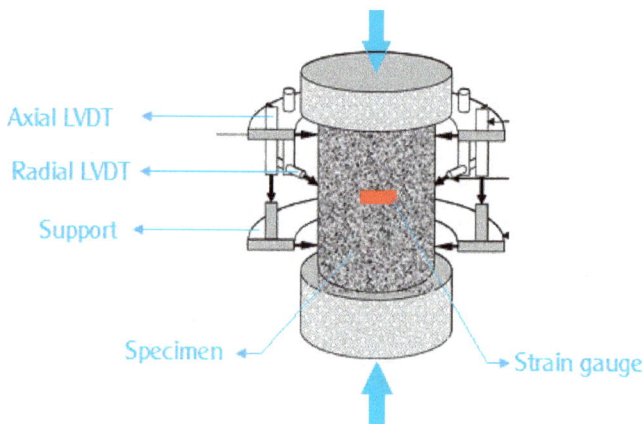

Figure 3. Test setup and instrumentation [50].

Table 1. Statistics for the experimental data.

	D (mm)	t (mm)	f'_{co} (MPa)	f'_{cc} (MPa)	ε_{co} (%)	ε_{cc} (%)	$\varepsilon_{h,rup}$ (%)	ρ_ε	E_f (GPa)	ρ_K
Min. value	51.00	0.09	19.70	31.40	0.20	0.23	0.10	0.29	10.50	0.01
Max. value	406.00	7.26	188.20	372.20	0.35	6.20	4.98	23.03	662.50	0.68
Average	158.46	0.89	46.52	83.88	0.24	1.68	1.12	4.80	183.09	0.07
Standard deviation	52.74	1.06	27.05	42.18	0.03	1.05	0.53	2.44	124.73	0.07

In the equations, f'_{co} is the unconfined concrete strength, ε_{co} is the corresponding axial strain of f'_{co}, E_f is the elastic modulus of the FRP wrap in the hoop direction, t is the total thickness of the FRP wrap, $\varepsilon_{h,rup}$ is the actual hoop rupture strain of the FRP wrap, and D is the diameter of the concrete core. The confinement ratio (f_l/f'_{co}) is a frequently used parameter in existing confinement models, which is equal to the product of ρ_K and ρ_ε. Several equations have been proposed for estimating the strain and strength of FRP-confined concrete cylinders which depend on ρ_K and ρ_ε [9,48]. According to Teng et al. [11], instead of the more approximate value of 0.002 for ε_{co}, it is assumed as follows:

$$\varepsilon_{co} = 9.37 \times 10^{-4} \sqrt[4]{f'_{co}}, \ f'_{co} \ in \ MPa \tag{8}$$

The data set is randomly grouped into two subsets; the first data set is adopted for training, and the second data set (20% of the whole database) is adopted for the testing stage. Before application of the RBNN, the training values of input and output are normalized between 0.2 and 0.8 as follows:

$$b_1 \frac{x_i - x_{min}}{x_{max} - x_{min}} + b_2 \tag{9}$$

In the equation, x_{max} and x_{min} are the maximum and minimum values of the training data. Here values of 0.6 and 0.2 are respectively assigned for b_1 and b_2, and the input data are normalized within a range of 0.2 to 0.8, as recommended in Cigizoglu [51]. According to that study, input parameters ranging from 0.2 to 0.8 allow the artificial neural network the flexibility to appraise beyond the training range.

The applied models are compared with the mean absolute relative error (MARE), root mean square error (RMSE) and determination coefficient (R^2). The definitions of statistical parameters are given as follows:

$$RMSE = \sqrt{\frac{1}{N}\sum_{i=1}^{N}(Xo_i - Xe_i)^2} \tag{10}$$

$$MARE = \frac{1}{N}\sum_{i=1}^{N}\frac{|Xo_i - Xe_i|}{Xo_i} \times 100 \tag{11}$$

where N, Xo_i and Xe_i are the number of samples, and the observed and estimated values, respectively.

3.1. Ultimate Strength Prediction

Testing and training results for the prediction of strength of the RBNN, ANFIS-SC, ANFIS-FCM and M5Tree models are listed in Table 2. The control parameter values of the optimal models are also provided in the second column. Different numbers of parameters and structures were tried for each method and the optimal ones were selected. Gaussian membership functions are used in ANFIS-SC and ANFIS-FCM models. The RBNN, ANFIS-SC and ANFIS-FCM methods can be easily obtained and applied by using new RBNN, genfis2 and genfis3 tools in MATLAB command windows. For the M5Tree method, code which is available for free online (http://www.cs.rtu.lv/jekabsons/regression.html) is used.

Table 2. Statistical performance of RBNN, ANFIS-SC, ANFIS-FCM and M5Tree models in strength predictions.

Method	Control Parameters	Training			Test		
		RMSE	MARE	R^2	RMSE	MARE	R^2
RBNN	0.8,15	0.32	11.6	0.880	0.27	10.5	0.899
ANFIS-SC	0.1	0.31	11.1	0.891	0.32	11.3	0.879
ANFIS-FCM	10	0.30	11.0	0.896	0.27	10.7	0.903
M5Tree		0.26	8.25	0.921	0.43	14.3	0.739

In the table, 0.8 and 15 indicate the spread value and the number of hidden layer neuron of the RBNN model, respectively, while 0.1 and 10 show the radii and cluster number of the ANFIS-SC and ANFIS-FCM models. A radii value of 0.1 in ANFIS-FCM corresponds to 15 clusters. This means that the ANFIS-FCM has fewer membership functions and parameters (10 Gaussian membership function each have 2 parameters, or 20 parameters in total) than those of ANFIS-SC. Table 2 implies that RBNN and ANFIS-FCM have almost the same accuracy, and they both are more efficient than the ANFIS-SC and M5Ttree models with respect to RMSE, MARE and R^2. It is interesting that the M5Tree approximated training data very well whereas its test results are worse than those of the other models. This implies that this method cannot adequately learn the investigated phenomenon. Different statistical indices were obtained for each of the methods. The main reason for this may be the fact that each method has different assumptions in developing models and their behaviors with the used data are distinct from each other. The rule base of the optimal ANFIS-FCM model is given in Table 3.

Table 3. Rule base of the optimal ANFIS-FCM in modeling strength.

1.	If (strain-ratio is in1cluster1) and (confinement-stiffness-ratio is in2cluster1) then (Strength is out1cluster1)
2.	If (strain-ratio is in1cluster2) and (confinement-stiffness-ratio is in2cluster2) then (Strength is out1cluster2)
3.	If (strain-ratio is in1cluster3) and (confinement-stiffness-ratio is in2cluster3) then (Strength is out1cluster3)
4.	If (strain-ratio is in1cluster4) and (confinement-stiffness-ratio is in2cluster4) then (Strength is out1cluster4)
5.	If (strain-ratio is in1cluster5) and (confinement-stiffness-ratio is in2cluster5) then (Strength is out1cluster5)
6.	If (strain-ratio is in1cluster6) and (confinement-stiffness-ratio is in2cluster6) then (Strength is out1cluster6)
7.	If (strain-ratio is in1cluster7) and (confinement-stiffness-ratio is in2cluster7) then (Strength is out1cluster7)
8.	If (strain-ratio is in1cluster8) and (confinement-stiffness-ratio is in2cluster8) then (Strength is out1cluster8)
9.	If (strain-ratio is in1cluster9) and (confinement-stiffness-ratio is in2cluster9) then (Strength is out1cluster9)
10.	If (strain-ratio is in1cluster10) and (confinement-stiffness-ratio is in2cluster10) then (Strength is out1cluster10)

Table 3 demonstrates that the model has 10 clusters and one rule for each of them. Figure 4 illustrates the strength estimates of the applied models in the forms of time variation and scatterplot.

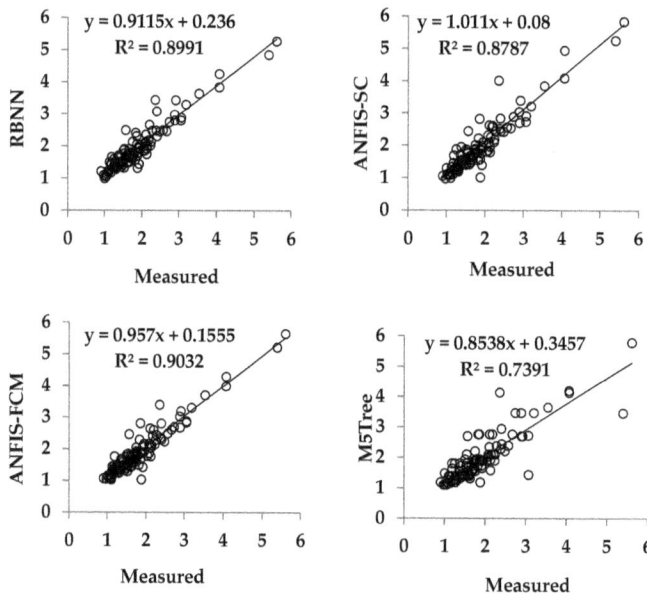

Figure 4. Strength estimates of the RBNN, ANFIS-SC, ANFIS-FCM and M5Tree models (dimensionless: f'_{cc}/f'_{co}).

From the figure, it is apparent that the RBNN and ANFIS-FCM models have less scattered assessments than other models, and the M5Tree model gave the most scattered estimates. This reveals that the investigated phenomenon is nonlinear, and therefore linear M5Tree cannot adequately simulate strength behavior. The variation of strength versus strain ratio and confinement stiffness ratio for the optimal ANFIS-FCM model is illustrated in Figure 5, in which it is clear that strength takes its maximum value when the strain ratio is also at its maximum.

The linear behavior for a confinement-stiffness-ratio greater than 0.3 is observed in Figure 5. This is because of the exponent of the confinement-stiffness-ratio in equations is usually 0.7; therefore, values greater than $0.3^{0.7}$ behave linearly.

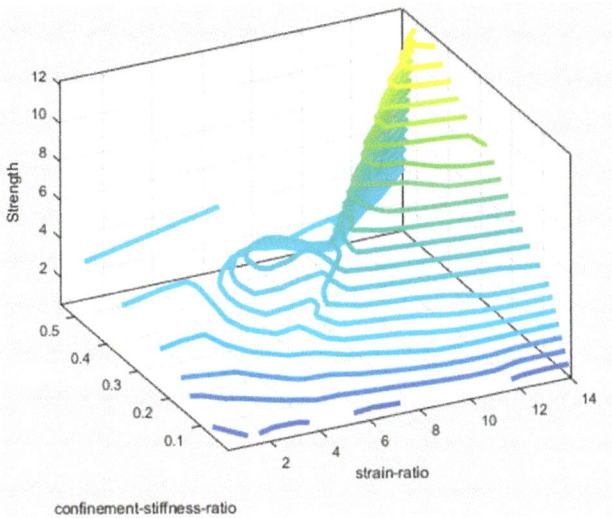

Figure 5. Variation of strength versus strain ratio and confinement stiffness ratio for the optimal ANFIS-FCM model.

3.2. Ultimate Strains Prediction

Strain predictions for the RMSE, MARE and R^2 values of the applied models are compared in Table 4.

Table 4. Statistical performance of RBNN, ANFIS-SC, ANFIS-FCM and M5Tree models in strain predictions.

Method	Control Parameters	Training			Test		
		RMSE	MARE	R^2	RMSE	MARE	R^2
RBNN	0.640	2.43	29.9	0.767	2.51	31.6	0.752
ANFIS-SC	0.1	2.31	29.4	0.790	2.47	30.5	0.766
ANFIS-FCM	6	2.33	31.3	0.786	2.57	31.9	0.742
M5Tree		1.90	21.8	0.858	2.72	33.3	0.711

This tables shows that the ANFIS-SC models outperform other models. Here also RBNN and ANFIS-FCM have similar accuracy, and they are slightly worse than the ANFIS-SC model. A radii value of 0.1 in ANFIS-SC corresponds to 13 clusters. Similar to the previous application, here too ANFIS-SC has many more membership functions and parameters compared to ANFIS-FCM. Table 5 gives the rule base of the optimal ANFIS-SC model. The estimated results of the optimal RBNN, ANFIS-FCM and ANFIS-SC models are shown in Figure 6.

This figure shows that the estimates of the ANFIS-SC model are closer to the corresponding measured values, especially for high strain values. The variation of strength versus strain ratio and confinement stiffness ratio for the optimal ANFIS-SC model is shown in Figure 7. It is clear from the figure that strain takes its maximum value when the strain ratio and confinement stiffness ratio are both at their maximum values. The linear relationship between strain and strain ratio is clearly seen in the range of confinement-stiffness-ratio between 0.4 and 0.6.

The effects of strain ratio and the confinement stiffness ratio inputs on strength and strain were also investigated using the ANFIS-SC method because it has less control parameters than RBNN. For the RBNN models, the optimal radii and hidden node numbers should be determined whereas, in obtaining ANFIS-SC model, the radii value which indicates the number of clusters is only distinguished. The simulation results of the ANFIS-SC models are reported in Table 6.

Table 5. Rule base of the optimal ANFIS-SC in modeling strain.

1.	If (strain-ratio is in1cluster1) and (confinement-stiffness-ratio is in2cluster1) then (Strength is out1cluster1)
2.	If (strain-ratio is in1cluster2) and (confinement-stiffness-ratio is in2cluster2) then (Strength is out1cluster2)
3.	If (strain-ratio is in1cluster3) and (confinement-stiffness-ratio is in2cluster3) then (Strength is out1cluster3)
4.	If (strain-ratio is in1cluster4) and (confinement-stiffness-ratio is in2cluster4) then (Strength is out1cluster4)
5.	If (strain-ratio is in1cluster5) and (confinement-stiffness-ratio is in2cluster5) then (Strength is out1cluster5)
6.	If (strain-ratio is in1cluster6) and (confinement-stiffness-ratio is in2cluster6) then (Strength is out1cluster6)
7.	If (strain-ratio is in1cluster7) and (confinement-stiffness-ratio is in2cluster7) then (Strength is out1cluster7)
8.	If (strain-ratio is in1cluster8) and (confinement-stiffness-ratio is in2cluster8) then (Strength is out1cluster8)
9.	If (strain-ratio is in1cluster9) and (confinement-stiffness-ratio is in2cluster9) then (Strength is out1cluster9)
10.	If (strain-ratio is in1cluster10) and (confinement-stiffness-ratio is in2cluster10) then (Strength is out1cluster10)
11.	If (strain-ratio is in1cluster11) and (confinement-stiffness-ratio is in2cluster11) then (Strength is out1cluster11)
12.	If (strain-ratio is in1cluster12) and (confinement-stiffness-ratio is in2cluster12) then (Strength is out1cluster12)
13.	If (strain-ratio is in1cluster13) and (confinement-stiffness-ratio is in2cluster13) then (Strength is out1cluster13)

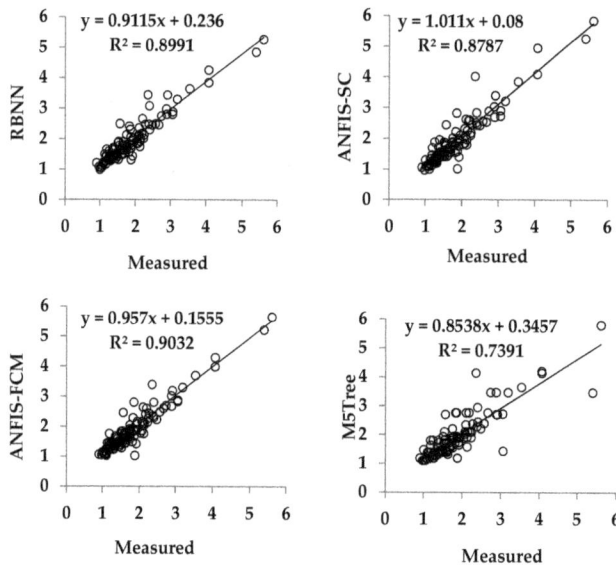

Figure 6. Strain estimates of the RBNN, ANFIS-SC, ANFIS-FCM and M5Tree models (dimensionless: $\varepsilon_{cc}/\varepsilon_{co}$).

Table 6. Effect of strain ratio (ρ_ε) and the confinement stiffness ratio (ρ_K) inputs on strength and strain with respect to ANFIS-SC.

Inputs	Control Parameters	Training			Test		
		RMSE	MARE	R^2	RMSE	MARE	R^2
		Strength					
ρ_ε	0.1	0.852	32.8	0.249	0.804	32.8	0.054
ρ_K	0.1	0.646	21.3	0.512	0.648	25.1	0.445
		Strain					
ρ_ε	0.2	4.081	63.6	0.343	4.406	59.2	0.244
ρ_K	0.1	4.453	71.8	0.218	4.330	63.2	0.265

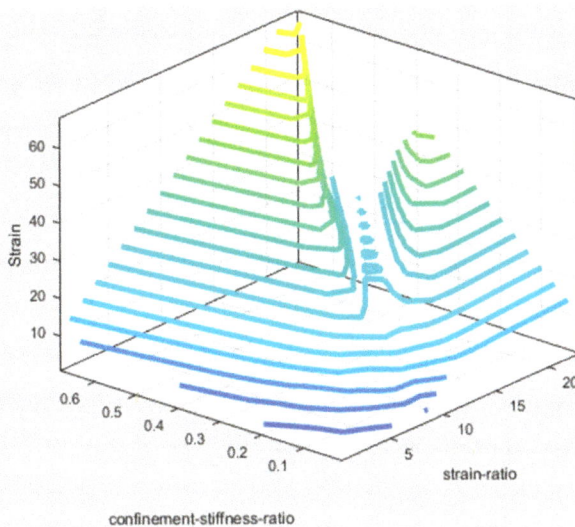

Figure 7. Variation of strain versus strain ratio and confinement stiffness ratio for the optimal ANFIS-SC model.

This table indicates that the effect of ρ_K is more significant than ρ_ϵ in strength and strain. The relative RMSE differences between ρ_K and ρ_ϵ based ANFIS-SC models are 19.4% and 1.7% for strength and strain, respectively.

It should be noted that providing explicit formulation for the RBNN, ANFIS-SC and ANFIS-FCM is impossible because they are black-box models. However, the regression tree of the optimal M5Tree models for the strain and strength modeling are provided in Appendix A.

4. Conclusions

In this paper, the ultimate strength and strain of cylindrical concrete specimens confined with FRP composites were studied using four soft computational methods. The optimal RBNN, ANFIS-SC, ANFIS-FCM and M5Tree models obtained by trying different control parameters were compared with respect to RMSE, MARE and R^2 statistics. RBNN and ANFIS-FCM provided almost the same level of accuracy, and they performed better than the other models in estimating strength of FRP-confined concrete cylindrical specimens by using the inputs of strain ratio and the confinement stiffness ratio. In estimating strain of FRPs, however, the ANFIS-SC model performed slightly better than the RBNN and ANFIS-FCM models. Among the applied models, the M5Tree model was the least accurate at estimating strength and strain of FRPs. The effects of strain ratio and the confinement stiffness ratio inputs on strength and strain were also examined using ANFIS-SC, and the confinement stiffness ratio was found to be more significant than strain ratio in strength and strain.

Acknowledgments: This research was supported by the Basic Science Research Program through the National Research Foundation of Korea (NRF), which is funded by the Ministry of Science, ICT & Future Planning of the Republic of Korea (2017R1A2B2010120).

Author Contributions: Pedram Sadeghian conceived and designed the database; Ozgur Kisi and Iman Mansouri analyzed the data; Jong Wan Hu and Iman Mansouri wrote the paper. Chang-Hwan Lee also analyzed the data and handled the final proofreading.

Conflicts of Interest: The authors declare no conflict of interest.

Appendix A. Regression Tree of the Optimal M5Tree Models for the Strain and Strength Modeling

Table A1. Regression tree obtained from M5Tree for strength modeling.

if x2 <= 0.065174	else if x2 <= 0.034551	else if x1 <= 6.6355
	if x2 <= 0.033937	if x2 <= 0.15871
if x2 <= 0.0255		
if x1 <= 2.9025	y = 1.71921529875 (4)	if x1 <= 3.3762
if x1 <= 2.505	else	if x2 <= 0.071835
y = 1.108064681 (10)	y = 1.92418515966667 (6)	y = 1.55785802183333 (6)
else	else if x2 <= 0.037414	else if x1 <= 2.6111
y = 0.891721241833333 (6)	y = 1.45295335707692 (13)	y = 1.8102727921 (20)
else if x2 <= 0.021267	else if x2 <= 0.044137	else
if x2 <= 0.010372	y = 1.73181401571429 (14)	y = 1.57794946244278
		+4.99230783389806*x2 (11)
y = 1.483590448 (4)	else if x2 <= 0.045434	else if x2 <= 0.092595
else if x2 <= 0.015328	y = 1.41987727225 (4)	if x2 <= 0.06805
y = 1.13560753725 (12)	else	if x2 <= 0.066642
else if x2 <= 0.018965	y = 1.587916394 (4)	y = 2.34944086616667 (6)
y = 1.25916023821429 (14)	else if x1 <= 3.7759	else
else	y = 1.64059121775 (4)	y = 2.7420408164 (5)
y = 1.1897478106 (5)	else if x2 <= 0.056561	else if x2 <= 0.074024
else if x1 <= 6.5213	if x2 <= 0.04728	y = 1.9024165688 (5)
if x1 <= 4.9928	y = 1.7582115855 (4)	else
y = 1.26759138458333 (12)	else if x2 <= 0.050345	y = 2.10530567306667 (15)
else	y = 2.23252585925 (4)	else if x1 <= 3.9987
y = 1.3995581252 (5)	else	y = 5.07504348861174
		-23.5904126567999*x2 (11)
else if x1 <= 7.387	y = -4.7981786875333	else if x2 <= 0.10281
	+124.346272222065*x2 (9)	
y = 1.654221567 (4)	else	y = 2.9481859308 (5)
else	y = 2.72437967584461 −	else
	0.199395562322418*x1 (13)	
y = 1.4358405485 (4)	else if x2 <= 0.049162	y = 2.70621362185714 (7)
	if x2 <= 0.037766	else if x1 <= 2.9764
else if x1 <= 7.0025		
if x1 <= 3.4344	if x1 <= 8.8613	y = 2.767153359 (19)
if x2 <= 0.029531	if x2 <= 0.033956	else if x2 <= 0.22102
y = 1.16075967033333 (6)	y = 1.7132418315 (12)	y = 3.47170636066667 (9)
else if x1 <= 1.3994	else	else
y = 1.179568198 (4)	y = 1.89220380857143 (7)	y = 4.1398758346 (5)
else if x2 <= 0.059018	else	else if x2 <= 0.13591
y = 1.39728594933333 (24)	y = 2.15360432216667 (6)	if x2 <= 0.068264
else	else	y = 2.72699884416667 (6)
y = 1.59811338542857 (7)	y = 2.21282843818182 (11)	else if x2 <= 0.1031
else if x2 <= 0.04595	else if x1 <= 7.8378	y = 3.65529982828571 (7)
if x2 <= 0.033094	y = 2.3919454476 (5)	else
if x1 <= 5.5058	else	y = 4.204802991 (4)
y = 1.323997823 (12)	y = 2.915657196 (7)	else
else		y = 1.45436334578442
		+23.9129270980071*x2 (10)
y = 1.56201983057143 (7)		

Table A2. Regression tree obtained from M5Tree for strain modeling.

if x1 <= 6.2548	else y = 4.69127474228571 (7)	else y = 6.0040160536 (5)

if x2 <= 0.06311
  if x1 <= 3.6413
  if x2 <= 0.020455
	else
	y = 6.3533113762 (5)
		else if x2 <= 0.20919
		  y = 2.39097472537744
		  +2.19876304626978*x1 (10)

if x1 <= 1.8874
  y = 1.1440491595 (6)
else
  y = 1.68687631466667 (9)
else if x1 <= 2.6807
	else if x1 <= 5.6305
	  if x1 <= 4.5915
	  if x2 <= 0.038697
	  y = 5.82153494575 (4)
	  else if x2 <= 0.055755
		else
		  y = 14.51223043 (8)
		else if x2 <= 0.05346
		  if x2 <= 0.022672
		  y = 3.41272976260672
		  +0.218464713423608*x1 (22)

if x1 <= 2.2516
  if x1 <= 1.7107
  y = 1.78844272172727 (11)
	y = 7.6093721018 (5)
	  else
	  y = 6.4901269452 (5)
		else if x1 <= 8.8613
		  if x2 <= 0.033956
		  y = 11.8427843384812
		  -146.164747121597*x2 (20)

else
  y = 3.15371412385714 (7)
else
  y = 1.764552396 (6)
else if x2 <= 0.047075
	else
	  y = 6.42120672495833 (24)
	else
	  y = 8.64814412742857 (7)
		else if x2 <= 0.035225
		  y = 12.096092072 (5)
		else if x2 <= 0.041997
		  if x1 <= 7.2822
		  y = 5.2413583345 (4)

if x2 <= 0.035327
  if x1 <= 3.4003
  y = 2.8827251381 (10)
  else
  y = 4.1709103866 (5)
  else
	else if x1 <= 2.8545
	  if x2 <= 0.14161
	  if x1 <= 0.94068
	  y = 1.6541689808 (5)
	  else if x2 <= 0.12452
	  if x2 <= 0.091846
	  y = 3.8282330848 (10)
		else
		  y = 8.69478460875 (4)
		else
		  y = 10.7816929114 (10)
		else
		  y = −4.15948380355076
		  +1.79907971888928*x1 (8)

y = 3.072717053875 (8)	else	else if x2 <= 0.14656
		  if x1 <= 7.7892
		  if x2 <= 0.067874
		  y = 11.774363725 (6)

else
  y = 4.18276414975 (12)
else if x2 <= 0.021073
  if x2 <= 0.018742
  y = 2.46492040855556 (9)
else
  y = 3.35411568575 (4)
else if x1 <= 4.0509
  if x2 <= 0.054126
  y = 4.0987042836 (15)
else
  y = 6.1883060684 (5)
	y = 5.62676665785714 (7)
	else
	  y = 3.269071034 (4)
	else
	  y = 7.55286884866667 (21)
	else if x2 <= 0.14818
	  if x2 <= 0.092595
	  if x1 <= 4.2573
	  if x1 <= 3.429
	  y = 7.9752687216 (5)
	  else
	  y = 10.9299345495055
	  -67.7789156702637*x2 (13)
		else
		  y = 15.9303924215 (8)
		else
		  y = 17.838992625 (18)
		else
		  y = 24.3989610371429 (7)
		  else if x2 <= 0.11185
		  y = 10.2661860479091 (11)
		  else if x1 <= 4.1077

else if x2 <= 0.033258
  if x1 <= 5.6609
  if x2 <= 0.024021
  y = 6.21866667766667 (6)
  else if x2 <= 0.029178
  y = 2.6673381094 (5)
	else if x1 <= 5.2858
	  if x1 <= 4.3562
	  y = 7.779023415 (4)
	  else
	  y = 9.292927947 (10)
		y = 6.97833482855556 (9)
		  else
		  y = 9.9430306388 (5)

References

1. Green, M.F.; Bisby, L.A.; Fam, A.Z.; Kodur, V.K.R. Frp confined concrete columns: Behaviour under extreme conditions. *Cem. Concr. Compos.* **2006**, *28*, 928–937. [CrossRef]
2. Mirmiran, A.; Shahawy, M.; Samaan, M.; El Echary, H.; Mastrapa, J.C.; Pico, O. Effect of column parameters on frp-confined concrete. *J. Compos. Constr.* **1998**, *2*, 175–185. [CrossRef]
3. Nanni, A.; Bradford, N.M. Frp jacketed concrete under uniaxial compression. *Constr. Build. Mater.* **1995**, *9*, 115–124. [CrossRef]
4. Pessiki, S.; Harries, K.A.; Kestner, J.T.; Sause, R.; Ricles, J.M. Axial behavior of reinforced concrete columns confined with frp jackets. *J. Compos. Constr.* **2001**, *5*, 237–245. [CrossRef]
5. Sadeghian, P.; Rahai, A.R.; Ehsani, M.R. Effect of fiber orientation on compressive behavior of cfrp-confined concrete columns. *J. Reinf. Plast. Compos.* **2010**, *29*, 1335–1346. [CrossRef]

6. Mansouri, I.; Gholampour, A.; Kisi, O.; Ozbakkaloglu, T. Evaluation of peak and residual conditions of actively confined concrete using neuro-fuzzy and neural computing techniques. *Neural Comput. Appl.* **2016**, 1–16. [CrossRef]
7. Mansouri, I.; Ozbakkaloglu, T.; Kisi, O.; Xie, T. Predicting behavior of frp-confined concrete using neuro fuzzy, neural network, multivariate adaptive regression splines and m5 model tree techniques. *Mater. Struct. Materiaux Constr.* **2016**, *49*, 4319–4334. [CrossRef]
8. Karbhari, V.M.; Gao, Y. Composite jacketed concrete under uniaxial compression—Verification of simple design equations. *J. Mater. Civ. Eng.* **1997**, *9*, 185–193. [CrossRef]
9. Lam, L.; Teng, J.G. Design-oriented stress-strain model for frp-confined concrete. *Constr. Build. Mater.* **2003**, *17*, 471–489. [CrossRef]
10. Samaan, M.; Mirmiran, A.; Shahawy, M. Model of concrete confined by fiber composites. *J. Struct. Eng.* **1998**, *124*, 1025–1031. [CrossRef]
11. Teng, J.G.; Jiang, T.; Lam, L.; Luo, Y.Z. Refinement of a design-oriented stress-strain model for frp-confined concrete. *J. Compos. Constr.* **2009**, *13*, 269–278. [CrossRef]
12. Toutanji, H.A. Stress-strain characteristics of concrete columns externally confined with advanced fiber composite sheets. *ACI Mater. J.* **1999**, *96*, 397–404.
13. Harries, K.A.; Kharel, G. Behavior and modeling of concrete subject to variable confining pressure. *ACI Mater. J.* **2002**, *99*, 180–189.
14. Spoelstra, M.R.; Monti, G. Frp-confined concrete model. *J. Compos. Constr.* **1999**, *3*, 143–150. [CrossRef]
15. Fam, A.Z.; Rizkalla, S.H. Confinement model for axially loaded concrete confined by circular fiber-reinforced polymer tubes. *ACI Struct. J.* **2001**, *98*, 451–461.
16. Pham, T.M.; Hadi, M.N.S. Predicting stress and strain of frp-confined square/rectangular columns using artificial neural networks. *J. Compos. Constr.* **2014**, *18*, 1–9. [CrossRef]
17. Berthet, J.F.; Ferrier, E.; Hamelin, P. Compressive behavior of concrete externally confined by composite jackets: Part b: Modeling. *Constr. Build. Mater.* **2006**, *20*, 338–347. [CrossRef]
18. Bisby, L.A.; Dent, A.J.S.; Green, M.F. Comparison of confinement models for fiber-reinforced polymer-wrapped concrete. *ACI Struct. J.* **2005**, *102*, 62–72.
19. Cascardi, A.; Micelli, F.; Aiello, M.A. Unified model for hollow columns externally confined by frp. *Eng. Struct.* **2016**, *111*, 119–130. [CrossRef]
20. Cascardi, A.; Micelli, F.; Aiello, M.A. An artificial neural networks model for the prediction of the compressive strength of frp-confined concrete circular columns. *Eng. Struct.* **2017**, *140*, 199–208. [CrossRef]
21. Faustino, P.; Chastre, C. Analysis of load–strain models for rc square columns confined with cfrp. *Compos. Part B Eng.* **2015**, *74*, 23–41. [CrossRef]
22. Ilki, A.; Kumbasar, N.; Koc, V. Low strength concrete members externally confined with frp sheets. *Struct. Eng. Mech.* **2004**, *18*, 167–194. [CrossRef]
23. Matthys, S.; Toutanji, H.; Audenaert, K.; Taerwe, L. Axial load behavior of large-scale columns confined with fiber-reinforced polymer composites. *ACI Struct. J.* **2005**, *102*, 258–267.
24. Naderpour, H.; Kheyroddin, A.; Amiri, G.G. Prediction of frp-confined compressive strength of concrete using artificial neural networks. *Compos. Struct.* **2010**, *92*, 2817–2829. [CrossRef]
25. Rousakis, T.C.; Tourtouras, I.S. Modeling of passive and active external confinement of rc columns with elastic material. *ZAMM J. Appl. Math. Mech.* **2015**, *95*, 1046–1057. [CrossRef]
26. Saiidi, M.S.; Kandasamy, S.; Claudia, P. Simple carbon-fiber-reinforced-plastic-confined concrete model for moment-curvature analysis. *J. Compos. Constr.* **2005**, *9*, 101–104. [CrossRef]
27. Sidney, A.G.; Lukito, G. Strengthening of reinforced concrete bridge columns with frp wrap. *Pract. Period. Struct. Des. Constr.* **2006**, *11*, 218–228.
28. Tamuzs, V.; Tepfers, R.; Sparnins, E. Behavior of concrete cylinders confined by carbon composite 2. Prediction of strength. *Mech. Compos. Mater.* **2006**, *42*, 109–118. [CrossRef]
29. Teng, J.G.; Lin, G.; Yu, T. Analysis-oriented stress-strain model for concrete under combined frp-steel confinement. *J. Compos. Constr.* **2015**, *19*. [CrossRef]
30. Lim, J.C.; Karakus, M.; Ozbakkaloglu, T. Evaluation of ultimate conditions of frp-confined concrete columns using genetic programming. *Comput. Struct.* **2016**, *162*, 28–37. [CrossRef]
31. Zounemat-kermani, M.; Kisi, O.; Rajaee, T. Performance of radial basis and lm-feed forward artificial neural networks for predicting daily watershed runoff. *Appl. Soft Comput.* **2013**, *13*, 4633–4644. [CrossRef]

32. Senthil Kumar, A.R.; Ojha, C.S.P.; Goyal, M.K.; Singh, R.D.; Swamee, P.K. Modeling of suspended sediment concentration at kasol in india using ann, fuzzy logic, and decision tree algorithms. *J. Hydrol. Eng.* **2012**, *17*, 394–404. [CrossRef]
33. Powell, M.J.D. Radial basis functions for multivariable interpolation: A review. In Proceedings of the IMA Conference on Algorithms for the Approximation of Functions and Data, RMCS, Shrivenham, UK, 1–30 July 1987; pp. 143–167.
34. Broomhead, D.S.; Lowe, D. Multivariable functional interpolation and adaptive networks. *Complex Syst.* **1988**, *2*, 321–355.
35. Rezaeian-Zadeh, M.; Zand-Parsa, S.; Abghari, H.; Zolghadr, M.; Singh, V.P. Hourly air temperature driven using multi-layer perceptron and radial basis function networks in arid and semi-arid regions. *Theor. Appl. Climatol.* **2012**, *109*, 519–528. [CrossRef]
36. Haykin, S. *Neural Networks and Learning Machines*, 3rd ed.; Prentice Hall: New Jersey, NJ, USA, 2008.
37. Deka, P.; Chandramouli, V. Fuzzy neural network model for hydrologic flow routing. *J. Hydrol. Eng.* **2005**, *10*, 302–314. [CrossRef]
38. Yilmaz, A.G.; Muttil, N. Runoff estimation by machine learning methods and application to the euphrates basin in Turkey. *J. Hydrol. Eng.* **2014**, *19*, 1015–1025. [CrossRef]
39. Jang, J.S.R. Anfis: Adaptive-network-based fuzzy inference system. *IEEE Transact. Syst. Man Cybern.* **1993**, *23*, 665–685. [CrossRef]
40. Jang, J.S.R.; Sun, C.T.; Mizutani, E. *Neuro-Fuzzy and Soft Computing: A Computational Approach to Learning and Machine Intelligence*; Prentice-Hall Upper Saddle River: New Jersey, NY, USA, 1997.
41. Mansouri, I.; Kisi, O. Prediction of debonding strength for masonry elements retrofitted with frp composites using neuro fuzzy and neural network approaches. *Compos. Part B Eng.* **2015**, *70*, 247–255. [CrossRef]
42. Drake, J.T. Communications Phase Synchronization Using the Adaptive Network Fuzzy Inference System. Ph.D. Thesis, New Mexico State University, Albuquerque, NM, USA, 2000.
43. Mansouri, I.; Shariati, M.; Safa, M.; Ibrahim, Z.; Tahir, M.M.; Petković, D. Analysis of influential factors for predicting the shear strength of a v-shaped angle shear connector in composite beams using an adaptive neuro-fuzzy technique. *J. Intell. Manuf.* **2017**, *28*, 1–11. [CrossRef]
44. Yager, R.R.; Filev, D.P. Approximate clustering via the mountain method. *IEEE Transact. Syst. Man Cybern.* **1994**, *24*, 1279–1284. [CrossRef]
45. Ayvaz, M.T.; Karahan, H.; Aral, M.M. Aquifer parameter and zone structure estimation using kernel-based fuzzy c-means clustering and genetic algorithm. *J. Hydrol.* **2007**, *343*, 240–253. [CrossRef]
46. Quinlan, J.R. Learning with continuous classes. In Proceedings of the 5th Australian Joint Conference on Artificial Intelligence (AI '92), Hobart, Australia, 16–18 October 1992; Adams, A., Sterling, L., Eds.; World Scientific Publishing: Singapore, 1992; pp. 343–348.
47. Wang, L.; Kisi, O.; Zounemat-Kermani, M.; Zhu, Z.; Gong, W.; Niu, Z.; Liu, H.; Liu, Z. Prediction of solar radiation in China using different adaptive neuro-fuzzy methods and m5 model tree. *Int. J. Climatol.* **2017**, *37*, 1141–1155. [CrossRef]
48. Sadeghian, P.; Fam, A. Improved design-oriented confinement models for frp-wrapped concrete cylinders based on statistical analyses. *Eng. Struct.* **2015**, *87*, 162–182. [CrossRef]
49. Sadeghian, P.; Fam, A. A rational approach toward strain efficiency factor of fiber-reinforced polymer-wrapped concrete columns. *ACI Struct. J.* **2014**, *111*, 135–144.
50. Berthet, J.F.; Ferrier, E.; Hamelin, P. Compressive behavior of concrete externally confined by composite jackets. Part a: Experimental study. *Constr. Build. Mater.* **2005**, *19*, 223–232. [CrossRef]
51. Cigizoglu, H.K. Estimation, forecasting and extrapolation of river flows by artificial neural networks. *Hydrol. Sci. J.* **2003**, *48*, 349–362. [CrossRef]

applied
sciences

MDPI

Article

Experimental Study on Robustness of an Eddy Current-Tuned Mass Damper

Junda Chen [1], Guangtao Lu [1,2,*], Yourong Li [1], Tao Wang [2], Wenxi Wang [3,4] and Gangbing Song [3,*]

[1] Key Laboratory for Metallurgical Equipment and Control of Ministry of Education,
   Wuhan University of Science and Technology, Wuhan 430081, China; chen_junda@yahoo.com (J.C.);
   liyourong@wust.edu.cn (Y.L.)
[2] Hubei Key Laboratory of Mechanical Transmission and Manufacturing Engineering,
   Wuhan University of Science and Technology, Wuhan 430081, China; wangtao77@wust.edu.cn
[3] Smart Materials and Structures Laboratory, Department of Mechanical Engineering, University of Houston,
   Houston, TX 77204, USA; wangwenxi_hnu@sina.com
[4] Key Laboratory for Bridge and Wind Engineering of Hunan Province, College of Civil Engineering,
   Hunan University, Changsha 410082, China
* Correspondence: luguangtao@wust.edu.cn (G.L.); gsong@uh.edu (G.S.)

Received: 3 August 2017; Accepted: 28 August 2017; Published: 1 September 2017

Abstract: In this paper, an eddy current tuned mass damper (ECTMD) is utilized to control the vibration of a cantilever beam. The robustness of the ECTMD against frequency detuning is experimentally studied in cases of both free vibration and forced vibration. The natural frequency of the cantilever beam can be adjusted by changing the location of a lumped mass. For purposes of comparison with the ECTMD, the robustness of a tuned mass damper (TMD) is also studied. The experimental results in the free vibration case indicate that the ECTMD works well both in tuned and detuned situations, and the equivalent damping ratio of the cantilever beam equipped with the ECTMD is 2.08~5.91 times that of the TMD. However, the TMD only suppresses the free vibration effectively in the tuned situation. With forced vibration, the experimental results also demonstrate the robustness of the ECTMD in vibration suppression in detuned cases. On the other hand, the cantilever beam with TMD experiences 1.63~2.99 times the peak vibration of that of the ECTMD control.

Keywords: eddy current tuned mass damper (ECTMD); tuned mass damper (TMD); vibration control; frequency detuning; robustness

1. Introduction

Vibration control is commonly used in civil engineering [1–5], and includes passive [6,7], semi-active [8–10] and active approaches [11–13]. In the passive approach, dampers, such as magneto-rheological dampers, fluid viscous dampers, particle dampers and tuned mass dampers (TMDs), have been utilized to suppress structural vibration and dissipate vibration energy of the controlled structures [14–24]. A typical TMD device mainly consists of a spring-mass system and a damper, with mainly the spring-mass-damper model usually being used to investigate the characteristics of TMDs [25–27]. When the frequency of the TMD is tuned to the same as that of the controlled structure, the vibration energy of the controlled structure is transferred to the TMD through the spring-mass system and is dissipated by the damper. Owing to their simplicity and effectiveness, TMDs are widely employed in vibration control for bridges [28–32], high-rise buildings [33–35], and so on. However, TMDs are sensitive to frequency detuning, i.e., when the natural frequency of the primary structure is changed, the control performance of TMD will be degraded.

Recently, eddy current dampers (ECDs) have been proposed to suppress structural vibration, and the control effectiveness of ECDs has been demonstrated both experimentally and numerically [36,37]. ECDs can generate an electromagnetic force to hinder structural movement and dissipate energy through eddy currents, which are induced when an electrical conductor moves through a stationary magnetic field or vice versa. The damping force provided by ECDs mainly depends on the relative moving speed between the conductor and the magnetic field, the strength of the magnetic field, and the gap between the magnet and conductor [38]. Various applications for the use of ECDs for controlling structural vibrations have been studied. Bae et al. [39] investigated the performance of an ECD on vibration control of a cantilever beam. A theoretical model of the ECD was established based on electromagnetic theory, and this model was validated by experiments. Zhang et al. [40] proposed a novel planar ECD that can provide damping forces in two different directions, and the analytical model for simulation of the proposed planar ECD was validated by experiments. Bae et al. [41] applied an ECD to provide additional damping for a tuned mass damper system on the vibration control of a large-scale beam structure, and the feasibility of applying the ECD in the vibration control of large structures was verified. Ao and Reynolds [42] used an ECD to control the vibration of a footbridge and the analytic results demonstrated that the damping provided by the ECD was enhanced under both harmonic and random inputs.

More recently, eddy current tuned mass dampers (ECTMDs), which combine a traditional TMD and an ECD, have been proposed to control undesirable vibrations. Based on the theories of mechanics and electromagnetism, the behaviors of ECTMDs have been studied by many researchers [37–41]. Niu et al. [43] applied an ECTMD on a bridge to suppress the wind-induced vibrations that occur in the bridge hangers. Lu et al. [44] investigated the vibration suppression performance of an ECTMD in Shanghai Center Tower, and the experimental results showed that the acceleration resulting from wind-induced vibration was reduced by 45–60%.

The feasibility and effectiveness of ECTMDs in vibration control has been demonstrated by the aforementioned studies. However, the robustness of ECTMDs has rarely been reported, especially in cases of experimental studies. In this paper, the robustness of an ECTMD against frequency detuning was studied experimentally in cases of both free vibration and forced vibration. Furthermore, the robustness of a traditional TMD was also studied to make a comparison with the ECTMD.

2. Vibration Control of a Cantilever Beam with an Eddy Current Tuned Mass Damper

In this paper, a cantilever beam is used as the primary structure, and an ECTMD is installed on the end of the primary cantilever beam, as shown in Figure 1, to control the vibration of the beam. The ECTMD consists of a small cantilever beam, a permanent magnet and a copper plate fixed on the primary cantilever beam. The small cantilever beam can provide stiffness for the ECTMD and the frequency of the ECTMD can be adjusted by changing the length of the beam. The permanent magnet is placed on the end of the small cantilever beam, and can be regarded as a lumped mass. In this study, the ECTMD or TMD is designed to control the first mode vibration of the primary cantilever beam. When the primary cantilever beam is excited by external forces, the vibration energy can be transferred through the small cantilever beam to the ECTMD.

To investigate the robustness of the ECTMD, the natural frequency of the cantilever beam is designed to be a variable, while the natural frequencies of the TMD and ECTMD are fixed. When the natural frequency of the cantilever beam is set close to the natural frequency of the ECTMD or the TMD, it is defined as a perfectly tuned situation; the opposite conditions are defined as a detuned situation. To study the detuned influence on the robustness of the ECTMD and TMD, the detuned ratio γ is defined by

$$\gamma = \frac{f_p - f_{damp}}{f_{damp}} \times 100\%$$

where f_p is the fundamental natural frequency of the cantilever beam, and f_{damp} is the natural frequency of the damper. When the detuned ratio γ is less than zero, this situation is defined as downward-detuned; and if the detuned ratio γ is larger than zero, it is defined as upward-detuned.

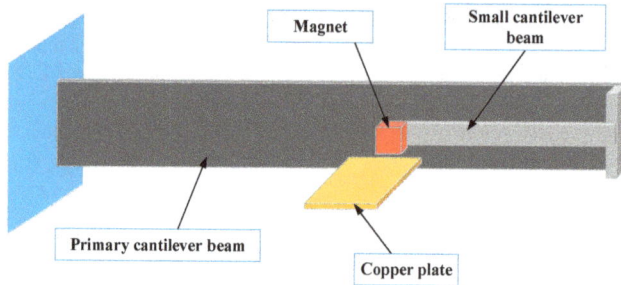

Figure 1. A cantilever beam controlled by an eddy current tuned mass damper (ECTMD).

3. Experimental Setup

The experimental setup is shown in Figure 2. A data acquisition system with 16-bit resolution and 1 MHz maximum repeat rate (IOtech WaveBook 516) is used to collect data and a laser displacement sensor with 0.6 μm repeatability, 150 mm reference distance and ±40 mm measurement range (Keyence LK-G155H) is used to record the displacement of the primary cantilever beam. A vibration exciting motor with an unbalanced mass (WS-25GA370R) is used to generate a harmonic excitation on the end of the cantilever beam. The sampling frequency is set to 1000 Hz.

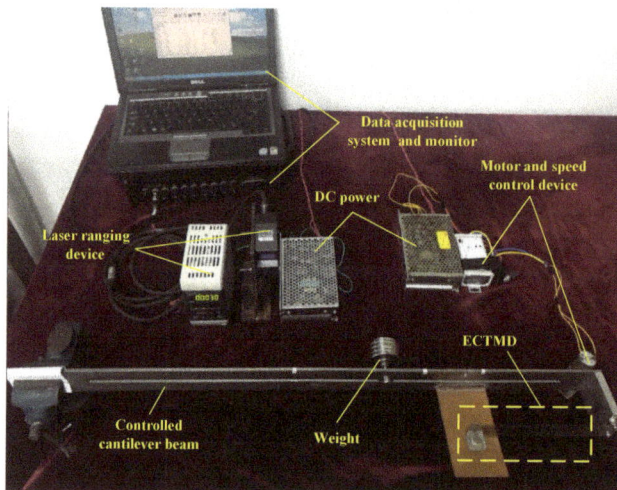

Figure 2. Experimental setup for robustness study.

The higher modes of vibration of the controlled structure will reduce quickly due to the higher damping ratio in higher modes. Therefore, the dominant vibration mode of the cantilever beam will be the first vibration mode. Therefore, the detuned characteristics under the first natural frequency are only considered in this paper. The fundamental frequency of the primary cantilever beam can be adjusted by changing the placement of a lumped mass (0.4 kg) on the cantilever beam, as shown in Figure 3. The primary cantilever beam is shown in Figure 3. When the movable lumped mass is set in the middle of the cantilever beam, the fundamental frequency of the beam is 3.53 Hz. When the weight is placed at the free end of the primary cantilever beam, the fundamental frequency of the cantilever beam is 3.05 Hz, and when it is moved close to the fixed end, the fundamental frequency is 4.06 Hz.

An electric motor with an unbalanced mass, which is used as a vibration exciter, is installed at the free end of the primary cantilever beam, as shown in Figure 3. The parameters of the primary cantilever beam are listed in Table 1.

Table 1. Parameters of the primary cantilever beam.

Parameter	Value
Material	Stainless steel 304
Length	750 mm
Width	40 mm
Thickness	5 mm
Adjustable frequency range	3.06 Hz~4.05 Hz

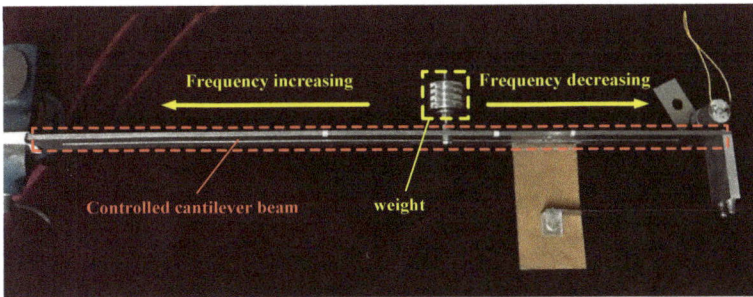

Figure 3. A moveable lumped weight installed on the cantilever beam.

The vibration suppression of the damper is much more effective when the damper is installed at the location where the maximum mode displacement happens [45]. In this study, the maximum mode displacement of the cantilever beam happens at the free end, and therefore the ECTMD or TMD is fixed at the free end.

The configuration of the designed ECTMD is shown in Figure 4. The ECTMD consists of a small cantilever beam, a copper plate and a cube of acrylic glass embedded with a disk magnet. One end of the small cantilever beam is fixed on the controlled cantilever beam, and the other end is attached to the cube acrylic glass. The copper plate is fixed on the primary cantilever beam. The gap between the magnet and the copper plate is 3 mm. When the copper plate is removed from the cantilever beam, the ECTMD becomes a TMD. Since the mass of the thin copper plate is relatively small, the fundamental frequency of the cantilever beam will not be affected.

Some parameters of the small cantilever beam, the copper plate and the cylindrical magnet are listed in Tables 2–4.

When the fundamental frequency of the cantilever beam is changed from 3.06 Hz to 4.05 Hz, dynamic properties of the ECTMD and TMD, such as the frequency, remain at a constant value, equal to 3.53 Hz.

Table 2. Parameters of the small cantilever beam.

Parameter	Value
Material	Stainless steel 304
Length	180 mm
Width	30 mm
Thickness	0.3 mm

Table 3. Parameters of the copper plate.

Parameter	Value
Material	Copper
Thickness	1 mm

Table 4. Parameters of the magnet.

Parameter	Value
Material	N35 (NdFeB)
Diameter	12 mm
Height	10 mm

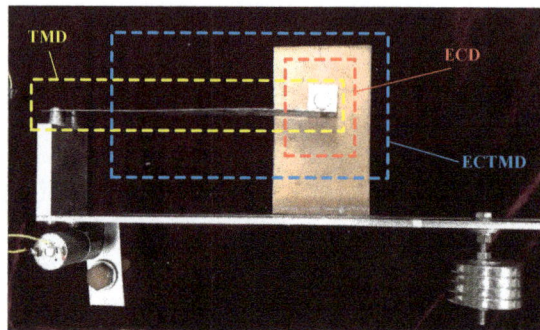

Figure 4. The configuration of the ECTMD.

4. Experimental Results

4.1. Case I: Free Vibration

To test the robustness of the ECTMD and the TMD in the free vibration case, an initial displacement is applied to the free end of the primary cantilever beam. The displacement values of the controlled beam are collected for detuned ratios γ of 0%, 14.7% and −13.3%. The free vibration displacements of the controlled structure in these three detuned situations are shown in Figure 5, and the frequency spectra of the free vibration displacements of the controlled structure in the tuned situation are plotted in Figure 6.

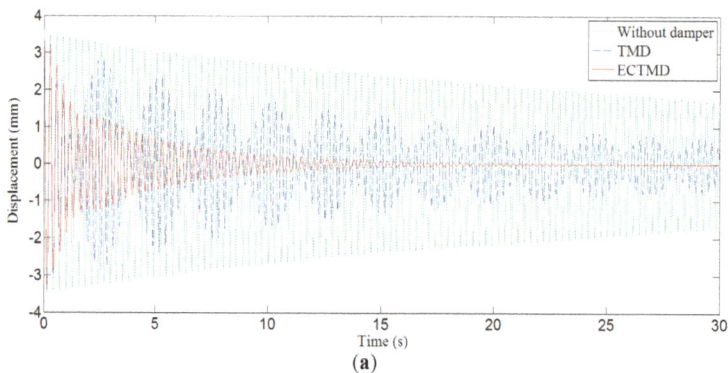

(a)

Figure 5. *Cont.*

(b)

(c)

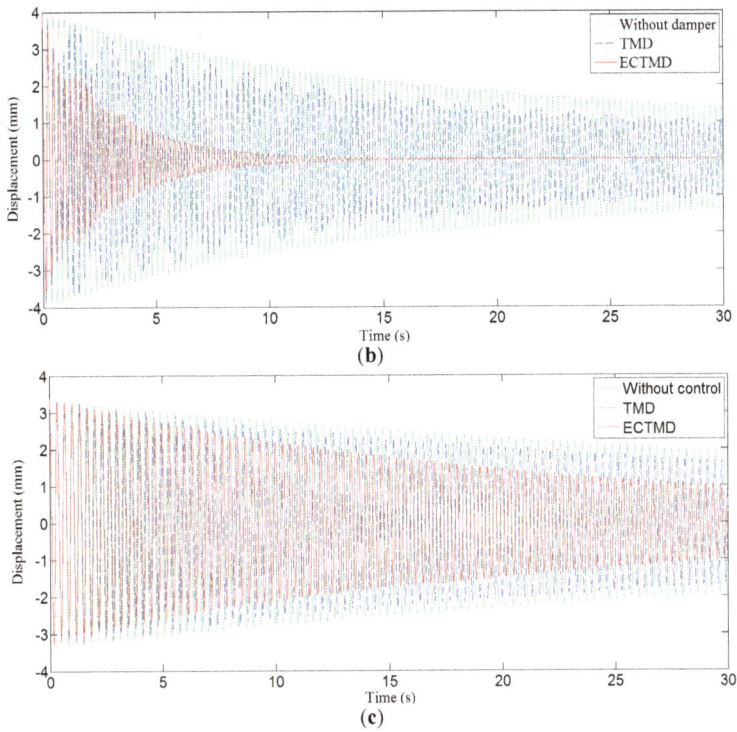

Figure 5. Free vibration responses of the cantilever beam in different situations. (**a**) Detuned ratio 0%; (**b**) Detuned ratio 14.7%; (**c**) Detuned ratio −13.3%.

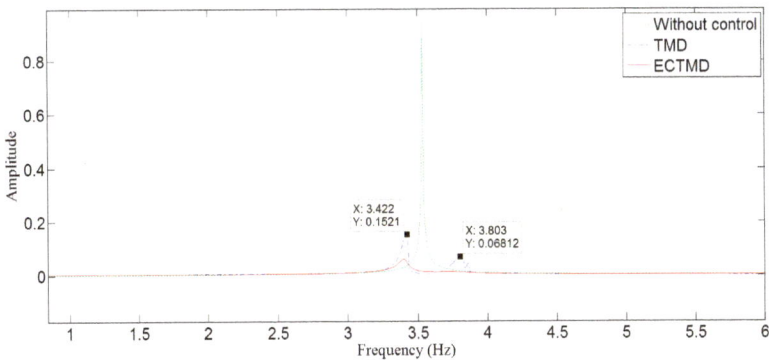

Figure 6. Frequency spectrum of the free vibration response of the cantilever beam in tuned situations.

As shown in Figure 6, there are two dominant frequencies when the controlled structure is with the TMD, and the difference between these two dominant frequencies is only 0.38 Hz, which indicates that a beat vibration [46] is excited in the structure and the vibration energy of the controlled structure is exchanged between the TMD and the controlled structure. Moreover, the beat vibration and vibration energy transformation can also be seen in Figure 5a,b. We noted that there is no beat vibration when the structure is controlled by the ECTMD.

Figure 5a shows that for the situation with a detuned ratio $\gamma = 0\%$, the damping ratios of the cantilever beam are 0.14% without control, 0.53% with the TMD and 1.39% with the ECTMD. Figure 5b shows that for the situation with a detuned ratio $\gamma = 14.7\%$, the damping ratios of the cantilever beam are 0.18% without control, 0.22% with the TMD and 1.3% with the ECTMD. Figure 5c shows that for the situation with a detuned ratio $\gamma = -13.3\%$, the damping ratios of the cantilever beam are 0.11% without control, 0.13% with the TMD and 0.27% with the ECTMD. The damping ratios of the cantilever beam in various cases are summarized in Table 5.

Table 5. Equivalent damping ratios in different situations.

Detuned Ratio	Without Control	With TMD	With ECTMD
$\gamma = 14.7\%$	0.18%	0.22%	1.3%
$\gamma = 0$	0.14%	0.53%	1.39%
$\gamma = -13.3\%$	0.11%	0.13%	0.27%

Both Figure 5 and Table 5 demonstrate that the TMD only works well in the perfectly tuned situation ($\gamma = 0$) while the ECTMD still has considerable effectiveness in both tuned and detuned situations ($\gamma = 0$, $\gamma = 14.7\%$ and $\gamma = -13.3\%$), with the equivalent damping ratio of the cantilever beam with the ECTMD being around 2.08~5.91 times that of the TMD control. In addition, it can be seen that the ECTMD works better in the upward-detuned situation ($\gamma = 14.7\%$) than in the downward-detuned situation ($\gamma = -13.3\%$). Therefore, from the results in Figure 5, it can be seen that the robustness of the ECTMD is better than that of the TMD in the free vibration case.

4.2. Case II: Forced Vibration

To study the robustness of the ECTMD in the forced vibration case, a motor with an unbalanced mass is used to generate the harmonic excitation on the free end of the cantilever beam. The frequency of the excitation load can be varied from 0 Hz to 5 Hz.

Figure 7 plots the steady-state response of the cantilever beam in a wide frequency domain under different detuned situation. The results show that, when the frequency of the excitation load is larger or smaller than that of the cantilever beam, some new resonance peaks are excited in both tuned and detuned situations ($\gamma = 0\%$, $\gamma = 14.7\%$ and $\gamma = -13.3\%$), and the displacement responses of the cantilever beam with the TMD near the resonance peaks are 1.63~2.99 times that with the ECTMD in both tuned and detuned ($\gamma = 0$, $\gamma = 14.7\%$ and $\gamma = -13.3\%$) situations. Therefore, the vibration performances plotted in Figure 7 demonstrate that the robustness of the ECTMD is much better than the TMD in the forced vibration case.

(a)

Figure 7. *Cont.*

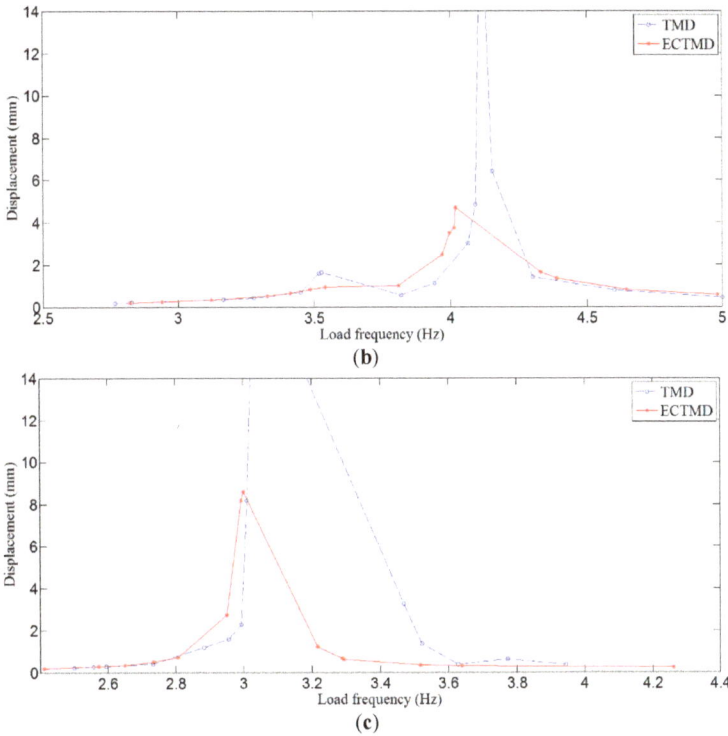

Figure 7. Frequency responses of the cantilever beam in different detuned situations. (a) Detuned ratio 0%; (b) Detuned ratio 14.7%; (c) Detuned ratio −13.3%.

Please note that in Figure 7c the curve (blue dashed line) for the vibration of the primary structure with TMD control is not continuous. The missing segment corresponds to the worst-case vibration of the primary stricture in the case of the TMD control, and was not recorded since the vibration of the primary structure was so severe that the primary structure starts to impact the mass of the TMD.

4.3. Discussion

The experimental results in the free vibration case clearly reveal that a beat vibration is excited in the structure with the TMD, and the TMD reduces the vibration of the controlled structure by transferring the energy to the TMD, and therefore the vibration reduction is effective only in the tuned situation. However, the ECTMD dissipates the vibration energy of the controlled structure by the damping force provided by the ECD and the damping force mainly depends on the relative motion between the damper and the main structure. Therefore, if there is a relative movement, the vibration energy of the controlled structure can be effectively dissipated by the ECTMD in both tuned and de-tuned situations, which results in the strong robustness of the ECTMD.

5. Conclusions

In this paper, the robustness of an ECTMD to control a cantilever beam is studied experimentally under different detuned cases. Moreover, to compare with the performance of the ECTMD, the robustness of a TMD is also studied. The experimental results show that the ECTMD is more robust in both the free vibration case and the forced vibration case. In addition, in the forced vibration case, some new resonance peaks are excited in both perfectly tuned and detuned situations, and the displacement responses of the cantilever beam with the ECTMD near the resonance peaks are much smaller than those with the TMD control in both the downward-detuned and upward-detuned cases. Though we have experimentally demonstrated the robustness of ECTMD in both free vibration and forced vibration cases, an in-depth theoretical analysis of the robustness of the ECTMD is yet to be undertaken, and will be carried out as a future work.

Acknowledgments: The authors are grateful for the financial support from National Natural Science Foundation of China (Grant No.: 51375354 and 51475339).

Author Contributions: All authors discussed and agreed upon the idea, and made scientific contributions. Gangbing Song and Yourong Li designed the experiments and wrote the paper, Junda Chen and Guangtao Lu collected and analyzed the experimental data, Tao Wang and Wenxi Wang designed the ECTMD and revised the paper.

Conflicts of Interest: The authors declare no conflict of interest.

References

1. Sahasrabudhe, S.S.; Nagarajaiah, S. Semi-active control of sliding isolated bridges using MR dampers: An experimental and numerical study. *Earthq. Eng. Struct. Dyn.* **2005**, *34*, 965–983. [CrossRef]
2. Ou, J.; Zhang, W. Analysis and design method for wind-induced vibration passive damping control of tall buildings. *J. Build. Struct.* **2003**, *6*, 4.
3. Basu, B.; Bursi, O.S.; Casciati, F.; Casciati, S.; Del Grosso, A.E.; Domaneschi, M.; Faravelli, L.; Holnicki-Szulc, J.; Irschik, H.; Krommer, M. A european association for the control of structures joint perspective. Recent studies in civil structural control across europe. *Struct. Control Health Monit.* **2014**, *21*, 1414–1436. [CrossRef]
4. Li, L.; Song, G.; Ou, J. Hybrid active mass damper (AMD) vibration suppression of nonlinear high-rise structure using fuzzy logic control algorithm under earthquake excitations. *Struct. Control Health Monit.* **2011**, *18*, 698–709. [CrossRef]
5. Spencer, B., Jr.; Nagarajaiah, S. State of the art of structural control. *J. Struct. Eng.* **2003**, *129*, 845–856. [CrossRef]
6. Casciati, F.; De Stefano, A.; Matta, E. Simulating a conical tuned liquid damper. *Simul. Model. Pract. Theory* **2003**, *11*, 353–370. [CrossRef]
7. Casciati, F.; Faravelli, L. A passive control device with sma components: From the prototype to the model. *Struct. Control Health Monit.* **2009**, *16*, 751–765. [CrossRef]
8. Casciati, F.; Rodellar, J.; Yildirim, U. Active and semi-active control of structures–theory and applications: A review of recent advances. *JIMSS* **2012**, *23*, 1181–1195. [CrossRef]
9. Nagarajaiah, S.; Sonmez, E. Structures with semiactive variable stiffness single/multiple tuned mass dampers. *J. Struct. Eng.* **2007**, *133*, 67–77. [CrossRef]
10. Casciati, F.; Domaneschi, M. Semi-active electro-inductive devices: Characterization and modelling. *JVC* **2007**, *13*, 815–838. [CrossRef]
11. Casciati, S.; Faravelli, L.; Chen, Z. A DC-motor based active-mass-damper with wireless feedback for structural control. In Proceedings of the International Symposium on Innovation & Sustainability of Structures in Civil Engineering, Xiamen, China, 28–30 October 2011; Southeast University Press: Xiamen, China, 2011; pp. 560–565.
12. Casciati, S.; Chen, Z. An active mass damper system for structural control using real-time wireless sensors. *Struct. Control Health Monit.* **2012**, *19*, 758–767. [CrossRef]
13. Li, L.; Song, G.; Ou, J. Adaptive fuzzy sliding mode based active vibration control of a smart beam with mass uncertainty. *Struct. Control Health Monit.* **2011**, *18*, 40–52. [CrossRef]
14. Yang, G.; Spencer, B.F., Jr.; Jung, H.-J.; Carlson, J.D. Dynamic modeling of large-scale magnetorheological damper systems for civil engineering applications. *J. Eng. Mech.* **2004**, *130*, 1107–1114. [CrossRef]
15. Lu, G.; Li, Y.; Song, G. Analysis of a magneto-rheological coupler with misalignment. *Smart Mater. Struct.* **2011**, *20*, 105028. [CrossRef]
16. Zhengqiang, Y.; Aiqun, L.; Youlin, X. Fluid viscous damper technology and its engineering application for structural vibration energy dissipation. *J. Southeast Univ.* **2002**, *32*, 466–473.

17. Isalgue, A.; Lovey, F.C.; Terriault, P.; Martorell, F.; Torra, R.M.; Torra, V. SMA for dampers in civil engineering. *Mater. Trans.* **2006**, *47*, 682–690. [CrossRef]

18. Demetriou, D.; Nikitas, N. A novel hybrid semi-active mass damper configuration for structural applications. *Appl. Sci.* **2016**, *6*, 397. [CrossRef]

19. Zhang, Z.; Ou, J.; Li, D.; Zhang, S. Optimization design of coupling beam metal damper in shear wall structures. *Appl. Sci.* **2017**, *7*, 137. [CrossRef]

20. Li, H.; Qian, H.; Song, G.; Gao, D. Type of shape memory alloy damper: Design, experiment and numerical simulation. *J. Vib. Eng.* **2008**, *21*, 179–184.

21. Song, G.; Zhang, P.; Li, L.; Singla, M.; Patil, D.; Li, H.; Mo, Y. Vibration control of a pipeline structure using pounding tuned mass damper. *J. Eng. Mech.* **2016**, *142*, 04016031. [CrossRef]

22. Lin, W.; Song, G.; Chen, S. PTMD control on a benchmark TV tower under earthquake and wind load excitations. *Appl. Sci.* **2017**, *7*, 425. [CrossRef]

23. Zhao, D.; Li, H.; Song, G.; Qian, H. Experiment investigation for a new type of piezoelectric friction damper. In Proceedings of the Earth and Space 2014, St. Louis, MO, USA, 27–29 October 2014; ASCE: Reston, VA, USA, 2014; pp. 642–647.

24. Yang, G.; Spencer, B.; Carlson, J.; Sain, M. Large-scale mr fluid dampers: Modeling and dynamic performance considerations. *Eng. Struct.* **2002**, *24*, 309–323. [CrossRef]

25. Cao, H.; Reinhorn, A.; Soong, T. Design of an active mass damper for a tall tv tower in Nanjing, China. *Eng. Struct.* **1998**, *20*, 134–143. [CrossRef]

26. Casciati, F.; Ubertini, F. Nonlinear vibration of shallow cables with semiactive tuned mass damper. *Nonlinear Dyn.* **2008**, *53*, 89–106. [CrossRef]

27. Rana, R.; Soong, T. Parametric study and simplified design of tuned mass dampers. *Eng. Struct.* **1998**, *20*, 193–204. [CrossRef]

28. Lin, Y.-Y.; Cheng, C.-M.; Lee, C.-H. A tuned mass damper for suppressing the coupled flexural and torsional buffeting response of long-span bridges. *Eng. Struct.* **2000**, *22*, 1195–1204. [CrossRef]

29. Tubino, F.; Piccardo, G. Tuned mass damper optimization for the mitigation of human-induced vibrations of pedestrian bridges. *Meccanica* **2015**, *50*, 809–824. [CrossRef]

30. Abdel-Rohman, M.; John, M.J. Control of wind-induced nonlinear oscillations in suspension bridges using a semi-active tuned mass damper. *J. Vib. Control* **2006**, *12*, 1049–1080. [CrossRef]

31. Lin, C.; Wang, J.-F.; Chen, B. Train-induced vibration control of high-speed railway bridges equipped with multiple tuned mass dampers. *J. Bridge Eng.* **2005**, *10*, 398–414. [CrossRef]

32. Jo, B.-W.; Tae, G.-H.; Lee, D.-W. Structural vibration of tuned mass damper-installed three-span steel box bridge. *Int. J. Press. Vessels. Pip.* **2001**, *78*, 667–675. [CrossRef]

33. Kawaguchi, A.; Teramura, A.; Omote, Y. Time history response of a tall building with a tuned mass damper under wind force. *J. Wind Eng. Ind. Aerodyn.* **1992**, *43*, 1949–1960. [CrossRef]

34. Kang, J.; Kim, H.S.; Lee, D.G. Mitigation of wind response of a tall building using semi-active tuned mass dampers. *Struct. Des. Tall Spec. Build.* **2011**, *20*, 552–565. [CrossRef]

35. Youssef, N. Supertall buildings with tuned mass damper. *Struct. Des. Tall Spec. Build.* **1994**, *3*, 1–12. [CrossRef]

36. Ebrahimi, B.; Khamesee, M.B.; Golnaraghi, F. A novel eddy current damper: Theory and experiment. *J. Phys. D Appl. Phys.* **2009**, *42*, 075001. [CrossRef]

37. Sodano, H.A.; Bae, J.-S.; Inman, D.J.; Belvin, W.K. Concept and model of eddy current damper for vibration suppression of a beam. *J. Sound Vib.* **2005**, *288*, 1177–1196. [CrossRef]

38. Van Beek, T.; Pluk, K.; Jansen, J.; Lomonova, E. Optimization and measurement of eddy current damping applied in a tuned mass damper. In Proceedings of the 2014 International Conference on Electrical Machines (ICEM), Berlin, Germany, 2–5 September 2014; IEEE: New York, NY, USA; pp. 609–615.

39. Bae, J.-S.; Kwak, M.K.; Inman, D.J. Vibration suppression of a cantilever beam using eddy current damper. *J. Sound Vib.* **2005**, *284*, 805–824. [CrossRef]

40. Zhang, H.; Kou, B.; Jin, Y.; Zhang, L.; Zhang, H.; Li, L. Modeling and analysis of a novel planar eddy current damper. *J. Appl. Phys.* **2014**, *115*, 17E709. [CrossRef]

41. Bae, J.-S.; Hwang, J.-H.; Kwag, D.-G.; Park, J.; Inman, D.J. Vibration suppression of a large beam structure using tuned mass damper and eddy current damping. *Shock Vib.* **2014**, *2014*, 893914. [CrossRef]

42. Ao, W.K.; Reynolds, P. Analytical and experimental study of eddy current damper for vibration suppression in a footbridge structure. In *Dynamics of Civil Structures*; Conference Proceedings of the Society for Experimental Mechanics Series; Springer: New York, NY, USA, 2017; pp. 131–138.

43. Niu, H.; Chen, Z.; Lei, X.; Hua, X. Development of eddy current turned mass damper for suppressing windinduced vibration of bridge hangers. In Proceedings of the 6th European and African Wind Engineering Conference, Cambridge, UK, 7–11 July 2013.

44. Lu, X.; Zhang, Q.; Weng, D.; Zhou, Z.; Wang, S.; Mahin, S.A.; Ding, S.; Qian, F. Improving performance of a super tall building using a new eddy-current tuned mass damper. *Struct. Control Health Monit.* **2017**, *24*, e1882. [CrossRef]

45. Li, H.; Zhang, P.; Song, G.; Patil, D.; Mo, Y. Robustness study of the pounding tuned mass damper for vibration control of subsea jumpers. *Smart Mater. Struct.* **2015**, *24*, 095001. [CrossRef]

46. Thomson, W.T.; Dahleh, M.D. *Theory of Vibration with Applications*, 5th ed.; Prentice Hall: Upper Saddle River, NJ, USA, 1998; pp. 131–135. ISBN 013651068X.

applied sciences

MDPI

Article

Experimental Study on Vibration Control of a Submerged Pipeline Model by Eddy Current Tuned Mass Damper

Wenxi Wang [1,2], Dakota Dalton [2], Xugang Hua [1], Xiuyong Wang [3,*], Zhengqing Chen [1] and Gangbing Song [2,*]

[1] Key Laboratory for Bridge and Wind Engineering of Hunan Province, College of Civil Engineering, Hunan University, Changsha 410082, China; wxwang@hnu.edu.cn (W.W.); cexghua@hnu.edu.cn (X.H.); zqchen@hnu.edu.cn (Z.C.)
[2] Department of Mechanical Engineering, University of Houston, Houston, TX 77204, USA; ddalton94@hotmail.com
[3] College of Civil Engineering, Hunan University of Science and Technology, Xiangtan 411201, China
* Correspondence: cexywang@hnust.edu.cn (X.W.); gsong@uh.edu (G.S.); Tel.: +1-832-606-1000 (G.S.)

Received: 4 September 2017; Accepted: 22 September 2017; Published: 25 September 2017

Abstract: Undesirable vibrations occurring in undersea pipeline structures due to ocean currents may shorten the lifecycle of pipeline structures and even lead to their failure. Therefore, it is desirable to find a feasible and effective device to suppress the subsea vibration. Eddy current tuned mass damper (ECTMD), which employs the damping force generated by the relative movement of a non-magnetic conductive metal (such as copper or aluminum) through a magnetic field, is demonstrated to be an efficient way in structural vibration control. However, the feasibility and effectiveness of ECTMD in a seawater environment has not been reported on before. In this paper, an experiment is conducted to validate the feasibility of an eddy current damper in a seawater environment. A submerged pipeline is used as the controlled structure to experimentally study the effectiveness of ECTMD. The dynamic properties of the submerged pipeline are obtained from dynamic tests and the finite element method (FEM). The optimum design of TMD with a linear spring-damper element for a damped primary structure is carried out through numerical optimization procedures and is used to determine the optimal frequency tuning ratio and damping ratio of ECTMD. In addition, the performance of ECTMD to control the submerged pipeline model is respectively studied in free vibration case and forced vibration case. The results show that the damping provided by eddy current in a seawater environment is only slightly varied compared to that in an air environment. With the optimal ECTMD control, vibration response of the submerged pipeline is significantly decreased.

Keywords: eddy current damping; tuned mass damper; submerged pipeline model; vibration control; seawater environment

1. Introduction

Many control devices have been proposed and successfully applied to control onshore civil structures in past decades [1–7]. Nowadays, thousands of kilometers of pipelines have been installed on the seabed to transport oil and gas. Undesirable vibrations of subsea pipelines subjected to the ocean current may cause the fatigue of pipeline structures and shorten their lifecycle [8]. The control devices used to suppress vibration in onshore structures may be unsuitable or may be too hard to implement in a seawater environment. Therefore, an effective control device which can be employed in seawater environments for vibration suppression of subsea pipelines should be developed. Tuned mass damper (TMD) [9] is a traditional passive control device that has been successfully applied in vibration control of civil structures [10–15] and machining processes [16–18], which consists of a spring-mass system and a damper. When the frequency of the TMD is tuned closely to that of the primary structure, the vibration energy of the primary structure will be transferred to the TMD through the stiffness force and partly dissipated by the damping force [19,20].

In the previous studies, conventional fluid dampers such as viscous oil dampers [21], Magneto rheological (MR) dampers [22] and impact dampers [23,24] are used to dissipate vibration energy in a TMD system. It should be noted that MR dampers are also employed to control the frequency of a semi-active TMD system in real-time [25,26]. Since the friction force always exists in the relative movement between the TMD and the fluid damper, the conventional TMD equipped with a fluid damper is only activated when the acceleration of the controlled structure exceeds a certain value [14,15]. Because the damping force generated by eddy currents is a non-contact force, no friction force exists in the relative movement between the TMD and eddy current damper. Moreover, compared with conventional oil damper, the eddy current damper is insensitive to the change of temperature. In the regard of those points, eddy current dampers have been proposed to provide the additional damping for a TMD system [27,28]. The damping force generated by the eddy currents is proportional to the relative velocity, which can be regarded as a linear viscous damping [27–31]. Some works on the vibration suppression using eddy current tuned mass damper (ECTMD) have been presented. Larose et al. [32] investigated the vibration control of a wind tunnel bridge model using an ECTMD. In their device, the damping ratio of the ECTMD can be adjusted by changing the clearance between the magnet and the aluminum conductive plate. Bae et al. [33] designed an ECTMD to control the vibration of a cantilever beam and proposed a relatively simple method to calculate the damping force provided by eddy currents numerically. For verifying that an ECTMD is an effective way to control a large-scale structure, a large scale ECTMD was designed and constructed to control a large beam structure [34]. Wen et al. [35] installed an ECTMD system on a footbridge to control the human-induced vibration, resulting in the large-amplitude vibrations being suppressed. Bourquin et al. [36] proposed two ECTMDs to control the first two modes of a bridge mock-up and found that the maximum achievable damping is obtained when a desirable viscous damping level is selected. However, the feasibility of using ECTMDs for vibration control in a seawater environment has not been studied before.

In this paper, an underwater experiment is conducted to validate the feasibility of eddy current damping in a seawater environment. A submerged pipeline is employed as the controlled structure to experimentally study the effectiveness of an ECTMD and the dynamic properties of the pipeline are obtained from dynamic tests and the finite element method (FEM). The optimum design of a TMD for the damped primary structure is carried out through numerical search procedures and is used to determine the optimal tuning ratio and damping ratio of the ECTMD. Finally, the effectiveness of an optimal ECTMD to control the submerged pipeline is experimentally demonstrated both in the free vibration case and forced vibration case.

2. Effectiveness of Eddy Current Damping in Seawater

2.1. Eddy Current Damping

Eddy current damping is generated by magnetic inductions. When a conductive plate moves through a stationary magnetic field, or vice versa, eddy currents will appear in the conductive plate to hinder the movement of the conductive plate [24–26], as shown in Figure 1. The strength of the damping depends upon the strength of the magnetic field, the surface area, and conductivity of the conductive plate, and the rate of change of the magnetic flux [29].

2.2. Free Vibration Tests of a Eddy Current Damper in Seawater Environment

As mentioned before, eddy current dampers have been successfully applied in TMD systems as a linear damping source. The key issue for applying ECTMD to control vibrations occurring in subsea pipelines is the validation of the damping effect of eddy currents in a seawater environment.

Studies worked on the application of eddy current damping in a seawater environment have not been reported on before. In this paper, to verify the feasibility of eddy current damping in a seawater environment, the free vibration responses of a cantilever beam are measured in four cases and the damping ratios of each case are calculated.

The experimental setup is shown in Figure 2a–c. A laser displacement sensor is used to collect the response of the cantilever beam and a seawater container is used to carry seawater to simulate the seawater environment. Moreover, an eddy current damping device, as shown in Figure 2d, with two permanent magnets is fixed on the bottom of the water container to provide extra damping force for the cantilever beam. The sample frequency of the laser displacement sensor is set to 100 Hz and the gap between the magnets and the aluminum plate is set to 5 mm.

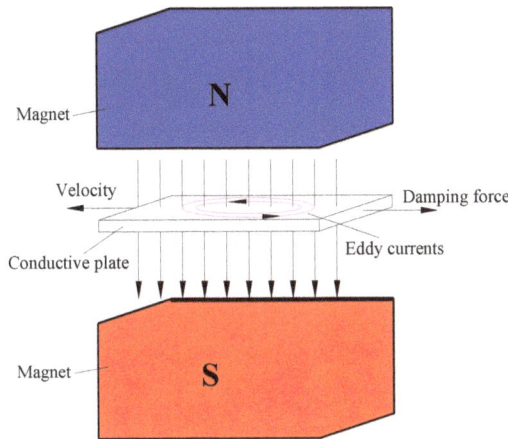

Figure 1. Generation of eddy current damping in a conductive plate.

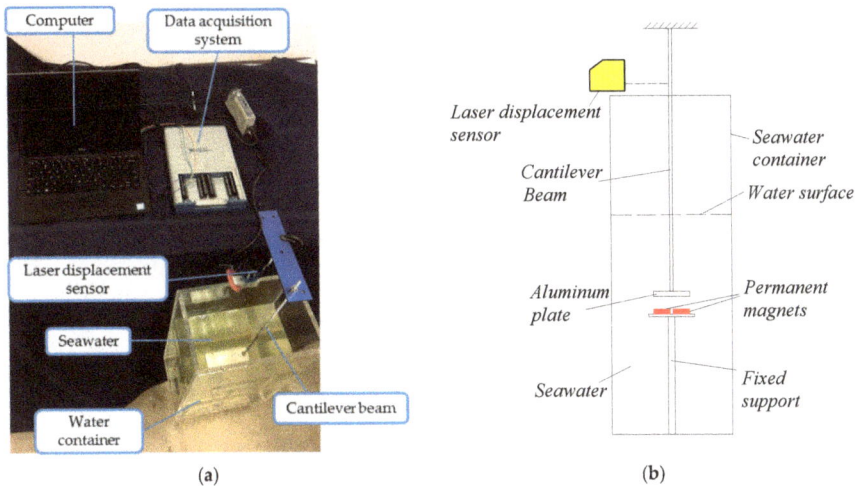

(a)

(b)

Figure 2. *Cont.*

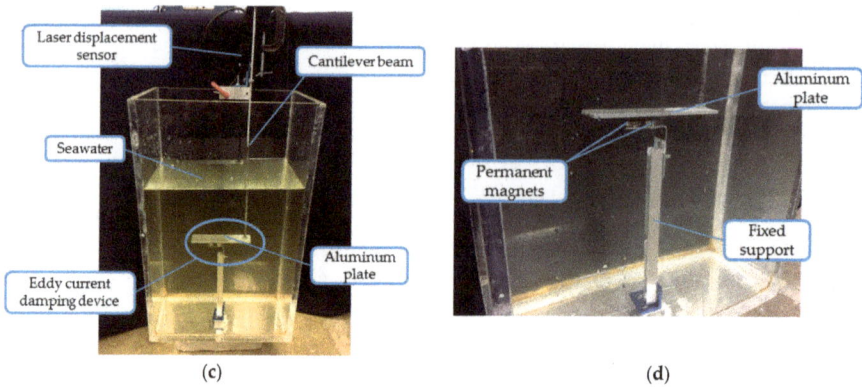

Figure 2. Experimental setup for free vibration tests: (**a**) global view; (**b**) side view of seawater container; (**c**) front view of seawater container; (**d**) eddy current damper.

2.3. Experimental Results

The free vibration of the cantilever beam is measured in four cases to study the additional damping provided by the eddy currents. Figure 3 plots the free vibration response of the cantilever beam in these four cases. The average damping ratios of the cantilever beam and descriptions of the four cases are given in Table 1. It can be seen that the original damping ratios of the cantilever beam without eddy current damping are 0.55% and 2.87% in the air and in the seawater environment, respectively. The damping ratios with eddy current damping are increased from the original values to 7.33% and 8.98%, respectively. The increased damping ratio provided by eddy current damping is 6.78% in air and 6.02% in the seawater environment. It is shown that the influence of the seawater environment on the eddy current damping is limited and the feasibility of using eddy current damping in seawater is validated.

Moreover, the damping ratio in the first five vibration cycles are obtained from the logarithmic decrement method and plotted in Figure 4. It can be seen that the damping ratio of the cantilever beam at first one or two cycles is larger than that at last few cycles. This phenomenon indicates that the structural damping is much higher during large-amplitude vibration.

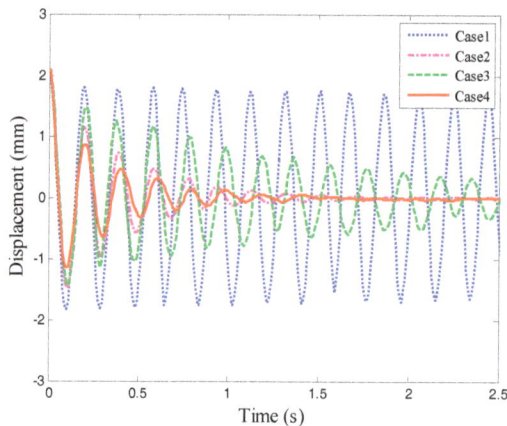

Figure 3. Free vibration response of cantilever beam in each case.

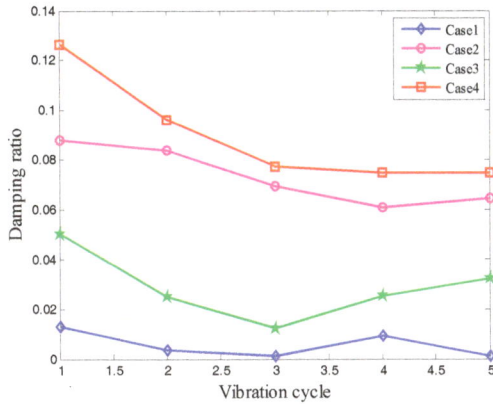

Figure 4. Damping ratio of cantilever beam in the first five vibration cycles.

Table 1. Average damping ratio for each case.

Experimental Case	Case Description	Average Damping Ratio	Increased Damping Ratio
Case1	Without eddy current damping in air	0.55%	In air: 6.78%
Case2	With eddy current damping in air	7.33%	
Case3	Without eddy current damping in seawater	2.87%	In seawater: 6.02%
Case4	With eddy current damping in seawater	8.98%	

3. Experimental Model of a Submerged Pipeline

A submerged pipeline built with polyvinyl chloride (PVC) pipe is used as the controlled structure to verify the effectiveness of an ECTMD in a seawater environment, as shown in Figure 5a. Moreover, the finite element (FE) model of the pipeline shown in Figure 5b is established in ANSYS (11.0, ANSYS Inc., Canonsburg, PA, USA, 2009) and the dynamic properties such as frequency, modal shape, and modal mass are computed. The first vibration mode of the pipeline is selected as the controlled target and the mode shape of this mode is plotted in Figure 5c. The free vibration test of the submerged pipeline is conducted by giving an initial displacement in the middle of the pipeline. Free vibration history is shown in Figure 6a and its Fast Fourier Transform (FFT) amplitude in the frequency domain is plotted in Figure 6b. The dynamic properties of the pipeline are listed in Table 2, which were obtained through experimental testing and FE computation. The modal frequency and modal mass are 3.2 Hz and 0.728 kg, respectively. The vertical damping ratio of the pipeline is 2.4%, which is a relatively high value compared with onshore civil structures.

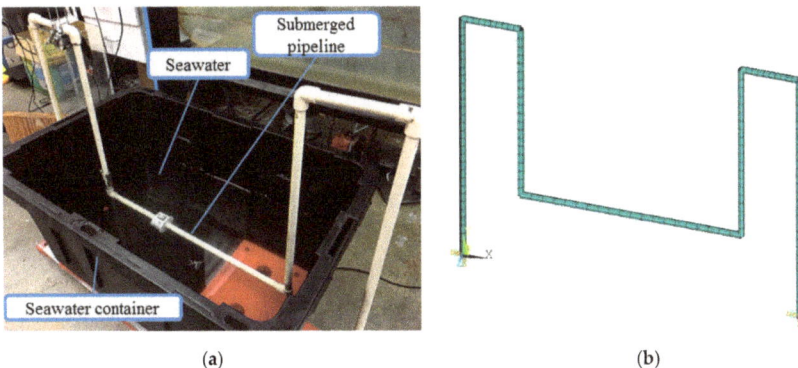

(a)

(b)

Figure 5. *Cont.*

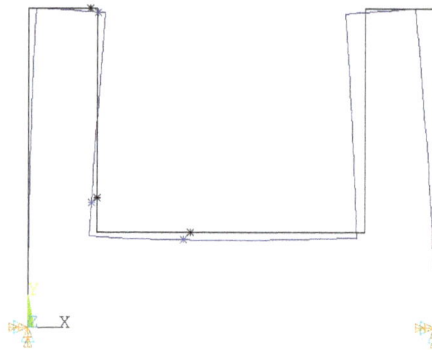

(c)

Figure 5. Submerged pipeline model and its finite element model: (**a**) polyvinyl chloride (PVC) pipeline in the seawater container; (**b**) finite element (FE) model; (**c**) first vibration mode shape.

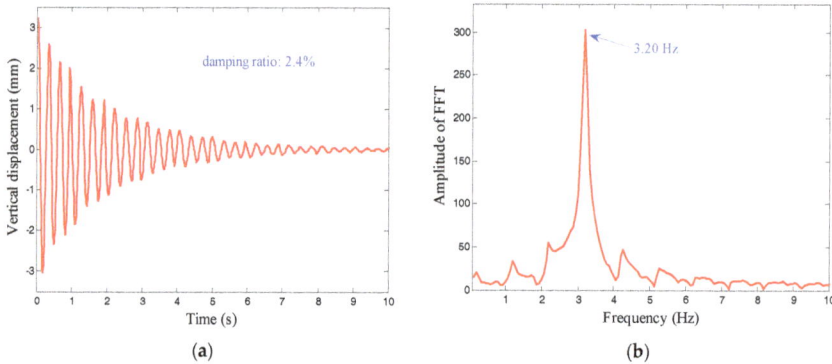

(a)

(b)

Figure 6. Vertical response and its Fast Fourier transform (FFT) amplitude of the submerged pipeline in seawater environment: (**a**) free vibration response; (**b**) FFT amplitude.

Table 2. Modal parameters of the pipeline structure.

Result	Frequency	Damping Ratio	Modal Mass
FE computation	3.26 Hz	-	0.728 kg
Experimental result	3.20 Hz	2.4%	-

4. Optimal Design of ECTMD for Damped Structure

If its tuning ratio and damping ratio is off-tuned, the ECTMD will lose its control efficiency. Due to the eddy current damping being linearly proportional to the relative velocity between ECTMD and the primary structure, the optimization procedures of TMD with a linear viscous damper can be employed to design the optimal parameters of the ECTMD. The optimal tuning ratio and damping ratio of a tuned mass damper (TMD) system for an undamped structure are given by [37],

$$r_{opt} = \frac{1}{1+\mu} \tag{1}$$

$$\zeta_{opt} = \sqrt{\frac{3\mu}{8(1+\mu)}} \tag{2}$$

where r_{opt} and ζ_{opt} are the optimal frequency tuning ratio and the optimal damping ratio of the TMD respectively, $\mu = m_{tmd}/m_s$ is the mass ratio of the TMD, m_{tmd} and m_s are the masses of the primary structure and the TMD, respectively.

The damping ratio of the primary structure before a TMD is attached can be calculated by $\zeta_s = c_s/(2m_s(k_s/m_s)^{1/2})$. For a primary structure with a low damping ratio, such as 0.5% or lower, Equations (1) and (2) can be used to design the optimal parameters of TMD. When the damping ratio of the controlled structure is higher than 2%, such as the tested submerged pipeline in the Section 3, the optimal tuning ratio and damping ratio calculated by Equations (1) and (2) may result in detuning to the ECTMD system, and reduce its optimal control ability [38,39]. An optimum design of a TMD for a damped controlled structure is proposed by Thompson [40] using the frequency focus method. This method for finding the optimal parameter of a TMD is relatively complicated. Thompson's work focused on highly damped structures, but the damping ratios of most civil structures are under 5%. Therefore, the optimal parameters of a TMD system for damped structures based on H_∞ optimization method are studied in this paper. The equation of motion of the damped structure with a linear TMD control, as shown in Figure 7, can be expressed as Equation (3),

$$\begin{cases} m_s\ddot{x}_s + c_s\dot{x}_s + k_sx_s + c_{tmd}(\dot{x}_s - \dot{x}_{tmd}) + k_{tmd}(x_s - x_{tmd}) = f_0\sin(\omega t) \\ m_{tmd}\ddot{x}_{tmd} + c_{tmd}(\dot{x}_{tmd} - \dot{x}_s) + k_{tmd}(x_{tmd} - x_s) = 0 \end{cases} \tag{3}$$

where m_s, c_s, k_s, x_s are the mass, damping coefficient, stiffness, and displacement of the controlled structure, and m_{tmd}, c_{tmd}, k_{tmd}, x_{tmd} are the mass, damping coefficient, stiffness, and displacement of the TMD. $f_0\sin(\omega t)$ is the external excitation applied on the controlled structure.

The dynamic magnification factor of the primary structure is defined as $DMF(\omega) = x_{s0}k_s/f_0$, and it can be expressed as Equations (4),

$$DMF(\omega) = \frac{x_{s0}}{f_0/k_s} = \frac{k_s}{\left|(k_s + i\omega c_s - m_s\omega^2) + k_{tmd} + i\omega c_{tmd} - \frac{(k_{tmd} + i\omega c_{tmd})^2}{k_{tmd} + i\omega c_{tmd} - m_{tmd}\omega^2}\right|} \tag{4}$$

where i is complex unit $(-1)^{1/2}$ and x_{s0} is the amplitude of steady-state response. The modal mass ratio, tuning ratio, and damping ratio of TMD are defined as $\mu = m_{tmd}/m_s$, $r = (k_{tmd}/m_{tmd})^{1/2}/(k_s/m_s)^{1/2}$, and $\zeta = c_{tmd}/(2m_{tmd}(k_{tmd}/m_{tmd})^{1/2})$, respectively. The optimization objective function based on H_∞ optimization method is written as Equation (5) [41,42]. It should be noted the objective function is aimed at minimizing the displacement response of the primary structure. The objective function targeted to minimize the acceleration of the primary structure can be found in reference [43].

$$\min_{r,\zeta} \max_{\omega} DMF(\omega) \tag{5}$$

Equation (5) is solved through a numerical optimization and the optimal parameters of the TMD are achieved. The optimal tuning ratio and damping ratio of a TMD for the damped structure from the numerical optimization are shown in Figure 8 along with the optimal parameters obtained from Equations (1) and (2). The mass ratio of TMD μ is varied from 0.01 to 0.1 with an interval of 0.01. It can be observed from Figure 8 that the optimal parameters obtained by the numerical optimization have the same values with the results calculated from Equations (1) and (2) when it is designed for an undamped structure. When the damping ratio of the controlled structure is increased, the optimal tuning ratio of TMD is lower than the value for the undamped structure, and the optimal damping ratio of TMD will increase slightly. The optimal parameters of a TMD for different mass ratios and various damping ratios of primary structures are listed in Table 3.

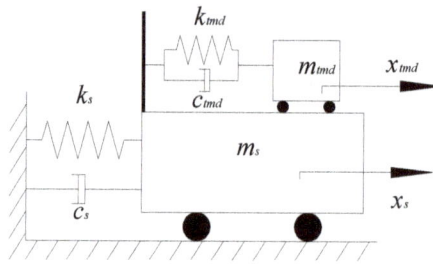

Figure 7. Mechanical model of damped structure with a tuned mass damper (TMD).

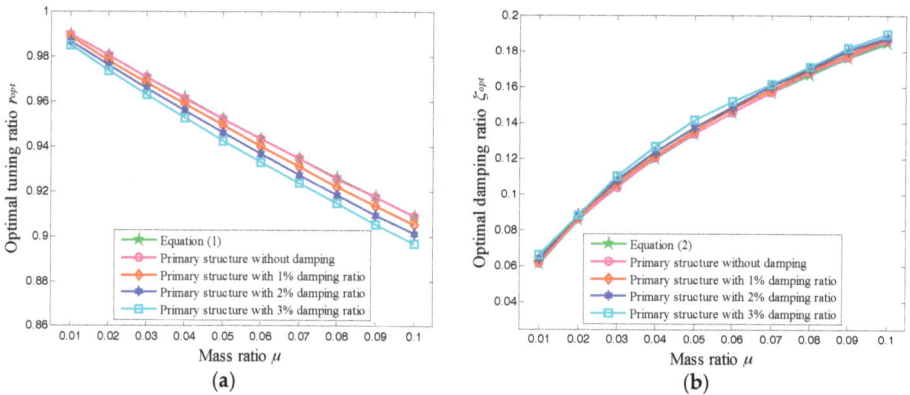

Figure 8. Optimal frequency tuning ratio r_{opt} and damping ζ_{opt} of a TMD for damped structure: (a) optimal frequency tuning ratio; (b) optimal damping ratio.

Table 3. Optimal parameters of a TMD for the damped structure.

Mass Ratio	$\zeta_s = 1\%$		$\zeta_s = 2\%$		$\zeta_s = 3\%$	
μ	r_{opt}	ζ_{opt}	r_{opt}	ζ_{opt}	r_{opt}	ζ_{opt}
$\mu = 0.01$	0.9886	0.0625	0.9869	0.0643	0.9851	0.0662
$\mu = 0.02$	0.9783	0.0874	0.9761	0.0886	0.9738	0.0883
$\mu = 0.03$	0.9684	0.1064	0.9658	0.1084	0.9630	0.1100
$\mu = 0.04$	0.9588	0.1213	0.9558	0.1233	0.9526	0.1268
$\mu = 0.05$	0.9493	0.1355	0.9461	0.1371	0.9426	0.1413
$\mu = 0.06$	0.9401	0.1473	0.9367	0.1488	0.9330	0.1518
$\mu = 0.07$	0.9311	0.1584	0.9274	0.1604	0.9237	0.1614
$\mu = 0.08$	0.9222	0.1691	0.9184	0.1702	0.9145	0.1710
$\mu = 0.09$	0.9135	0.1784	0.9095	0.1805	0.9053	0.1817
$\mu = 0.1$	0.9051	0.1869	0.9010	0.1877	0.8967	0.1896

5. Effectiveness of ECTMD to Control a Submerged Pipeline

In general, the mass ratio of a TMD cannot be too large. Therefore, the mass of the designed ECTMD added on the experimental pipeline structure is 32 g, which means the mass ratio μ is 4.4%. Considering the mass ratio $\mu = 4.4\%$ and the primary damping ratio of the pipeline structure $\zeta_s = 2.4\%$, the optimal tuning ratio and damping ratio of the ECTMD are obtained from Table 3 to be 0.9508 and 12.63%, respectively. For the implementation of the ECTMD, two tiny cantilever beams are used to provide stiffness for the ECTMD and the eddy current damper are utilized to provide the additional damping for the ECTMD. The natural frequency and damping ratio of the ECTMD can be adjusted through changing the length of the cantilever beams and the gap between the aluminum plates and the magnets. The configuration of the ECTMD and its eddy current damper device for the submerged pipeline are shown in Figure 9. The eddy current damper device consists of two aluminum plates with a thickness of 3 mm and permanent magnets. The ECTMD is installed in the mid-span of the pipeline.

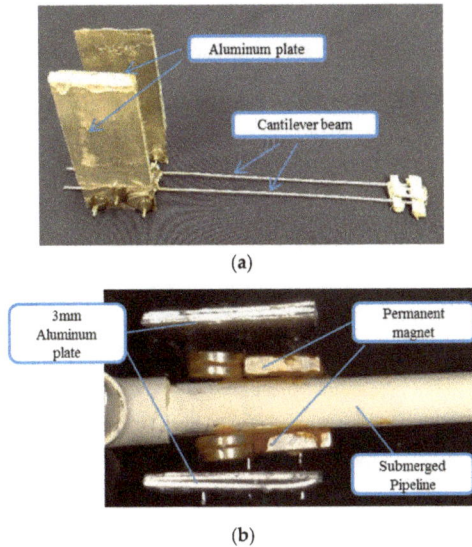

(a)

(b)

Figure 9. Implementation of the eddy current tuned mass damper (ECTMD) on the pipeline model: (a) configuration of ECTMD; (b) eddy current damper device.

To obtain the fundamental natural frequency and damping ratio of the ECTMD, free vibration tests of ECTMD are conducted in the seawater environment. When the length of the cantilever beam is set to 13.5 cm and the gap between magnets and aluminum plates is set to 4.2 mm, the frequency and the damping ratio is quite close to the optimal values. Figure 10 shows the vertical free vibration history of the ECTMD device with and without magnets, and the frequency responses are also shown in Figure 10. It can be seen that the natural frequency and the damping ratio of the ECTMD without magnets are respectively, 3.07 Hz and 6.1%, in the seawater environment. However, when the permanent magnets are added and eddy current damping is activated, the damping ratio of the ECTMD is sharply increased from 6.1% to 13.9%, which helps the ECTMD to achieve the optimal damping. The parameters of the ECTMD in the experiment are summarized in Table 4.

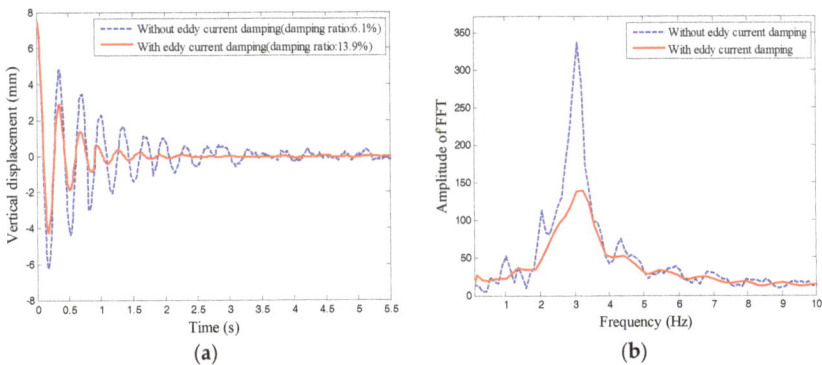

(a)

(b)

Figure 10. Vertical free vibration history and its response in frequency domain of the ECTMD with and without eddy current damping in seawater: (**a**) free vibration; (**b**) FFT amplitude.

Table 4. Dynamic property of ECTMD.

Case	Mass Ratio	Frequency	Tuning Ratio	Damping Ratio
Target value	4.4%	3.04 Hz	0.9508	12.63%
Experimental value		3.07 Hz	0.9594	13.9%

Figure 11 shows the experimental setup for the testing of the submerged pipeline model. A seawater container is employed to simulate the seawater environment. The pipeline model is submerged 10 cm beneath the surface of the seawater, which is a sufficient depth for the full submersion of the ECTMD during vibration tests. A laser displacement sensor is installed to measure the vertical displacement of the submerged pipeline in the mid-span location. Two different cases, namely structure without control and structure controlled by ECTMD, are tested both in the free vibration and forced vibration case. A motor with unbalanced mass is mounted on the pipeline structure as a harmonic force generator, which can apply harmonic loads in different frequencies by adjusting the input voltage.

Figure 12 plots the vertical response of the pipeline model in free vibration and resonant response. It can be observed that the damping ratio of the vertical vibration is significantly increased from 2.4% to 5.5% when the ECTMD is installed on the pipeline model and the free vibration reduces to zero within 10 periods. The resonant response is reduced by 80.5% while the peak of the no control case is 6.20 mm and the peak with ECTMD control is 1.21 mm. Additionally, the steady-state response in a large frequency range is tested and the vertical response amplitudes is plotted in Figure 13. The results show that the ECTMD can suppress the large amplitude vibration of the submerged pipeline model in a wide frequency range.

The purpose of the current work is aimed to reduce large amplitude vibrations occurred on the subsea pipeline structures. To apply the proposed ECTMD in vibration control of a real environmental setting, the modal parameters such as frequency and modal mass of the vibration mode should be obtained from the finite element model or field test in the design stage. According to the proposed optimization method, the optimal frequency tuning ratio and the optimal damping ratio of the ECTMD can be designed when the modal parameters of a subsea pipeline are already known. In the implementation stage, the configuration of the ECTMD used in the control experiment is worth to be considered. The magnet can be fixed on the pipeline and the aluminum plates can be used as mass. The frequency of the ECTMD will be easily tuned by changing the length of the supporting cantilever beam.

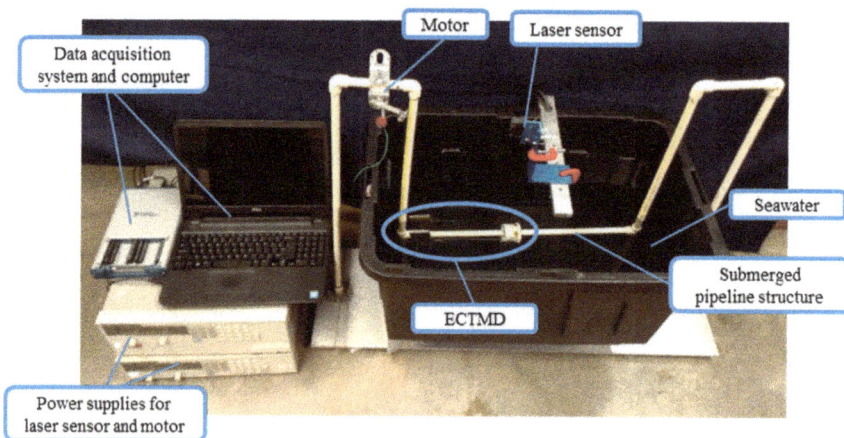

Figure 11. Experimental setup for the testing of the submerged pipeline.

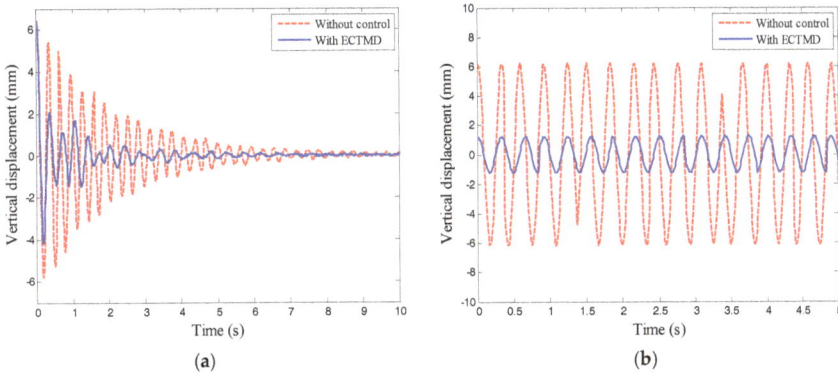

Figure 12. Vertical response of the pipeline model in seawater environments: (**a**) Free vibration; (**b**) resonant vibration.

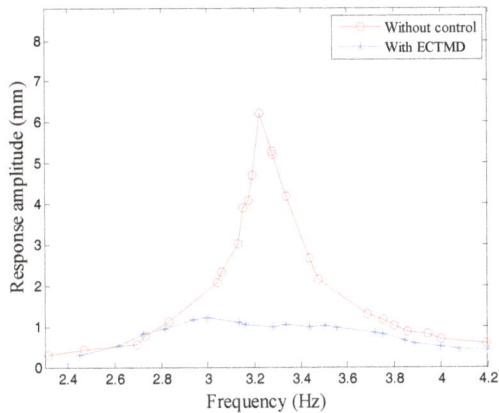

Figure 13. Vertical response of the pipeline model in a large frequency domain.

6. Conclusions

In the present work, a conception of using an ECTMD to control a subsea pipeline structure was introduced. The feasibility experiment was performed to demonstrate that eddy current damping can be used in a seawater environment. To obtain the optimal parameters of a TMD for a damped structure, a numerical optimization method was adopted. Then the attenuation experiments were carried out to show the performance of the ECTMD in a seawater environment.

The results of this study show that eddy current damping can be employed in a seawater environment. The additional damping by eddy currents in the seawater environment is almost the same as in the air environment. As shown from the vibration suppression experiment in the seawater environment, the damping of the submerged pipeline model is increased from 2.4% to 5.5% and the ECTMD can significantly reduce the peak response of the pipeline in a wide frequency range.

Acknowledgments: This research work was supported by the National Key Basic Research Program of China ('973 Project') (Grant Number 2015CB057702) and the Intergovernmental Innovation and Corporation Program of China (Grant Number 2016YFE0127900). The support from the National Natural Foundation of China (Grant Numbers 51422806; 51378203) is also greatly acknowledged. In addition, the first author's research stay at the University of Houston is sponsored by the China Scholarship Council (CSC) and the second author is sponsored by the Summer Undergraduate Research Fellowship at the University of Houston.

Author Contributions: All authors discussed and agreed upon the idea, and made scientific contributions. Wenxi Wang, Zhengqing Chen and Gangbing Song designed the experiments and wrote the paper, Dakota Dalton and Wenxi Wang collected the experimental and analyzed the data, Xugang Hua and Xiuyong Wang performed the optimum design of ECTMD and revised the paper.

Conflicts of Interest: The authors declare no conflict of interest.

References

1. Lu, Z.; Masri, S.F.; Lu, X. Parametric studies of the performance of particle dampers under harmonic excitation. *Struct. Control Health Monit.* **2011**, *18*, 79–98. [CrossRef]
2. Lu, Z.; Masri, S.F.; Lu, X. Studies of the performance of particle dampers attached to a two-degrees-of-freedom system under random excitation. *J. Vib. Control* **2011**, *17*, 1454–1471.
3. Tian, L.; Rong, K.; Zhang, P.; Liu, Y. Vibration control of a power transmission tower with pounding tuned mass damper under multi-component seismic excitations. *Appl. Sci.* **2017**, *7*, 477. [CrossRef]
4. Li, H.; Liu, M.; Li, J.; Guan, X.; Ou, J. Vibration control of stay cables of the shandong binzhou yellow river highway bridge using magnetorheological fluid dampers. *J. Bridge Eng.* **2007**, *12*, 401–409. [CrossRef]
5. Cai, C.S.; Wu, W.J.; Araujo, M. Cable vibration control with a TMD-MR damper system: Experimental exploration. *J. Struct. Eng.* **2007**, *133*, 629–637. [CrossRef]
6. Jiang, J.; Zhang, P.; Patil, D.; Li, H.; Song, G. Experimental studies on the effectiveness and robustness of a pounding tuned mass damper for vibration suppression of a submerged cylindrical pipe. *Struct. Control Health Monit.* **2017**, e2027. [CrossRef]
7. Fu, W.; Zhang, C.; Sun, L.; Askari, M.; Samali, B.; Chung, K.L.; Sharafi, P. Experimental investigation of a base isolation system incorporating MR dampers with the high-order single step control algorithm. *Appl. Sci.* **2017**, *7*, 344. [CrossRef]
8. Biswas, S.K.; Ahmed, N.U. Optimal control of flow-induced vibration of pipeline. *Dyn. Control* **2001**, *11*, 187–201. [CrossRef]
9. Frahm, H. Device for Damping Vibrations of Bodies. U.S. Patent 989958, 18 April 1911.
10. Tao, T.; Wang, H.; Yao, C.; He, X. Parametric sensitivity analysis on the buffeting control of a long-span triple-tower suspension bridge with MTMD. *Appl. Sci.* **2017**, *7*, 395. [CrossRef]
11. Li, C. Performance of multiple tuned mass dampers for attenuating undesirable oscillations of structures under the ground acceleration. *Earthq. Eng. Struct. Dyn.* **2000**, *29*, 1405–1421. [CrossRef]
12. Varadarajan, N.; Nagarajaiah, S. Wind response control of building with variable stiffness tuned mass damper using empirical mode decomposition/Hilbert transform. *J. Eng. Mech.* **2004**, *130*, 451–458. [CrossRef]
13. Setareh, M.; Hanson, R.D. Tuned mass dampers to control floor vibration from humans. *J. Struct. Eng.* **1992**, *118*, 741–762. [CrossRef]
14. Caetano, E.; Cunha, Á.; Magalhães, F.; Moutinho, C. Studies for controlling human-induced vibration of the Pedro e Inês footbridge, Portugal. Part 1: Assessment of dynamic behaviour. *Eng. Struct.* **2010**, *32*, 1069–1081. [CrossRef]
15. Caetano, E.; Cunha, Á.; Moutinho, C.; Magalhães, F. Studies for controlling human-induced vibration of the Pedro e Inês footbridge, Portugal. Part 2: Implementation of tuned mass dampers. *Eng. Struct.* **2010**, *32*, 1082–1091. [CrossRef]
16. Sims, N.D. Vibration absorbers for chatter suppression: A new analytical tuning methodology. *J. Sound Vib.* **2007**, *301*, 592–607. [CrossRef]
17. Yang, Y.; Munoa, J.; Altintas, Y. Optimization of multiple tuned mass dampers to suppress machine tool chatter. *Int. J. Mach. Tools Manuf.* **2010**, *50*, 834–842. [CrossRef]
18. Wang, M.; Zan, T.; Yang, Y.; Fei, R. Design and implementation of nonlinear TMD for chatter suppression: An application in turning processes. *Int. J. Mach. Tools Manuf.* **2010**, *50*, 474–479. [CrossRef]
19. Kim, S.M.; Wang, S.; Brennan, M.J. Optimal and robust modal control of a flexible structure using an active dynamic vibration absorber. *Smart Mater. Struct.* **2011**, *20*, 045003. [CrossRef]
20. Weber, F. Semi-active vibration absorber based on real-time controlled MR damper. *Mech. Syst. Signal Process.* **2014**, *46*, 272–288. [CrossRef]
21. Kwok, K.C.S.; Samali, B. Performance of tuned mass dampers under wind loads. *Eng. Struct.* **1995**, *17*, 655–667. [CrossRef]

22. Hong, S.R.; Wang, G.; Hu, W.; Wereley, N.M. Liquid spring shock absorber with controllable magnetorheological damping. *Proc. Inst. Mech. Eng. Part D* **2006**, *220*, 1019–1029. [CrossRef]

23. Wang, W.; Hua, X.; Wang, X.; Chen, Z.; Song, G. Optimum design of a novel pounding tuned mass damper under harmonic excitation. *Smart Mater. Struct.* **2017**, *26*, 055024. [CrossRef]

24. Lu, Z.; Chen, X.; Zhang, D.; Dai, K. Experimental and analytical study on the performance of particle tuned mass dampers under seismic excitation. *Earthq. Eng. Struct. Dyn.* **2017**, *46*, 697–714. [CrossRef]

25. Koo, J.H.; Ahmadian, M.; Setareh, M. Experimental robustness analysis of magneto-rheological tuned vibration absorbers subject to mass off-tuning. *J. Vib. Acoust.* **2006**, *128*, 126–131. [CrossRef]

26. Weber, F.; Maślanka, M. Frequency and damping adaptation of a TMD with controlled MR damper. *Smart Mater. Struct.* **2012**, *21*, 055011. [CrossRef]

27. Sodano, H.A.; Bae, J.S.; Inman, D.J.; Belvin, W.K. Improved concept and model of eddy current damper. *J. Vib. Acoust.* **2006**, *128*, 294–302. [CrossRef]

28. Sodano, H.A.; Inman, D.J.; Belvin, W.K. Development of a new passive-active magnetic damper for vibration suppression. *J. Vib. Acoust.* **2006**, *128*, 318–327. [CrossRef]

29. Sodano, H.A.; Bae, J.S.; Inman, D.J.; Belvin, W.K. Concept and model of eddy current damper for vibration suppression of a beam. *J. Sound Vib.* **2005**, *288*, 1177–1196. [CrossRef]

30. Bae, J.S.; Kwak, M.K.; Inman, D.J. Vibration suppression of a cantilever beam using eddy current damper. *J. Sound Vib.* **2005**, *284*, 805–824. [CrossRef]

31. Wang, Z.; Chen, Z.; Wang, J. Feasibility study of a large-scale tuned mass damper with eddy current damping mechanism. *Earthq. Eng. Eng. Vib.* **2012**, *11*, 391–401. [CrossRef]

32. Larose, G.L.; Larsen, A.; Svensson, E. Modelling of tuned mass dampers for wind-tunnel tests on a full-bridge aeroelastic model. *J. Wind Eng. Ind. Aerodyn.* **1995**, *54*, 427–437. [CrossRef]

33. Bae, J.S.; Hwang, J.H.; Roh, J.H.; Kim, J.H.; Yi, M.S.; Lim, J.H. Vibration suppression of a cantilever beam using magnetically tuned-mass-damper. *J. Sound Vib.* **2012**, *331*, 5669–5684. [CrossRef]

34. Bae, J.S.; Hwang, J.H.; Kwag, D.G.; Park, J.; Inman, D.J. Vibration suppression of a large beam structure using tuned mass damper and eddy current damping. *Shock Vib.* **2014**, *2014*, 89314. [CrossRef]

35. Wen, Q.; Hua, X.G.; Chen, Z.Q.; Yang, Y.; Niu, H.W. Control of Human-Induced Vibrations of a Curved Cable-Stayed Bridge: Design, Implementation, and Field Validation. *J. Bridge Eng.* **2016**, *21*, 04016028. [CrossRef]

36. Bourquin, F.; Caruso, G.; Peigney, M.; Siegert, D. Magnetically tuned mass dampers for optimal vibration damping of large structures. *Smart Mater. Struct.* **2014**, *23*, 085009. [CrossRef]

37. Den Hartog, J.P. *Mechanical Vibrations*; Courier Corporation: North Chelmsford, MA, USA, 1985.

38. Rana, R.; Soong, T.T. Parametric study and simplified design of tuned mass dampers. *Eng. Struct.* **1998**, *20*, 193–204. [CrossRef]

39. Hazra, B.; Sadhu, A.; Lourenco, R.; Narasimhan, S. Re-tuning tuned mass dampers using ambient vibration measurements. *Smart Mater. Struct.* **2010**, *19*, 115002. [CrossRef]

40. Thompson, A.G. Optimum tuning and damping of a dynamic vibration absorber applied to a force excited and damped primary system. *J. Sound Vib.* **1981**, *77*, 403–415. [CrossRef]

41. Du, D.; Gu, X.J.; Chu, D.Y.; Hua, H.X. Performance and parametric study of infinite-multiple TMDs for structures under ground acceleration by H_∞ optimization. *J. Sound Vib.* **2007**, *305*, 843–853. [CrossRef]

42. Zuo, L.; Nayfeh, S.A. Minimax optimization of multi-degree-of-freedom tuned-mass dampers. *J. Sound Vib.* **2004**, *272*, 893–908. [CrossRef]

43. Asami, T.; Nishihara, O. Closed-form exact solution to H_∞ optimization of dynamic vibration absorbers (application to different transfer functions and damping systems). *J. Vib. Acoust.* **2003**, *125*, 398–405. [CrossRef]

applied
sciences

MDPI

Article

Experimental Study on the Performance of Polyurethane-Steel Sandwich Structure under Debris Flow

Peizhen Li [1,2], Shutong Liu [2] and Zheng Lu [1,2,*]

1 State Key Laboratory of Disaster Reduction in Civil Engineering, Tongji University, Shanghai 200092, China; lipeizh@tongji.edu.cn
2 Research Institute of Structural Engineering and Disaster Reduction, Tongji University, Shanghai 200092, China; liushutong@tongji.edu.cn
* Correspondence: luzheng111@tongji.edu.cn; Tel.: +86-21-6598-6186

Received: 7 September 2017; Accepted: 29 September 2017; Published: 2 October 2017

Abstract: Polyurethane-steel sandwich structure, which creatively uses the polyurethane-steel sandwich composite as a structural material, is proposed to strengthen the impact resistance of buildings under debris flow. The impact resistance of polyurethane-steel sandwich structure under debris flow is investigated by a series of impact loading tests, compared with that of traditional steel frame structures. Additionally, further discussions regarding the hidden mechanism are performed. During the whole impact process, as for steel frame structure, the impacted column appeared obvious local deformation both at its column base and on the impact surface, leading to remarkable decrease of its impact resistance; while the stress and strain of polyurethane-steel sandwich structure develops more uniformly and distribute further in the whole structure, maintaining excellent integrity and impact transmission capability. The impact loading tests confirm that polyurethane-steel sandwich structure possesses superior impact resistance under debris flow. This is of great practical significance for the prevention and reduction of geological disasters.

Keywords: polyurethane-steel sandwich structure; debris flow; impact loading tests; impact resistance

1. Introduction

Debris flows are gravity-driven surges of roughly equal volumes of water and poorly sorted sediment, thoroughly mixed and agitated [1], characterized as being large-scale, unpredictable, and extremely destructive. The mountainous areas with low vegetation coverage are highly prone to debris flow hazards [2]. Many catastrophic debris flow disasters, resulting in heavy casualties and massive property losses, have been reported across the world (debris flows on 7 August 2010 in Zhouqu [2], 1998 volcaniclastic debris flows in the Sarno area [3], 2013 Wulipo landslide, and the resulting debris flow in Dujiangyan City [4]). Compared with strong structural facilities, such as railway bridges, common residential buildings are more easily damaged by debris flows. Moreover, most deaths and property losses are associated with the damage to buildings. For this reason, it is an urgent issue to study the dynamic response of structures under debris flow and take effective measures to increase their resistance to such disasters.

Impact force is the most important index and foundation for engineering design and risk assessment to buildings by debris flow. Additionally, it is the theoretical basis of the study on the impact of debris flow on structures [5–11]. In recent decades, several simplified analytical methods have been provided in the published literature to investigate the impact force of debris flow. Chen et al. [12] simplified the debris flow as two-phase fluid, establishing method to calculate velocities of solid phase and liquid phase of debris flow. Hungr et al. [13] provided a quantitative approach to calculate the impact force of debris flow, by means of simplifying the collision between stone and the retaining structure as the impact of rock on a cantilever beam. Moriguchi et al. [14] conducted small-scale laboratory physical modeling and corresponding numerical simulations in the Eulerian framework to estimate the impact force generated by granular flow. Some scholars simulated the flow behavior of debris flows by the smoothed particle hydrodynamics (SPH) method and have made promising progress [15–17]. Dai et al. [18] put forward a fluid–structure coupled numerical model based on SPH to predict the propagation and impact force of debris flows.

The time-frequency analysis of the impact fluctuation signals of debris flow would provide a meaningful basis for the study of debris flow load. A few field tests and field observations have also been reported. For example, Hu et al. [19] measured the first time a long-duration series of impact force at different flow depths, founding that the peak grain impacts at different depths were non-synchronous within the debris flows. Chen et al. [20] conducted debris flow tests of different solid and particle size combinations, and finally obtained the energy intensity of shock loading in different frequency bands using wavelet time frequency analysis.

In addition, analyzing potential damage of residential buildings to debris flow hazards is essential to study the impact of debris flow on structures. Although some progress has been made in quantifying the vulnerability of buildings to debris flows in many countries [21–23], field observations, experimental studies, and numerical simulations on the failure mechanism and impact resistance of engineering structures under the impact of debris flow are highly necessary. Zeng et al. [24] classified the failure models of reinforced concrete columns based on field investigation, and studied collapse mechanism for columns damaged by debris flow. Zhang et al. [25] carried out experimental research on reinforced concrete buildings struck by debris flow, collecting and analyzing experimental data. Hu et al. [2] distinguished the characteristics and patterns of damage to buildings by debris flows on 7 August 2010 in Zhouqu. Zanchetta et al. [3] assessed the relationship between the debris flow impact and structural damage in 1998 Sarno volcaniclastic debris flows.

To attenuate the dynamic impact force generated by debris flow and mitigate the subsequent destructive effects, many debris flow prevention and control engineering, such as debris flow barriers, debris racks and fences, debris-flow retention structures and debris breakers [26–30], are applied. Nevertheless, present design of these structures usually refers to empirical models [31]. As a result, the structures are often destroyed during the debris flow disasters. A sudden breakage of the protective structures would increase the destructive power of the debris flow and result in a more serious disaster, thus relying solely on the protective structure is inadvisable, and it is particularly critical to improve the impact resistance of the structure itself, which is the last line of defense against the impact of debris flow. Some innovative methods and concepts have been presented. For example, Li et al. [32] presented a novel masonry structure with strong resistance to debris flow; Lu et al. [33] proposed a shock absorption buffer to reduce the vulnerability of building structures under debris flows.

The impact resistance of traditional building material, like steel and masonry, could not fully meet the demand of resisting debris flow impact. With the progress of materials science and manufacturing technology, and the diversification of engineering requirements for materials, a polyurethane-steel sandwich composite emerges, in which the core material (polyurethane elastomer) and the steel plate are bonded by strong glue, or the polyurethane elastomer is directly poured into the hollow steel plate component. The polyurethane-steel sandwich composite is not only characterized by high strength and stiffness, but also has the advantages of light weight, easy processing, fire resistance, impact resistance, and fatigue endurance [34,35]. At present, the polyurethane-steel sandwich composite is widely used in ship engineering, sports venue construction, bridge pier anti-collision, as well as the bridge deck laying and repairing [36–40].

Although the polyurethane-steel sandwich composite has been applied in engineering in recent years, it is still rare to be used as a structural material for buildings. Moreover, it has a bright engineering application prospect to be widely used as the building material in debris flow prone areas due to its excellent impact resistance. In this paper, polyurethane-steel sandwich structure is proposed by columns that are filled with polyurethane elastomer to improve the impact resistance of traditional steel frame structures. The main objectives are to further assess the dynamic impact response and the impact resistance of building structure which uses the polyurethane-steel sandwich composite as structural material under the impact of debris flow. Based on field tests, the performance analysis and comparison of the polyurethane-steel sandwich structure to an ordinary steel frame structure under the impact of simulated debris flow are conducted, and the hidden mechanisms are also discussed.

2. Test Setup and Procedure

2.1. Test Model

The experiment is carried out by two test models fixed up on the platform, which are a two-story steel frame structure (Model A) and a polyurethane-steel sandwich structure (Model B), respectively. Compared with steel frame structure, polyurethane-steel sandwich structure's columns in the impact direction of debris flow, are filled with polyurethane elastomer, as shown in Figure 1a–c. In order to make the polyurethane elastomer a solid and bond the steel column together, the polyurethane-steel sandwich column is heated after the liquid polyurethane elastomer is poured in the steel pipe column. The height of first floor is 1.0 m and that of the second floor is 0.75 m. The test model in the present study is scaled down from full-scale buildings based on similitude principles. The model's scale is 1/4. Table 1 lists the specific parameters of various members. Figure 1d,e show the structural layout of Model B and photo of test model, respectively.

Table 1. Specific parameters of members.

Member Name	Material	Member Dimension (mm)
GKZ1		□100 × 100 × 3 × 3
GKZ2	Q235	□100 × 100 × 3 × 3
GKL1		H125 × 50 × 2 × 3
GL1		H100 × 50 × 2 × 3

2.2. Test Loading Schedule

In this test, referring to other experts' experience [25,41], the load of debris flow is simulated by an impact force caused by a steel ball rolling down in the slope track. The effective sliding height of the steel ball is chosen as 2.5 m, and the loading height is set to 1/2 floor height above ground [42]. A/2 column is the column directly subjected to the impact loading, as shown in Figure 1d. The different diameters of steel balls are used to simulate the impact loads, for the sake of hierarchical loading. The complete loading sequences are summarized in Table 2, which consists of six loading levels that are labeled from case G1 to G6. Figure 1f–h displays the elevation of loading device, photo of the test site and 3D schematic of the entire structure, respectively.

Table 2. Test loading schedule.

Loading Case	Diameter of the Steel Ball/mm	Mass of the Steel Ball/kg	Impact Velocity/m/s	Impact Energy/J
G1	200	33	7	808.5
G2	300	111	7	2791.5
G3–G6	350	180	7	4410

2.3. Instrumentation Arrangement

As depicted in Figure 2a,b, strain gauges are used to measure the structural strain, among which there are fifteen one-way strain gauges and ten three-way strain gauges. Additionally, nine displacement meters are used to measure the displacement at the impact position and the displacement at the top of each floor of the impacted column and columns adjacent to the impacted column (A/1,

A/3, and C/2 column). At the same time, the acceleration response at the impact position and the acceleration response at the top of each floor of A/3 column are measured by three accelerometers. Figure 2c displays the arrangement of displacement meters and accelerometers.

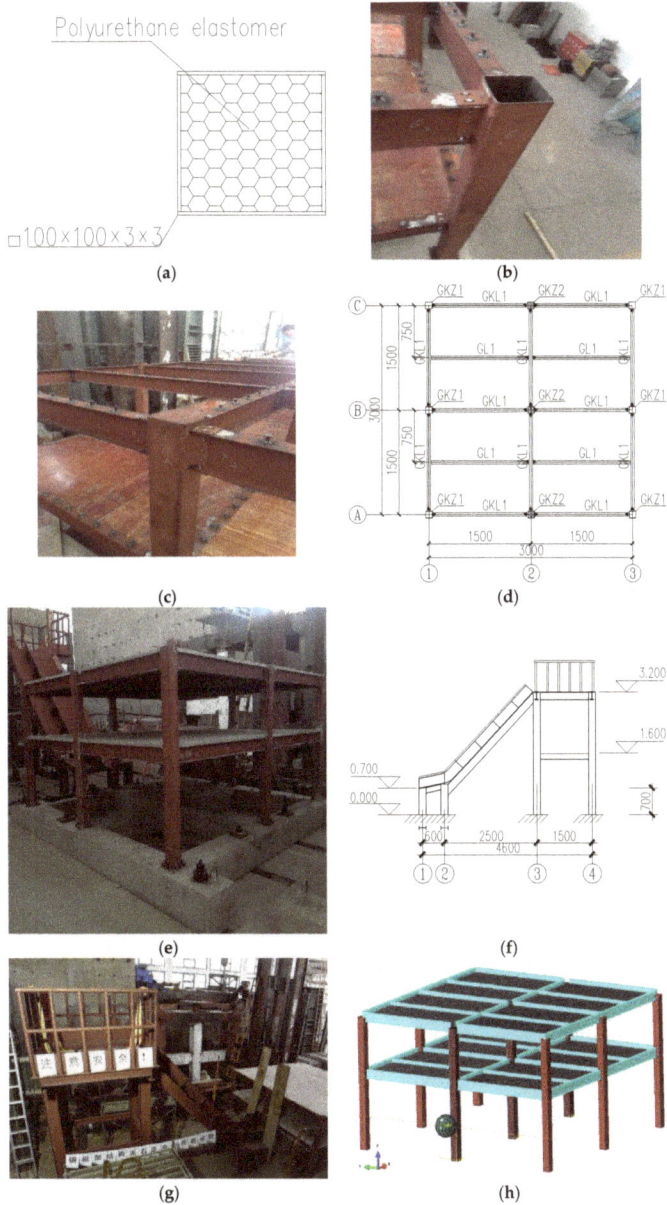

Figure 1. (a) Schematic diagram of the polyurethane-steel sandwich column; (b) photo of steel frame column; (c) photo of polyurethane-steel sandwich column; (d) the structural layout of Model B; (e) photo of test model; (f) elevation of loading device; (g) photo of the test site; and (h) a 3D schematic of the entire structure.

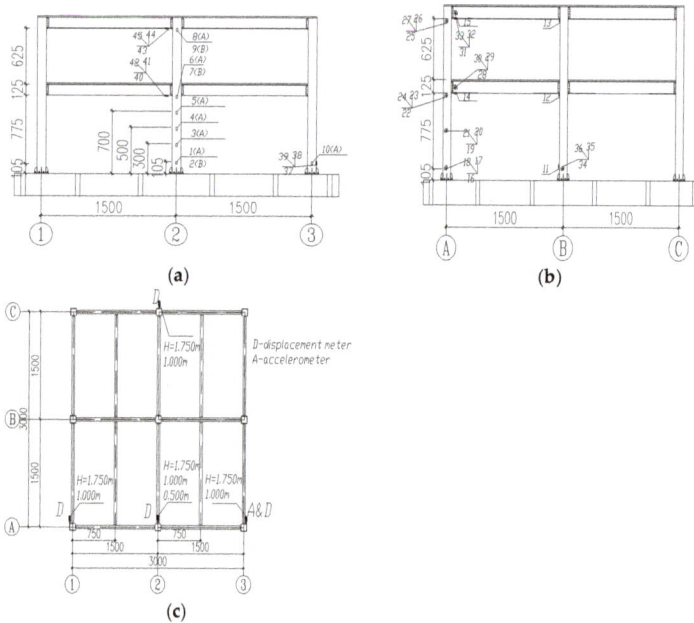

Figure 2. Instrumentation arrangement. (a) Strain gauge arrangement of axis A; (b) strain gauge arrangement of axis 2; and (c) displacement meter and accelerometer arrangement.

2.4. Polyurethane Property Test

Six uniaxial compression specimens are casted when the frame columns are filled, and its uniaxial compression tests are conducted. The stress-strain relationship of the polyurethane elastomer is shown in Figure 3.

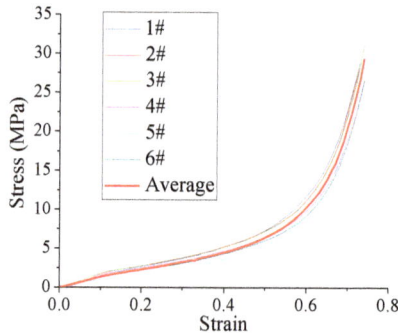

Figure 3. The stress-strain relationship of polyurethane elastomer in the test.

3. Test Phenomena and Analysis

In this section, through the test data and test phenomena, the impact responses of the polyurethane-steel sandwich structure and steel frame structure are analyzed and compared, from the impacted column to the overall structure, and some valuable conclusions are drawn. In consideration of debris flows' features of high velocity, extreme impact force, and severe damage, this study is focused on peak response.

3.1. Response of the Impacted Column

After the end of each impact, the local deformation of steel frame structure is significantly larger than that of the polyurethane-steel sandwich structure. The column's impacted surface of steel frame structure appeared deep dent and its both sides warped badly. Its local bending deformation is quite large. As shown in Figure 4, the column surface near the impact position of the steel frame structure is slightly cracked after case G3.

According to Figure 5a, it is noteworthy that the total value and incremental value of Model A's dent-depth are far greater than those of Model B's. This is because the columns of the polyurethane-steel sandwich structure are filled with a polyurethane elastomer that has low compression characteristics, which strengthens the structural integrity, hence, it can coordinate the whole force and deformation in the column. The polyurethane elastomer is bonded with the steel column to ensure a reliable connection, and it could transmit the shear force further and ensure the integrity of column against the impact force. Meanwhile, it also acts as a buffer against the impact force.

Figure 5b illustrates the displacement time-history of the impact position's back surface under case G6. The results clearly demonstrate that the displacement of Model B at that position is much larger than that of Model A. The reason for this phenomenon is that polyurethane elastomers are fully extruded, and the impact force is transmitted to the column's back surface by polyurethane elastomer. As to steel frame structure, its column completely relied on the side plate to transmit the impact force to its back surface, and its load transferring path is not direct, forming quite a large dent on the impact surface and a relatively small displacement on its back surface.

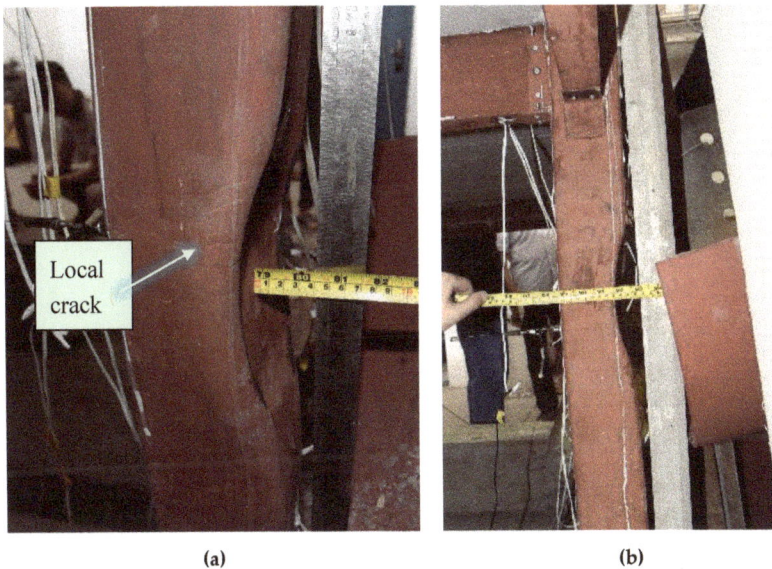

(a)   (b)

Figure 4. Photo of the impacted column after case G3: (a) Model A; and (b) Model B.

By comparing Figure 5a,b, at the end of case G6, it is evident that the displacement of the impact surface of Model A is relatively large while that of its back surface is small. However, the law of Model B is opposite to Model A. Accordingly, after case G6, the distance between the impact position and its back surface of Model A column is smaller than that of Model B's column, and the difference between the two models is about 30 mm. The decrease of distance between the impact position and its back surface means the decrease of section's bending arm and bending modulus. Consequently, the bending resistance of steel frame structure's column is evidently reduced. Furthermore, under the impact loading of debris flow, the midspan of the column is the key position to resisting the bending moment. The increase of the horizontal displacement at the midspan of column would also cause the second-order effect of gravity load to increase remarkably, even triggering the overall collapse of the structure, causing serious economic losses and casualties.

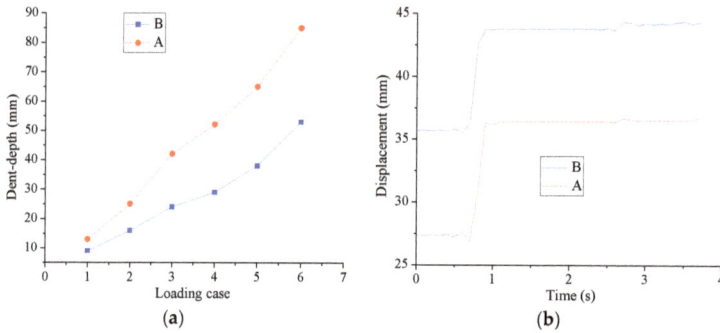

Figure 5. (**a**) Dent-depth at the impact position; and (**b**) the displacement time-history of the impact position's back surface under case G6.

Figure 6 shows the maximum normal strain at back surface of the impact position and the maximum shear strain at the corresponding side plate of the impacted column. It can be seen that the maximum normal strain of Model B at the back surface of the impact position is larger than that of Model A; for the maximum shear strain of the side plate, that of Model A is relatively larger. According to the analysis of the previous test phenomenon and the measured displacement, owing to the low compressibility of polyurethane elastomer and the reliable bonding between the column and polyurethane elastomer, the impact force on the impacted column could be transmitted directly to the opposite plate, thus the overall mechanical performance is effectively improved. However, the impact force of steel frame structure is directly transferred to the side plate, hence, its shear strain is relatively large.

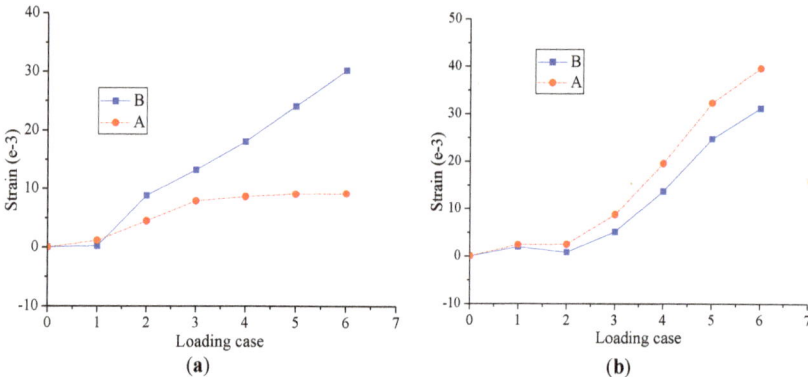

Figure 6. Strain curves. (**a**) Maximum normal strain at back surface of the impact position; and (**b**) maximum shear strain at the corresponding side plate of the impact position.

3.2. Local Deformation of Joint

Figure 7 shows the deformation of the impacted column at column base after case G6, it can be found that the screws of Model A's impacted column at the column base joints have bent. This is due to the fact that the bending resistance of the steel frame structure decreases and the catenary action increases sharply, which would increase the axial force, leading to the screws of Model A's impacted column at the column base joints being subjected to the interaction of the axial force, shear force, and bending moment; however, the screws of Model B's impacted column at the column base joints are mainly subjected to shear force and bending moment.

Figure 8 shows the damage of second floor slab around the impacted column at the end of case G6. It is shown that the slab damage at second floor slab around the impacted column of ordinary steel frame structure is worse than that of polyurethane-steel sandwich structure, and its rivets on the second floor slab around the impact column have been exposed. Similar to Figure 7, the reason for this phenomenon is that the impacted column of steel frame structure is seriously damaged, and its axial force increases sharply caused by the catenary action, which results in the serious damage of the second floor slab around the impacted column. This implies that the polyurethane-steel sandwich structure could maintain superior overall mechanical performance and transfer the impact force to the support and the frame beam stably.

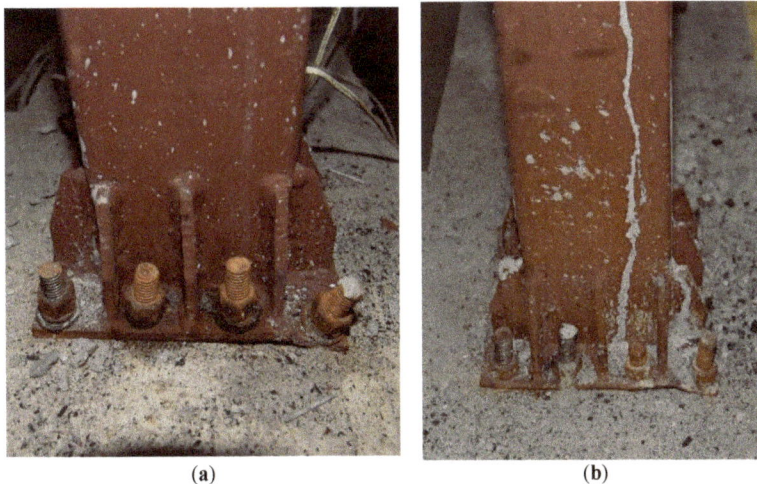

(a)   (b)

Figure 7. The deformation of the impacted column at column base. (a) Model A; and (b) Model B.

Figure 9a,b shows the maximum normal strain at the impacted column base joint and the maximum shear strain of its side plate under each impact, respectively. Figure 10a,b plots the maximum normal strain at the top of first floor impacted column and the maximum shear strain of its side plate under each impact, respectively. From these figures, some interesting phenomena are observed:

(1) As the impact progresses, for steel frame structure, its maximum normal strain at the impacted column base joint and the top of first floor impacted column gradually increases and exceeds the corresponding strain at the corresponding position of polyurethane-steel sandwich structure. This is because, with the mass of steel ball and the impact force increase, the local deformation of steel frame structure increases and the bending resistance of the column midspan is substantially weakened, hence, the contribution of catenary action to resist the impact increases gradually, which requires the column to provide great tension.

(2) However, for the polyurethane-steel sandwich structure, its maximum shear strain at the impacted column base joint and the top of first floor impacted column is much greater than that of the corresponding strain at the corresponding position of steel frame structure, which indirectly proved that the shear force and bending moment of the polyurethane-steel sandwich structure at the joint

area is much greater than that of steel frame structure, indicating that polyurethane-steel sandwich structure has better integrity and excellent impact transmission capability.

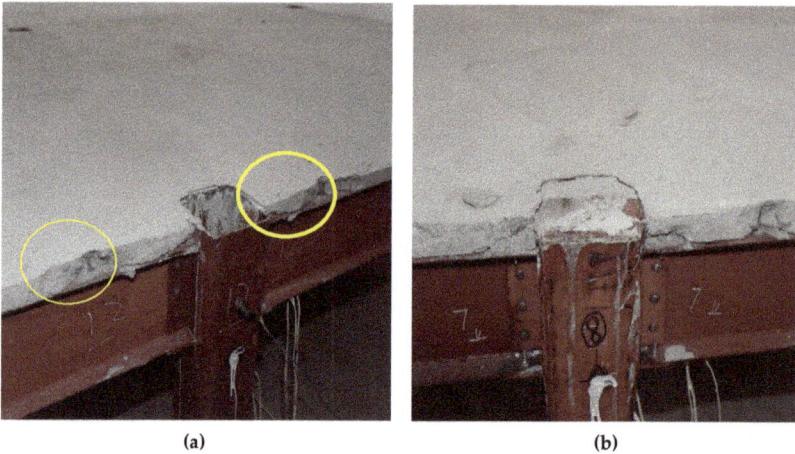

Figure 8. The damage of second floor slab around the impacted column at the end of case G6. (**a**) Model A; and (**b**) Model B.

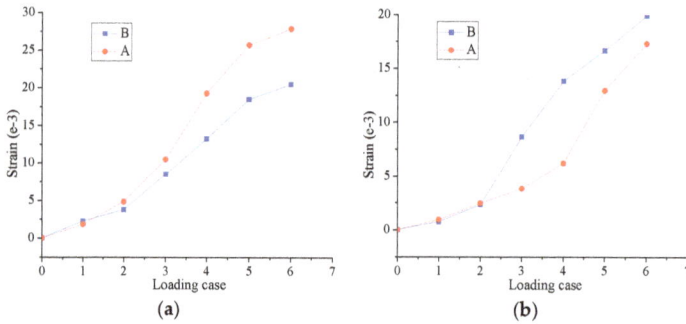

Figure 9. Strain curves at the impacted column base joint. (**a**) The maximum normal strain; and (**b**) the maximum shear strain of its side plate.

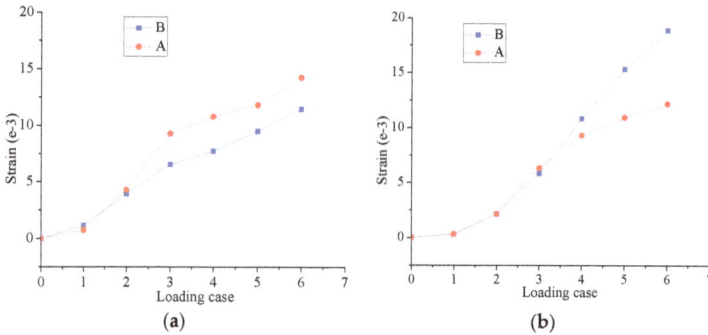

Figure 10. Strain curves at the top of first floor impacted column. (**a**) The maximum normal strain; and (**b**) the maximum shear strain of its side plate.

3.3. Overall Structural Deformation

In Figure 11, the overall deformation of the impacted column and its connected members at the end of case G6 are shown. It clearly demonstrates that the rotation angle between the impacted column and its connected beam of Model A is larger than that of Model B. Combining the analysis in Section 3.1, since the excessive local dent of Model A's impacted column leads to the decrease of bending resistance capacity, the catenary action of column plays a more important role in resisting the impact. At the same time, the constraint capacity of Model A's column on its beam is weak. Therefore, the angle between the beams and columns is relatively large.

As depicted in Figure 12, the vertical deformations of Model A and Model B at the end of all impact cases are measured as approximately 14 mm and 6 mm, respectively. This is because, as for the steel frame structure, the excessive local dent of the impacted column makes the catenary action more obvious, which results in increased column tension, and the displacement of the upper structure increases under the action of tension. However, the catenary action of polyurethane-steel sandwich structure is alleviated, and the displacement is mainly caused by the bending deformation.

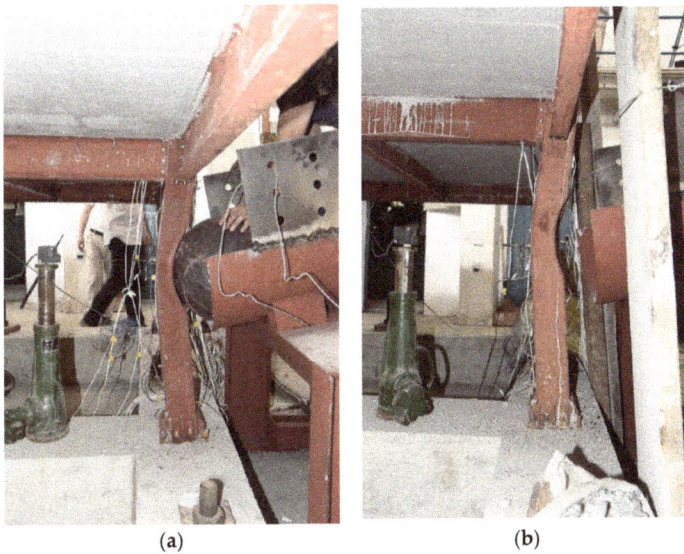

(a)   (b)

Figure 11. Photo of the impacted column after case G6. (**a**) Model A; and (**b**) Model B.

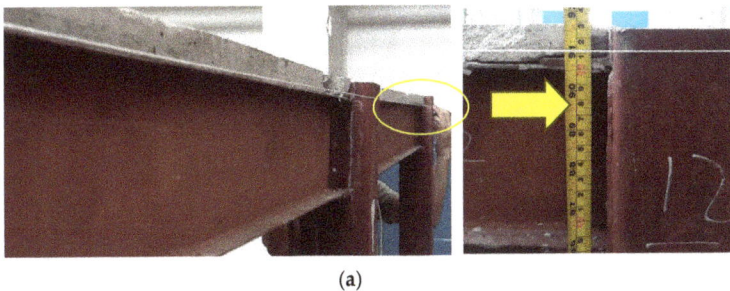

(a)

Figure 12. *Cont.*

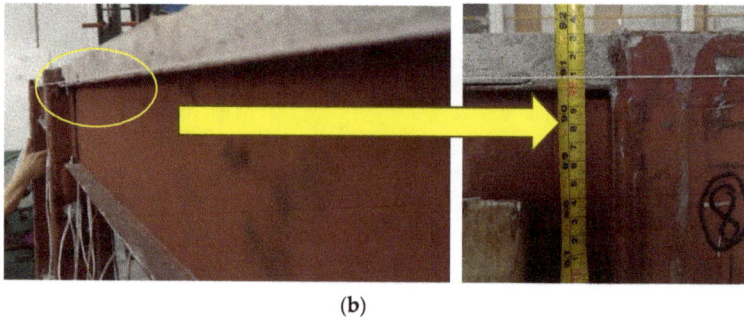

(b)

Figure 12. The vertical deformations at the end of case G6. (**a**) Model A; and (**b**) Model B.

Figure 13a,b displays the maximum normal strain of the B/2 and A/3 column base, which is adjacent to the impacted column. It is noted that because of the polyurethane elastomer's role in transmitting force and enhancing integrity, the maximum normal strain at these positions of the steel frame structure is smaller than that of polyurethane-steel sandwich structure. Correspondingly, the normal stress at these positions of the steel frame structure is relatively small. Figure 13c shows the maximum normal strain at the beam bottom flange near the beam-column joint, which is directly connected with the impacted column. This demonstrates that, under each impact, for the polyurethane-steel sandwich structure, the absolute increase of normal strain at that position is larger than that of steel frame structure (the negative strain represents the compression of beam). In other words, the larger the impact force, the better the performance of polyurethane-steel sandwich structure, which is obviously beneficial to practical engineering applications. These results mainly occur because the column of polyurethane-steel sandwich structure has a greater ability to restrain the beam, resulting in a greater bending moment of the beam, and the bending moment here would cause the compression of the bottom flange.

Figure 14 shows the displacement, velocity and acceleration time-history at the top of each floor of the A/3 column under case G2. From these figures, it can be seen that for polyurethane-steel sandwich structure (Model B), both the displacement, velocity, and acceleration time-history at the top of each floor of the A/3 column are greater than the corresponding time-history at the corresponding position of steel frame structure (Model A), indicating that polyurethane-steel sandwich structure has better integrity and excellent impact transmission capability. This is consistent with other phenomena in the paper, such as Figure 9a,b.

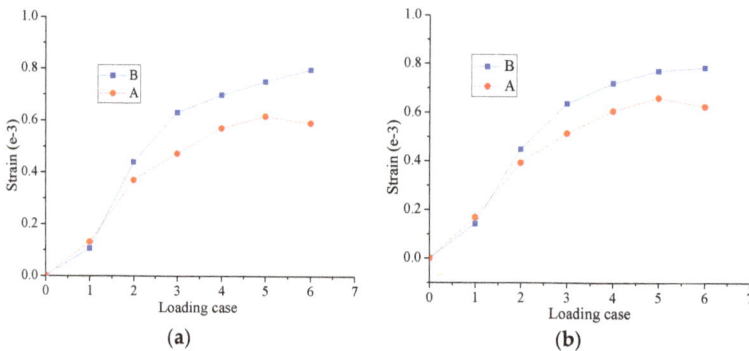

Figure 13. *Cont.*

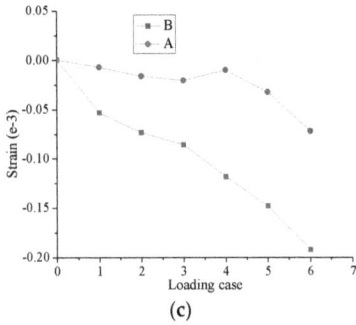

(c)

Figure 13. The maximum normal strain under each impact. (**a**) B/2 column base; (**b**) A/3 column base; and (**c**) the A end of the 2/A-B beam at the first floor.

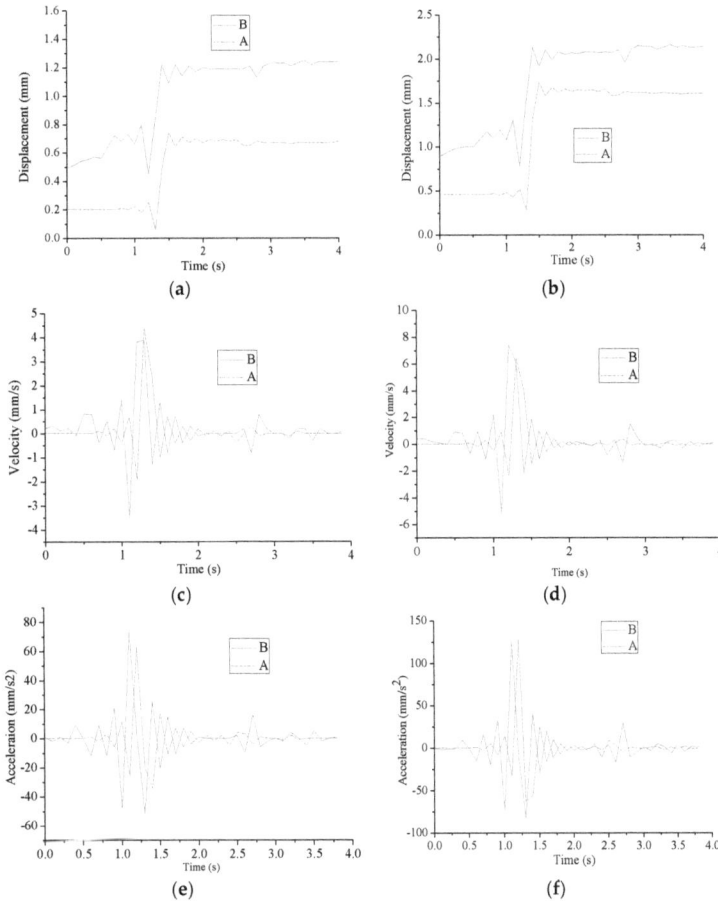

Figure 14. Response of the A/3 column in case G2: (**a**) displacement time-history at the top of the first floor; (**b**) displacement time-history at the top of the second floor; (**c**) velocity time-history at the top of the first floor; (**d**) velocity time-history at the top of the second floor; (**e**) acceleration time-history at the top of first floor; and (**f**) acceleration time-history at the top of the second floor.

Appl. Sci. **2017**, *7*, 1018

4. Conclusions

This paper proposes a creative polyurethane-steel sandwich composite structure to strengthen the impact resistance of buildings under debris flow. To make a comparative experimental study of the dynamic behaviors of such a polyurethane-steel sandwich structure and steel frame structure under the impact of simulated debris flow, a series of impact loading tests are designed and carried out. By analyzing the test results, the following conclusions can be drawn, which can provide a reference for future engineering applications:

(1) The steel frame structure mainly depends on the impacted column to resist the impact loading.

For steel frame structure, large local damage occurs under high or continuous impact loading; the local bending resistance of the impacted column decreases sharply, and instability may occur; furthermore, the mechanical behaviors and structural load-transferring path change, and the catenary action on the impacted column is quite obvious; the impact force is primarily borne by the impacted column, and the deformation is mainly concentrated on the impacted column.

(2) Under debris flow impact, the polyurethane-steel sandwich structure exhibits superior performance to resist the impact loading.

Under lower debris flow impact intensity, its deformation damage is minor, and does not affect its serviceability; under large and continuous impact loading, depending on the low compression performance of the polyurethane elastomer filled in column, it can coordinate the deformation of the column so as not to cause a great dent in local parts; it can ensure the bending resistance and vertical bearing capacity of column not decrease significantly, transferring the impact force more reliably to the whole structure; also, the dissipation of impact energy depends on overall force and deformation; meanwhile, the stress and strain of polyurethane-steel sandwich structure develop more uniformly.

In conclusion, the experimental study indicates that the polyurethane-steel sandwich structure can relieve the impact effect in time and space, and remarkably strengthens the overall mechanical performance of the whole structure under debris flow, compared with traditional steel frame structure. For the newly-built structures, the polyurethane-steel sandwich structures are favorable, which has bright prospects and deserves further study for its application in civil engineering. However one should be aware of that this is still the preliminary stage of the engineering design. During the next stages a more detailed parametric analysis for design would be necessary.

Acknowledgments: Financial support from the National Key Technology R&D Program through grant 2014BAL05B01 is highly appreciated.

Author Contributions: Peizhen Li conceived the new structural system, designed the work and revised the paper; Shutong Liu performed the experimental study and theoretical analysis, drafting the paper; Zheng Lu proposed the method, analyzed the data and revised the paper critically for important intellectual content. All authors approve the final version to be published and have agreement to be accountable for all aspects of the work in ensuring that questions related to the accuracy or integrity of any part of the work are appropriately investigated and resolved.

Conflicts of Interest: The authors declare no conflict of interest.

References

1. Iverson, R.M. The physics of debris flow. *Rev. Geophys.* **1997**, *35*, 245–296. [CrossRef]
2. Hu, K.H.; Cui, P.; Zhang, J.Q. Characteristics of damage to buildings by debris flows on 7 August 2010 in Zhouqu, Western China. *Nat. Hazards Earth Syst. Sci.* **2012**, *12*, 2209–2217. [CrossRef]
3. Zanchetta, G.; Sulpizio, R.; Pareschi, M.T.; Leoni, F.M.; Santacroce, R. Characteristics of May 5ˆ6, 1998 volcaniclastic debris flows in the Sarno area (Campania, southern Italy): Relationships to structural damage and hazard zonation. *J. Volcanol. Geotherm. Res.* **2004**, *133*, 377–393. [CrossRef]
4. Chen, X.Z.; Cui, Y.F. The formation of the Wulipo landslide and the resulting debris flow in Dujiangyan City, China. *J. Mt. Sci.* **2017**, *14*, 1100–1112. [CrossRef]
5. Wang, W.X.; Hua, X.G.; Wang, X.Y.; Chen, Z.Q.; Song, G.B. Optimum design of a novel pounding tuned mass damper under harmonic excitation. *Smart Mater. Struct.* **2017**, *26*. [CrossRef]

6. Song, G.B.; Zhang, P.; Li, L.Y.; Singla, M.; Patil, D.; Li, H.N.; Mo, Y.L. Vibration Control of a Pipeline Structure Using Pounding Tuned Mass Damper. *J. Eng. Mech.* **2016**, *142*. [CrossRef]

7. Zhang, P.; Song, G.B.; Li, H.N.; Lin, Y.X. Seismic Control of Power Transmission Tower Using Pounding TMD. *J. Eng. Mech.* **2013**, *139*, 1395–1406. [CrossRef]

8. Lu, Z.; Wang, D.C.; Masri, S.F.; Lu, X.L. An experimental study of vibration control of wind-excited high-rise buildings using particle tuned mass dampers. *Smart Struct. Syst.* **2016**, *18*, 93–115. [CrossRef]

9. Lu, Z.; Chen, X.Y.; Zhang, D.C.; Dai, K.S. Experimental and analytical study on the performance of particle tuned mass dampers under seismic excitation. *Earthq. Eng. Struct. Dyn.* **2017**, *46*, 697–714. [CrossRef]

10. Lu, Z.; Lu, X.L.; Jiang, H.J.; Masri, S.F. Discrete element method simulation and experimental validation of particle damper system. *Eng. Comput.* **2014**, *31*, 810–823. [CrossRef]

11. Lu, Z.; Chen, X.Y.; Lu, X.L.; Yang, Z. Shaking table test and numerical simulation of an RC frame-core tube structure for earthquake-induced collapse. *Earthq. Eng. Struct. Dyn.* **2016**, *45*, 1537–1556. [CrossRef]

12. Chen, H.K.; Tang, H.M.; Chen, Y.Y. Research on method to calculate velocities of solid phase and liquid phase in debris flow. *Appl. Math. Mech.* **2006**, *27*, 399–408. [CrossRef]

13. Hungr, O.; Morgan, G.C.; Kellerhals, R. Quantitative analysis of debris torrent hazards for design of remedial measures. *Can. Geotech. J.* **2011**, *21*, 663–677. [CrossRef]

14. Moriguchi, S.; Borja, R.I.; Yashima, A.; Sawada, K. Estimating the impact force generated by granular flow on a rigid obstruction. *Acta. Geotech.* **2009**, *4*, 57–71. [CrossRef]

15. Laigle, D.; Lachamp, P.; Naaim, M. SPH-based numerical investigation of mudflow and other complex fluid flow interactions with structures. *Comput. Geosci.* **2007**, *11*, 297–306. [CrossRef]

16. Pasculli, A.; Minatti, L.; Sciarra, N.; Paris, E. SPH modeling of fast muddy debris flow:Numerical and experimental comparison of certain commonly utilized approaches. *Ital. J. Geosci.* **2013**, *132*, 350–365. [CrossRef]

17. Wang, W.; Chen, G.; Han, Z.; Zhang, H.; Jing, P.D. 3D numerical simulation of debris-flow motion using SPH method incorporating non-Newtonian fluid behavior. *Nat. Hazards* **2016**, *81*, 1–18. [CrossRef]

18. Dai, Z.; Huang, Y.; Cheng, H.; Xu, Q. SPH model for fluid–structure interaction and its application to debris flow impact estimation. *Landslides* **2016**, *14*, 1–12. [CrossRef]

19. Hu, K.; Wei, F.; Li, Y. Real-time measurement and preliminary analysis of debris-flow impact force at Jiangjia Ravine, China. *Earth Surface Process. Landf.* **2011**, *36*, 1268–1278. [CrossRef]

20. Chen, H.K.; Xian, X.F.; Tang, H.M.; Zhang, Y.P.; He, X.Y.; Wen, G.J.; Tang, L. Energy distribution in spectrum of shock signal for non-viscous debris flow. *J. Vib. Shock.* **2012**, *31*, 56–59.

21. Spence, R.J.S.; Baxter, P.J.; Zuccaro, G. Building vulnerability and human casualty estimation for a pyroclastic flow: a model and its application to Vesuvius. *J. Volcanol. Geother. Res.* **2004**, *133*, 321–343. [CrossRef]

22. Luna, B.Q.; Blahut, J.; Westen, C.J.V.; Asch, T.W.J.V.; Akbas, S.O. The application of numerical debris flow modelling for the generation of physical vulnerability curves. *Nat. Hazards Earth Syst. Sci.* **2011**, *11*, 2047–2060. [CrossRef]

23. Kang, H.S.; Kim, Y.T. The physical vulnerability of different types of building structure to debris flow events. *Nat. Hazards* **2016**, *80*, 1–19. [CrossRef]

24. Zeng, C.; Cui, P.; Su, Z.; Lei, Y.; Chen, R. Failure modes of reinforced concrete columns of buildings under debris flow impact. *Landslides* **2015**, *12*, 561–571. [CrossRef]

25. Zhang, Y.; Wei, F.; Wang, Q. Experimental Research of Reinforced Concrete Buildings Struck by Debris Flow in Mountain Areas of Western China. *Wuhan Univ. J. Nat. Sci.* **2007**, *12*, 645–650. [CrossRef]

26. Brunkal, H.; Santi, P. Exploration of design parameters for a dewatering structure for debris flow mitigation. *Eng. Geol.* **2016**, *208*, 81–92. [CrossRef]

27. Wang, F.; Chen, X.; Chen, J.; You, Y. Experimental study on a debris-flow drainage channel with different types of energy dissipation baffles. *Eng. Geol.* **2017**, *220*, 43–51. [CrossRef]

28. Leonardi, A.; Wittel, F.K.; Mendoza, M.; Vetter, R.; Herrmann, H.J. Particle–Fluid–Structure Interaction for Debris Flow Impact on Flexible Barriers. *Comput.-Aided Civ. Infrastruct. Eng.* **2014**, *31*, 323–333. [CrossRef]

29. Jiang, Y.J.; Towhata, I. Experimental Study of Dry Granular Flow and Impact Behavior Against a Rigid Retaining Wall. *Rock Mech. Rock Eng.* **2013**, *46*, 713–729. [CrossRef]

30. Johnson, P.A.; Mccuen, R.H. Slit Dam Design for Debris Flow Mitigation. *J. Hydraul. Eng.* **1989**, *115*, 1293–1296. [CrossRef]

31. Huang, H.P.; Yang, K.C.; Lai, S.W. Impact force of debris flow on filter dam. *Eur. Geosci. Union Gen. Assem.* **2007**, *9*, 1–32.
32. Li, P.; Li, T.Z.H.; Lu, Z.; Li, J. Study on Dynamic Response of Novel Masonry Struct. Impacted by Debris Flow. *Sustainability* **2017**, *9*, 1122. [CrossRef]
33. Lu, Z.; Yang, Y.L.; Lu, X.L.; Liu, C.Q. Preliminary Study on the Damping Effect of a Lateral Damping Buffer under a Debris Flow Load. *Appl. Sci.* **2017**, *7*, 201. [CrossRef]
34. Datta, J.; Haponiuk, J. Advanced coating of interior of tanks for rising environmental safety-novel applications of polyurethanes. *Polish Marit. Res.* **2008**, *15*, 8–13. [CrossRef]
35. Sharma, S.C.; Krishna, M.; Murthy, H.N.N.; Sathyamoorthy, M.; Bhattacharya, D. Fatigue studies of polyurethane sandwich structures. *J. Mater. Eng. Perform.* **2004**, *13*, 637–641. [CrossRef]
36. Szarnik, A.; Kuryłko, A. Application of steel sandwich panels to hull structure of two-segment inland navigation passenger ship. *Polish Marit. Res.* **2006**, *S2*, 85–87.
37. Alia, C.; Arenas, J.M.; Suárez, J.C.; Pinilla, P. Mechanical behavior of polyurethane adhesive joints used in laminated materials for marine structures. *Ocean Eng.* **2016**, *113*, 64–74. [CrossRef]
38. Xiao, Y.K.; Ji, W.F.; Chang, K.S.; Hsu, K.T.; Yeh, J.M.; Liu, W.R. Sandwich-structured rGO/PVDF/PU multilayer coatings for anti-corrosion application. *Rsc. Adv.* **2017**, *7*, 33829–33836. [CrossRef]
39. Xu, X.J.; Shan, C.L. Impact analysis of the bridge pier anti-collision floating box sets made by sandwich structure with curved-shaped. *J. Hunan Univ. Nat. Sci.* **2015**, *42*, 106–111.
40. Harris, D.K.; Cousins, T.; Sotelino, E.D.; Murray, T.M. Flexural lateral load distribution characteristics of sandwich plate system bridges: parametric investigation. *J. Bridge Eng.* **2010**, *15*, 684–694. [CrossRef]
41. Onoue, K.; Tamai, H.; Suseno, H. Shock-absorbing capability of lightweight concrete utilizing volcanic pumice aggregate. *Construct. Build. Mater.* **2015**, *83*, 261–274. [CrossRef]
42. He, N.; Chen, N.; Zeng, C. Current Situation and Tendencies of Debris Flow Initiation Mechanism. *J. Catastrophol.* **2013**, *28*, 121–125.

*applied
sciences*

MDPI

Article

Soil–Structure–Equipment Interaction and Influence Factors in an Underground Electrical Substation under Seismic Loads

Bo Wen [1,*], Lu Zhang [1], Ditao Niu [1] and Muhua Zhang [2]

[1] School of Civil Engineering, Xi'an University of Architecture & Technology, Xi'an 710055, China;
    15202419379@163.com (L.Z.); niuditao@163.com (D.N.)
[2] School of Humanities, Xi'an Polytechnic University, Xi'an 710048, China; dr.longbow@163.com
*   Correspondence: wenbo_mail@163.com

Received: 25 July 2017; Accepted: 25 September 2017; Published: 12 October 2017

Abstract: Underground electrical substations play an increasingly significant role in urban economic development for the power supply of subways. However, in recent years, there have been few studies on the seismic performance of underground electrical substations involving the interaction of soil–structure–equipment. To conduct the study, three-dimensional finite element models of an underground substation are established. The implicit dynamic numerical simulation analysis is performed by changing earthquake input motions, soil characteristics, electrical equipment type and structure depths. According to a seismic response analysis, acceleration amplification coefficients, displacements, stresses and internal forces are obtained and analyzed. It is found that (1) as a boundary condition of soil–structure, the coupling boundary is feasible in the seismic response of an underground substation; (2) the seismic response of an underground substation is sensitive to burial depth and elastic modulus; (3) the oblique incidence of input motion has a slight influence on the horizontal seismic response, but has a significant impact on the vertical seismic response; and (4) the bottom of the side wall is the seismic weak part of an underground substation, so it is necessary to increase the stiffness of this area.

Keywords: underground electrical substation; seismic analysis; influence factor; boundary condition

1. Introduction

In order to meet the requirements of the rapid growth of urban power load under the constraint of densely populated urban land, the number of underground electrical substations has been increasing in recent years. The disaster prevention and mitigation issue of underground structures, including with regard to underground electrical substations, is receiving increasing attention. Therefore, many scholars have focused on the interaction of structure–equipment and the seismic performance of underground structures.

Suarezn and Singh (1989) [1] first developed the interaction principle of primary–secondary structure and obtained response spectrum curves based on the different floors by a simple model involving equipment–structure interaction. Pires (1996) [2] established a simulation model of a San Francisco substation for seismic reliability analysis and compared the numerical substation damages with the real damages caused by Loma Prieta earthquake in 1989. Bazán-Zurita (2009) [3] illustrated concepts interpreting and supplementing the seismic provisions of Manual 113 (Kempner, 2008 [4]) which defined seismic design spectra as simultaneous occurrence of ground motion in three mutually perpendicular directions and estimation of deflections under seismic loading. Poor seismic performance of these system components could result in extended power outages that might propagate far beyond the local epicentral region. Knight and Kempner, Jr. (2009) [5]

proposed a Seismic Vulnerability Reduction Program (SVRP) design method to improve the seismic weak parts of substation and present several suggestions on the seismic design of a transformer substation. The international atomic energy agency (IAEA, [6]) provided seismic evaluation criteria on the interaction of structure–equipment in a nuclear power plant. Furthermore, from the study of the interaction between the main plant and electrical equipment (EE) in transformer substations, it was found that the seismic response of the main structure and equipment was not usually synchronized and equipment had a negative effect on the seismic ability of the substation (Wen and Niu, 2013 [7]; Wen et al., 2015 [8]).

To date, many researchers have focused on the interaction of soil–structure. Assuming that interaction force is transformed by a spring and damper, Yuan (1970, [9]) introduced the soil–structure interaction principle. Rodriguez (1985, [10]) analyzed the static interaction between soil and the tunnel lining and found that the force between soil and structure had a significant relationship with contact surface roughness. Penzien (2000, [11]) developed an analytic method involving shear interaction between soil and structure. Based on Penzien's studies, Huo (2000, [12]) established a hysteretic non-linear soil model to investigate the load transfer mechanisms between underground structure and surrounding soil. Therefore, dynamic finite element analyses were conducted to investigate the load transfer mechanisms between underground structure and surrounding soil. Gong (2002, [13]) studied the responses of an underground structure under vertical seismic excitation through the shaking table test, analyzing the dynamic interaction of soil and underground structure. Based on the background of metro construction in Nanjing city, China, a shaking table test was conducted to model the interaction of soil–metro structure with a liquefiable soil effect. Chen and Zhuang (2006, 2007, [14,15]) studied the dynamic damage characteristics of underground structures in far-field earthquake motions and near-field earthquake motions. According to the static and dynamic coupling effects on the underground structure and surrounding rock system, Du (2009, [16]) proposed stress viscoelastic artificial boundary conditions to establish an underground structure–soil computational model. Tao (2012, [17]) calculated the side wall deformation and structural internal force of a single-story frame of Zhijing line subway in Beijing with different geological conditions and cover depths. In order to understand the distribution of internal force of the underground shield tunnel under earthquake motions, Geng (2013, [18]) investigated the seismic response of underground tunnels in different site types using a quasi-static method and indicated that such a method was more suitable to underground tunnels in soft foundation. Debiasi (2013, [19]) proposed a nonlinear static analysis of a rectangular tunnel by using several different simplified numerical methods, and the qualitative relationship between seismic dynamic strength and structure geometry, overburden thickness and soil stiffness were obtained. Liu (2014, [20]) put forward a Pushover analysis method with good accuracy and reliability, which applied to substructure seismic analysis and design. Kang (2014, [21]) studied the influence of foundation liquefaction on the substructure and conducted an experimental study and numerical analysis on the seismic response of a circular subsurface structure using a modified simple calculating method. Kyriazis (2014, [22]) performed a numerical analysis on an arched tunnel using finite element software and found that some given parameters such as the sectional area of arch, and the thickness of surrounding rock and soil characteristic, had a significant influence on the tunnel seismic response.

From the above research, it is known that underground structure types become more complicated and there are few studies on the seismic performance of underground substations involving the interaction of soil–structure–equipment. In past studies, such interaction in underground substations was usually ignored, and the seismic response caused by characteristics of electrical equipment and soil were neglected. For these reasons, some positive research is proposed in this paper. A three-dimensional finite element model of an underground substation is established to conduct dynamic numerical simulation analysis. By changing some parameters such as ground motions, properties of surrounding soils, electrical equipment and structure depth, the dynamic results of acceleration amplification coefficients, displacements, stresses and internal forces under different earthquake motions are obtained and analyzed.

2. Dynamic Interaction Mechanism of Soil–Structure–Equipment

It is well known that the dynamic equation of structure under earthquake motions can be shown as Equation (1) (Clough and Penzien, 1993, [23]).

$$[M]\left\{\ddot{u}\right\} + [C]\left\{\dot{u}\right\} + [K]\{u\} = \{F\} \tag{1}$$

where M is the mass matrix, $\ddot{u}$ is the acceleration vector, C is the damping matrix, $\dot{u}$ is the velocity vector, K is the stiffness matrix, u is the displacement vector, and F is the load vector.

For the underground substation model, it usually includes three parts: soil, structure and electrical equipment. Similar to Equation (1), the dynamic equation of an underground substation can be described as Equation (2), in which subscripts s, c, and j are denoted as soil, structure, and electrical equipment respectively.

$$\begin{bmatrix} M_s & & \\ & M_c & \\ & & M_j \end{bmatrix}\left\{\ddot{u}\right\} + \begin{bmatrix} C_{ss} & C_{sc} & C_{sj} \\ C_{cs} & C_{cc} & C_{cj} \\ C_{js} & C_{jc} & C_{jj} \end{bmatrix}\left\{\dot{u}\right\} + \begin{bmatrix} K_{ss} & K_{sc} & K_{sj} \\ K_{cs} & K_{cc} & K_{cj} \\ K_{js} & K_{jc} & K_{jj} \end{bmatrix}\{u\} = \left\{\begin{array}{c} F_s \\ F_c \\ F_j \end{array}\right\} \tag{2}$$

The approximate solution is solved by the Newton Method [24], as shown in Equations (3)–(5).

$$F^N\left(u^M\right) = 0 \tag{3}$$

$$F^N\left(u_i{}^M + c_{i+1}^M\right) = 0 \tag{4}$$

$$u_{i+1}^M = u_i^M + c_{i+1}^M \tag{5}$$

where u^M is the exact solution of Equation (2) and $u_i{}^M$ is the approximate solution of Equation (2). The dynamic equation numerical solution is solved when a small enough number c_{i+1}^M exists in Equation (5).

3. Numerical Simulations

3.1. Finite Element Models

The prototype substation used in this study is selected from Urban underground substation design rule (2005, China) [25]. The main plant of the underground substation is a three-story, four-bay reinforced concrete shear wall structure, as shown in Figures 1–4. The dimensions of the main plant are 50 m by 25 m, and floor heights from the bottom to top are 4.4 m, 4.8 m, and 5.2 m respectively.

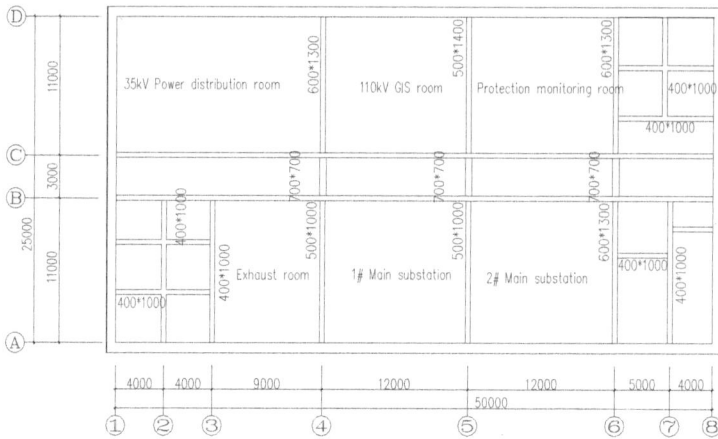

Figure 1. Plan of the second floor.

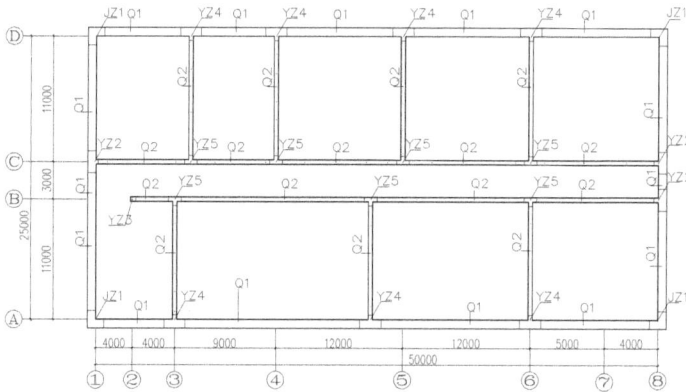

Figure 2. Dimensions of the floor and wall.

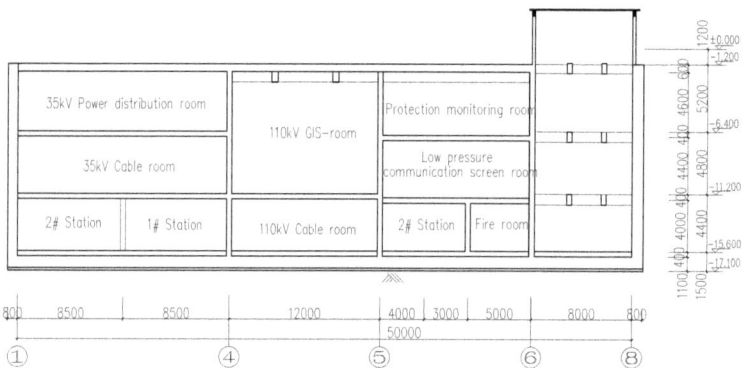

Figure 3. Long cross-section.

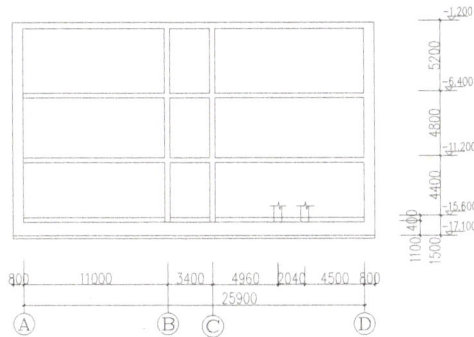

Figure 4. Short cross-section.

The exterior wall of the main plant structure is made of a reinforced concrete wall, 0.45 m in width. The dimensions of the main beams are 600 × 1300 mm^2, 500 × 1400 mm^2, 500 × 1000 mm^2 separately, and the cross-sectional dimension of the transverse beams is 400 × 1000 mm^2. The thicknesses of the first, second, third, and roof slabs are 900 mm, 400 mm, 400 mm, and 600 mm respectively. The total weight of the plant is approximately 27,530 kN.

Ten sets of electrical equipment, shown in Figure 5, are installed on the second floor, the dimensions of which are 6.0 m (length), 2.0 m (width) and 3.5 m (height). The total weight of the equipment is approximately 964 kN, which corresponds to 3.5% of the plant weight.

The plant was designed by Chinese loading Code for Design of Building Structures (2012, [26]), in which gravity loads are selected as 1.8 kN/m^2 for floors and 3.5 kN/m^2 for the roof. Live loads are selected as 2.5 kN/m^2 for floors and 2.0 kN/m^2 for the roof. The gravity load of equipment is 8.0 kN/m^2. The seismic design of the substation is allowed according to Chinese Code for Seismic Design of Buildings (2010, [27]) and Chinese Code for Seismic Design of Electrical Installations (2013, [28]). The underground substation is assumed to be located in Xi'an, China, where the design basis acceleration associated with 10% probability exceedance in 50 years is 0.2 g (g denotes the coefficient of gravitational acceleration). The compressive strength of concrete and the yield strength of steel are taken as f_c = 30 MPa, and f_y = 400 MPa, respectively. According to the first eigen-frequency, the factor β is shown in Equation (6).

$$\beta = \frac{2\xi_1}{\omega_1} \qquad (6)$$

Based on a natural frequency extraction analysis of the interaction system, the first eigen-frequency is 0.39 rad/s. Therefore, β is 0.179 s.

Figure 5. Electrical equipment.

The models were developed and analyzed in ABAQUS (2006, [29]). The beams and columns are modeled in three-dimensional beam elements, the floors and concrete walls are modeled in shell elements, and the equipment is modeled in pipe elements, as shown in Figure 6. The three-dimensional finite element electrical equipment model is shown in Figure 7. The concrete plastic-damage model is used for the constitutive properties of concrete material and the double linear dynamic reinforcement model is used for rebar. Druker–Prager, a generalized Mises yield criterion, is adopted in soil dilation. The effect of hydrostatic pressure on yield and strength is used as the constitutive model of soil (Liu et al., 2006, [30]). The physical and mechanical parameters of soil, as determined by the penetration test and consolidation test, are listed in Table 1. Based on the artificial constrained boundary, the boundary element is used to simulate the infinite field of soil, and the horizontal reaction forces on the lateral boundary are determined by static analysis. According to the literature (Liu et al., 2006 [31]; Huang et al., 2010 [32]), the coupling analysis method is used in the soil boundary condition. Meanwhile, it is suggested from literature (Lou et al., 1999 [33]; Lou et al., 2000 [34]) that five times the width of the structure can be defined as the width of the computational area of soil in order to reduce the motion reflection on the boundary. Following this rule, five times the width of the structure is selected as the soil width in the finite element model. Therefore, the soil size of the finite element is 250 m × 132.5 m × 50 m and the size of the entire soil is 500 m × 265 m × 50 m. The arbitrary Lagrangian Eulerian (ALE) adaptive meshing method is used to maintain a high-qualified meshing system for soil, which allows the meshes to move independently when large deformation of liquefied soils occurs. The soil element meshes are shown in Figure 8, in which the finite element of soil is simulated by unit C3D8R. In order to simulate the most realistic condition of soil and simplify some less salient factors in this paper, the following basic assumptions are advised in analysis (Chen [35], 2015; Novak [36], 1972).

(1) Each layer of soil is homogeneous, and extends infinitely in the horizontal direction;
(2) There is no relative sliding among different layers of soil;
(3) Earthquake excitations are supposed from the bottom of the foundation;
(4) The movement of each point on the bottom of the foundation is consistent without a traveling motion effect.

Table 1. The physical and mechanical soil parameters.

Number	Name of Soil	Density (kg/m³)	Elastic Modulus (MPa)	Poisson Ratio	Cohesion (kPa)	Internal Friction Angle (°)	Soil Depth (m)
1	Plain fill	1720	5.0	0.39	22	15	4.30
2	New loess	1600	12.0	0.30	28	18	6.20
3	Silty clay	1950	15.0	0.30	45	15	1.70
4	Pebble bed	2250	55.0	0.15	0	45	2.10
5	Silty clay	1950	15.0	0.30	45	15	27.0
6	Coarse ands	1920	40.0	0.26	0	35	4.50

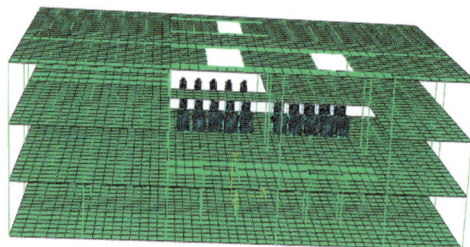

Figure 6. Finite element model of a substation.

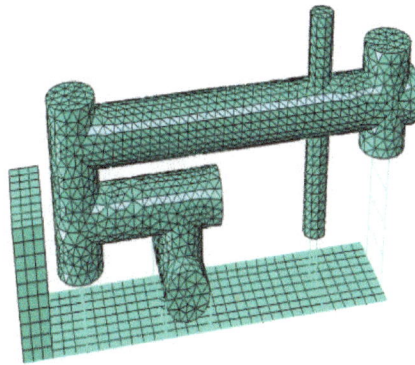

Figure 7. Finite element model of equipment.

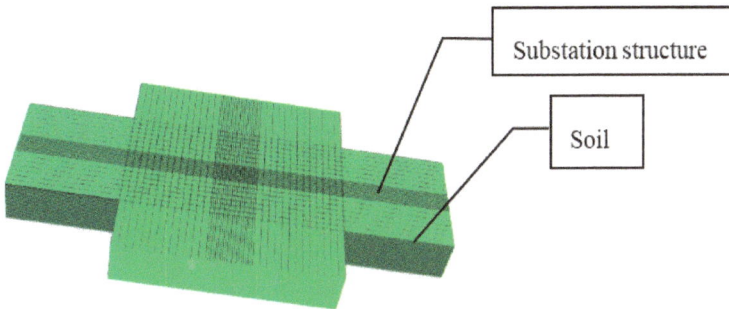

Figure 8. Finite element model of soil.

According to the classification of site type and seismic fortification intensity, three earthquake records—El-Centro record, Taft record and Lanzhou record—were derived from the registered seismic data. Moreover, in order to obtain earthquake motions, three earthquake records were back-calculated to the bottom of the foundation level by the SHAKE91 procedure (Idriss et al., 1992 [37]). According to the finite element division of the soil layer, soil properties such as density, shear velocity, shear modulus of elasticity, etc., were taken into account in earthquake motion inversion. Time-histories and spectra of these records after inversion are shown in Figures 9–14.

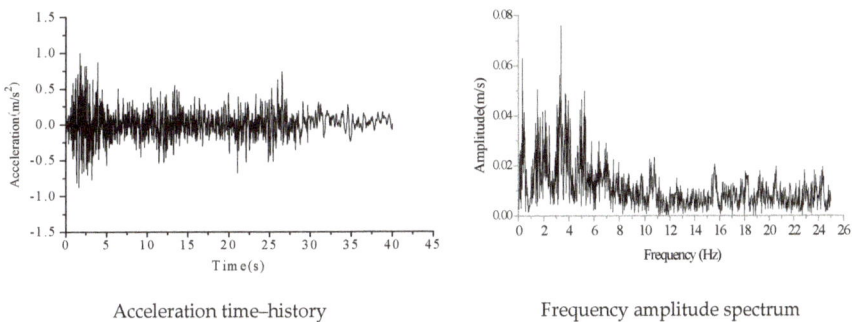

Acceleration time–history Frequency amplitude spectrum

Figure 9. El-Centro motion in the horizontal direction after inversion.

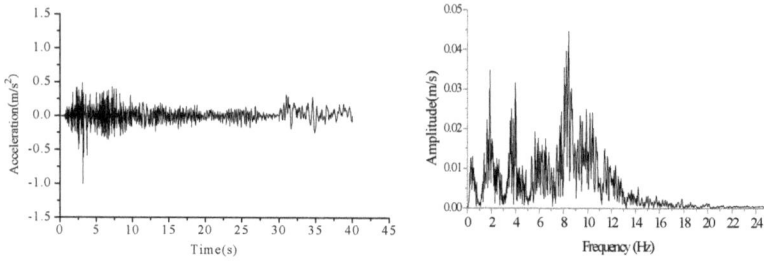

Acceleration time–history   Frequency amplitude spectrum

Figure 10. El-Centro motion in the vertical direction after inversion.

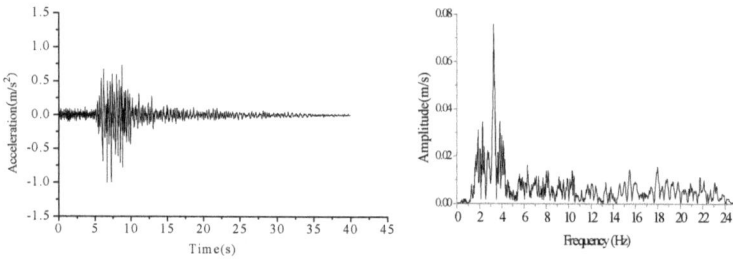

Acceleration time–history   Frequency amplitude spectrum

Figure 11. Taft motion in the horizontal direction after inversion.

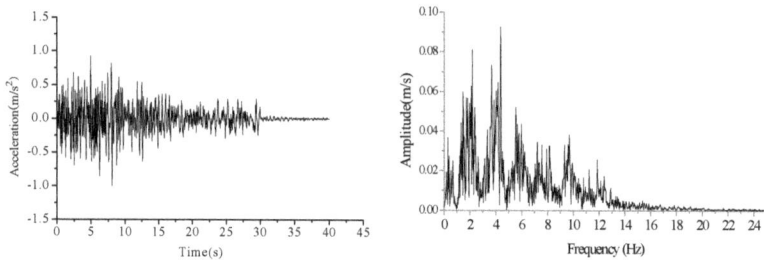

Acceleration time–history   Frequency amplitude spectrum

Figure 12. Taft motion in the vertical direction after inversion.

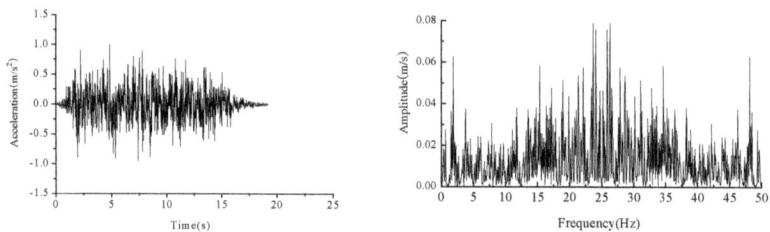

Acceleration time–history   Frequency amplitude spectrum

Figure 13. Lanzhou motion in the horizontal direction after inversion.

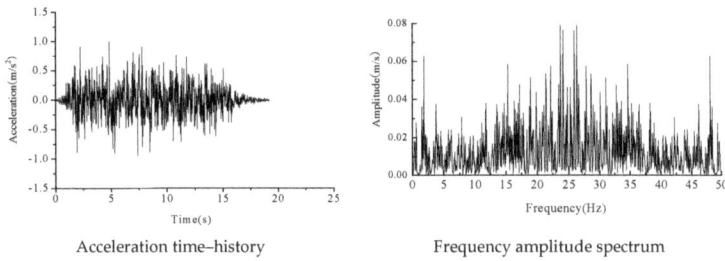

Acceleration time–history

Frequency amplitude spectrum

Figure 14. Lanzhou motion in the vertical direction after inversion.

3.2. Accuracy Verification of the Finite Element Model

The soil boundary condition of the coupling analysis method of the finite element and infinite element, which reasonably reflects the real boundary condition, gives a unified solving format, and makes the infinite element of the soil section become part of the finite element model (Liang et al., 2010 [38]; Meng et al., 2012 [39]). Compared with the viscoelastic boundary condition, the infinite element boundary condition can not only simulate the far field boundary to absorb seismic energy, but also correctly simulate the boundary conditions whose infinity displacement is zero. Therefore, the coupling analysis method of the finite element and infinite element is used as the soil boundary condition to simulate semi-infinite domain soil.

For the three-dimensional finite element model, the distance value r is first calculated from the random boundary node to the structure center. Secondly, normal spring stiffness, normal damping coefficient, tangential spring stiffness, and tangential damping coefficient of the node on the boundary are calculated by Equations (7) and (8). Finally, all of the boundary nodes are calculated cyclically in order to obtain a viscoelastic artificial boundary.

$$K_{li} = \frac{\lambda + 2G}{r} \quad C_{li} = \rho c_P \tag{7}$$

$$K_{li} = \frac{G}{r} \quad C_{li} = \rho c_S \tag{8}$$

where c_p is the longitudinal velocity of the medium, c_s is the longitudinal velocity of the medium, ρ is density, λ and G are lame constants, and r is the distance from the artificial boundary point to the scattering source.

The soil surface is supposed to be a free boundary, and the horizontal reaction forces at the nodes on the lateral boundary and bottom boundary are calculated according to the artificially constrained boundary assumption (Liu et al., 2006 [31]; Huang et al., 2010 [32]). Therefore, the fixed boundary model, viscoelastic boundary model and coupling analysis model are set up to verify different soil boundary conditions, which are simplified as follows:

Model 1 (M1): Coupling boundary between finite elements and infinite elements
Model 2 (M2): Viscoelastic boundary
Model 3 (M3): Fixed boundary

The acceleration time–history curves at the soil surface midpoint are shown in Figure 15.

From Figure 15, it can be inferred that the curves are almost the same in three different conditions before 4 s. After 4 s, only curves of Model 1 and Model 2 are similar to each other, while the acceleration values of Model 3 gradually increase compared to those of the other two curves. The reason is that seismic energy is absorbed and the seismic wave is rarely reflected when the motions pass through the infinite element boundary or viscoelastic artificial boundary, while seismic energy is greatly reflected when the motions pass through the fixed boundary, which leads to the accumulation of acceleration values on the free boundary. Therefore, compared with the viscoelastic boundary condition, the infinite element boundary condition does not need to be involved in the analytical solution expression in ABAQUS software. For this reason, Model 1 is selected as the soil boundary condition in this paper.

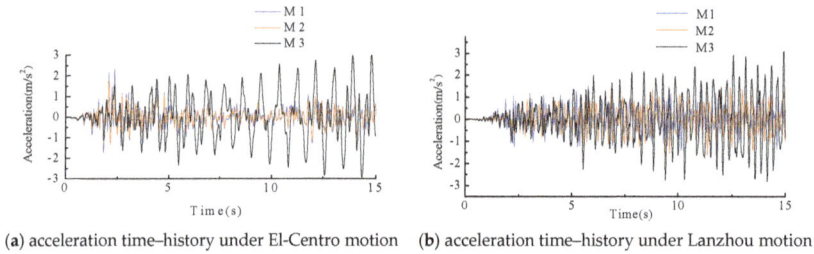

(**a**) acceleration time–history under El-Centro motion   (**b**) acceleration time–history under Lanzhou motion

Figure 15. Time–history curves of the midpoint's acceleration of soil.

4. Numerical Simulation Analysis

The underground substation model involving interaction of soil–structure–equipment is established for numerical simulation analysis. Based on finite element analysis, the influence of some parameters on seismic response such as oblique incidence of input motion, properties of surrounding soils, and burial depth of the underground substation are analyzed.

4.1. The Seismic Response under Different Ground Motions

According to Chinese Code for Seismic Design of Buildings (2010, China [27]), the peak ground acceleration (PGA) is given in Table 2, in which the scales of PGA in X, Y and Z directions are adjusted to 0.20 g, 0.17 g, and 0.13 g, respectively.

Table 2. The conditions of peak ground acceleration (PGA).

Model	Direction	Acceleration Input Value
M1	Horizontal	0.2 g:0.17 g (X:Y)
M2	Vertical	0.13 g (Z)
M3	Coupling of horizontal and vertical	0.2 g:0.17 g:0.13 g (X:Y:Z)

4.1.1. Acceleration Responses

The acceleration amplification factors under 0.20 g are shown in Table 3, in which the acceleration amplification factor is defined as the ratio of the peak acceleration of the layer over the peak acceleration of input motion.

Table 3. Acceleration amplification factors under 0.20 g.

Section		EL-Centro Motion			Lanzhou Motion		
		M1	M2	M3	M1	M2	M3
Roof	X-direction	0.671	/	0.683	0.683	/	0.714
	Y-direction	0.468	/	0.480	0.489	/	0.502
	Z-direction	/	1.476	1.526	/	1.108	1.163
Floor 3	X-direction	0.677	/	0.695	0.699	/	0.726
	Y-direction	0.573	/	0.582	0.503	/	0.520
	Z-direction	/	1.476	1.525	/	1.108	1.162
Floor 2	X-direction	0.679	/	0.719	0.722	/	0.742
	Y-direction	0.595	/	0.623	0.536	/	0.565
	Z-direction	/	1.475	1.523	/	1.016	1.159
Floor 1	X-direction	0.708	/	0.721	0.746	/	0.760
	Y-direction	0.681	/	0.692	0.568	/	0.590
	Z-direction	/	1.475	1.522	/	1.105	1.158

The following can be inferred from Table 3:

(1)   The acceleration amplification factors in the X-direction are larger than those in the Y-direction in M1 and M3, which means that the seismic response of the underground substation in the X-direction is more severe than that in the Y-direction. The seismic response including acceleration responses, displacement responses and internal force responses of 0.17 g PAG in the X-direction and 0.2 g PAG in the Y-direction would be reduced but the response rule is uniform.

(2)   The acceleration amplification factors in the Z-direction of M3 are larger than those in M2. The reason is that the seismic response is magnified under multi-direction earthquake motions compared with that under the one-direction earthquake input motion and the horizontal earthquake motions have a reverse effect on the seismic response in the Z-direction.

(3)   The peak vertical and horizontal acceleration amplification factors are 1.526 and 0.760, respectively, which means that the oblique incidence of input motion has a slight influence on the horizontal seismic response, but has a significant impact on the vertical seismic response.

4.1.2. Displacement Responses

The horizontal deformation of the side wall causes eccentric compression and a large bending moment, which can lead to damage of the side wall. Therefore, the horizontal layer drift of the underground substation is the major factor of side wall damage, and it is necessary to analyze the horizontal relative deformation of the underground substation under different ground motions.

The relative displacement time–history curves of the top floor under El-Centro motion and Lanzhou motion are shown in Figure 16, in which the relative displacement is defined as horizontal displacement values, and layer drifts in M1 and M3 under 0.20 g are shown in Table 4.

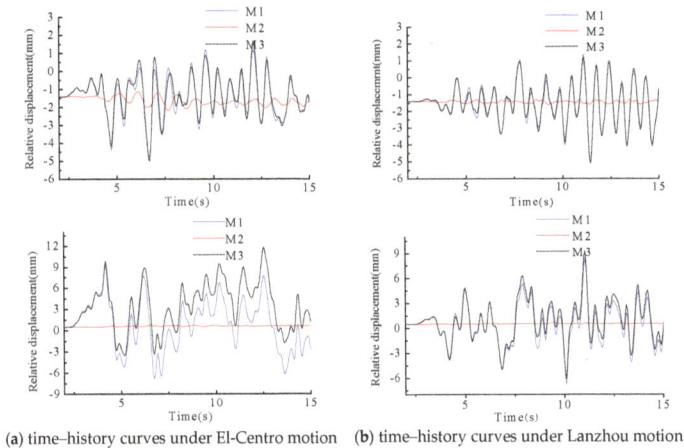

(a) time–history curves under El-Centro motion   (b) time–history curves under Lanzhou motion

Figure 16. Relative displacement curves of the top floor.

Table 4. Layer drifts (mm).

Section	El-Centro Motion				Lanzhou Motion			
	M1		M3		M1		M3	
	X-Direction	Y-Direction	X-Direction	Y-Direction	X-Direction	Y-Direction	X-Direction	Y-Direction
Floor 3	1.65	2.47	1.76	4.03	1.62	2.18	1.77	3.14
Floor 2	1.56	2.30	1.65	3.81	1.53	2.00	1.65	2.98
Floor 1	1.70	2.50	1.79	3.83	1.68	2.10	1.77	2.99

It can be seen from Figure 16 and Table 4 that layer drifts in the Y-direction are larger than those in the X-direction, which coincides with the feature that stiffness in X-direction is larger than that in the Y-direction. There is only a 2% difference in the value of the drifts between one-way input motions and multi-way input motions in the X-direction, which indicates that there is no obvious difference in torsion coupling in the X-direction. On the other hand, in the Y-direction, the horizontal drifts with multi-way input motions are about 1.4 times larger than those with one-way input earthquake motion. Therefore, the drifts of the corner column under multi-way input motions are larger than those under one-way input motions, which leads to severe damage of the corner columns. It is essential to improve the constraints of corner columns and increase the number of stirrups in these columns, which can strengthen the deformation ability of such components.

4.1.3. Internal Force Responses

The shear wall is an important component under seismic action. As the first defense of energy dissipation under earthquake motions, coupling beams cannot bear the internal force value owing to the limited sectional dimension. Thus, the main internal force value is borne by the shear wall after internal force redistribution. There are six edges of concrete walls represented by Q-2, Q-7, Q-12, Q-17, Q-22 and Q-27 on surface-6, as shown in Figure 17. Furthermore, there is a similar seismic response among different sections. Therefore, the surface-6 section is selected to analyze the internal force response of the side walls. The peak internal forces in each section of surface-6 under 0.20 g are given in Table 5 and Figure 18.

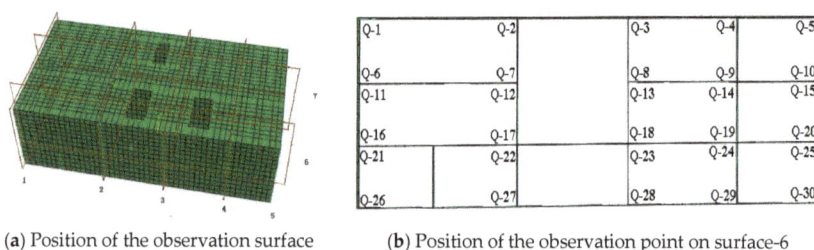

(a) Position of the observation surface   (b) Position of the observation point on surface-6

Figure 17. Position of the observation surface and point.

Table 5. The peak internal forces of each section on surface-6.

Section		Vertical Pressure (10^5 N/m)			Shear Force (10^5 N/m)			Bending Moment (10^3 N·m/m)		
		M1	M2	M3	M1	M2	M3	M1	M2	M3
Floor 3	El-Centro Top	3.74	4.32	4.43	5.41	4.74	5.42	7.19	8.22	8.28
	El-Centro Bottom	4.87	5.20	5.47	6.15	4.82	6.17	6.54	6.94	7.16
	Lanzhou Top	3.75	3.93	4.01	5.61	4.36	5.64	7.30	7.87	8.01
	Lanzhou Bottom	4.51	4.68	4.86	6.14	4.57	6.23	6.90	7.21	7.47
Floor 2	El-Centro Top	6.26	6.76	7.14	7.49	5.12	7.50	6.33	7.67	7.85
	El-Centro Bottom	9.39	10.13	10.31	7.83	4.86	7.92	9.41	11.50	11.62
	Lanzhou Top	5.75	6.07	6.24	7.04	4.82	7.22	6.33	6.97	6.97
	Lanzhou Bottom	8.69	9.05	9.44	7.00	4.53	7.35	9.08	10.07	10.13
Floor 1	El-Centro Top	11.39	12.52	12.77	6.62	3.14	6.64	6.32	7.95	8.19
	El-Centro Bottom	31.42	36.66	36.70	10.2	5.76	10.3	5.23	5.70	5.81
	Lanzhou Top	10.03	11.16	11.55	5.65	3.00	5.86	6.51	6.90	6.91
	Lanzhou Bottom	30.20	32.66	33.11	8.89	5.39	9.22	5.45	5.93	6.00

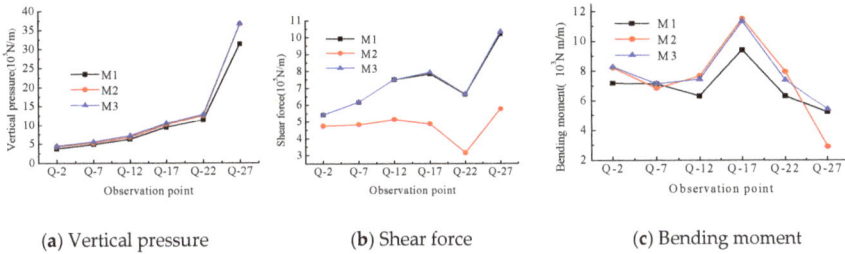

(a) Vertical pressure (b) Shear force (c) Bending moment

Figure 18. The peak internal forces of surface-6 under El-Centro motion.

The following can be concluded from Figure 18 and Table 5:

(1) Different input motions have different influences on the internal force of side walls. The shear forces of components are similar between horizontal direction motion and coupling input motion. Furthermore, the vertical pressure and bending moments of components under vertical seismic motion are similar to those in coupling input motion. Oblique incidence motion makes the vertical pressure and bending moments increase obviously.

(2) In comparing the peak value of internal force in different sections, it can be seen that the shear force and vertical pressure of the bottom of the side wall on the third floor are, at most, about 1.03×106 N/m and 3.67×106 N/m respectively, values which are significantly greater than those on the other floors. So, the bottom of the side walls is the seismic weak part of the structure.

(3) The bending moments at joint parts of the side wall on the second floor are larger than the rest of the structure, because the extra inertia force caused by the electrical equipment leads to the second floor being in a complex bend torsion condition. The bending moments of the second floor under coupling seismic action increase by 20% compared to those under horizontal seismic action. So, it can be speculated that the vertical pressure and bending moments increase in vertical seismic motion.

4.2. Soil Properties in the Seismic Response of the Underground Substation

In order to analyze the seismic response of soil and the underground substation, M1 (1 time soil elasticity modulus model), M2 (2 times soil elasticity modulus model) and M3 (3 times soil elasticity modulus model) were set up separately, the seismic response results of which are shown in Table 6.

Table 6. Soil elasticity modulus (MPa).

Number	Name of Soil	M1	M2	M3
1	Plain fill	5.0	10.0	15.0
2	New loess	12.0	24.0	36.0
3	Silty clay	15.0	30.0	45.0
4	Pebble bed	55.0	110.0	165.0
5	Silty clay	15.0	30.0	45.0

4.2.1. Displacement Responses

The relative displacements under 0.20 g are shown in Figure 19. The layer drifts and layer drift angles under 0.20 g are shown in Table 7.

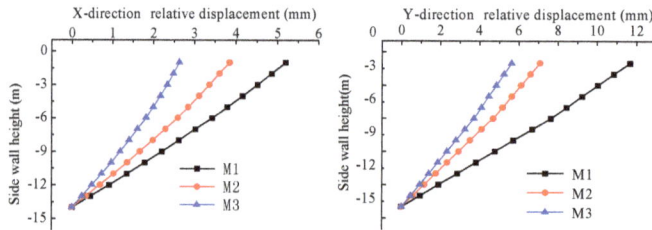

(a) Relative horizontal displacement under El-Centro motion

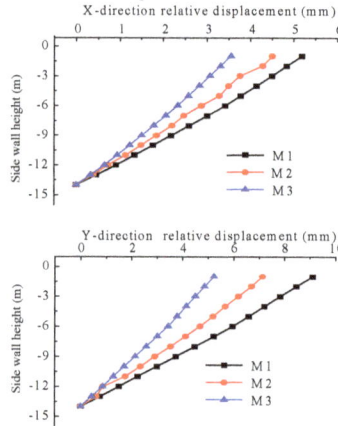

(b) Relative horizontal displacement under Lanzhou motion

Figure 19. Relative horizontal displacement of the side wall.

Table 7. Layer drifts (mm) and layer drift angles (rad).

Section		El-Centro Motion			Lanzhou Motion		
		M1	M2	M3	M1	M2	M3
Floor 3	layer drift in the X-direction	1.76	1.26	0.8	1.77	1.15	1.04
	layer drift angle in the X-direction	1/2898	1/4048	1/6375	1/2881	1/4435	1/4904
	layer drift in the Y-direction	4.03	2.37	1.91	3.14	1.83	0.76
	layer drift angle in the Y-direction	1/1266	1/2025	1/2513	1/1529	1/2623	1/6316
Floor 2	layer drift in the X-direction	1.65	1.23	0.86	1.65	1.12	0.78
	layer drift angle in the X-direction	1/2909	1/3902	1/5581	1/2909	1/4286	1/6154
	layer drift in the Y-direction	3.81	2.34	1.85	2.98	1.76	0.68
	layer drift angle in the Y-direction	1/1260	1/2051	1/2595	1/1611	1/2727	1/7059
Floor 1	layer drift in the X-direction	1.79	1.36	0.96	1.77	1.23	1.23
	layer drift angle in the X-direction	1/2682	1/3529	1/5000	1/2712	1/3902	1/3902
	layer drift in the Y-direction	3.83	2.36	1.87	2.99	1.75	0.93
	layer drift angle in the Y-direction	1/1253	1/2034	1/2567	1/1605	1/2743	1/5161

(1) From Figure 19, it can be inferred that the displacements of the side wall decrease with the increasing of the elastic modulus of soil and the horizontal drift diminishes mostly at the top of the side wall, and the horizontal relative displacement in the Y-direction decreases by about 73.9% in Lanzhou motion. The reason is that, with the increasing of the soil elastic modulus, the structure constraint increases and structural deformation decreases, which usually lead to the diminishing of horizontal drift.

(2) From Table 7, the drift angle is inversely proportional to the elastic modulus of surrounding soil. With the increase in the soil elastic modulus, the drift angle of each layer significantly reduces;

the minimum value decreases sharply from 1/1611 to 1/7059. Therefore, the soil elastic modulus has a very important influence on the deformation of the underground substation.

4.2.2. Stress Responses

The stress responses of the underground substation are similar under El-Centro, Taft and Lanzhou earthquakes; therefore, the stress response of the structure under Lanzhou earthquake is taken as an example. The stress nephograms of the underground substation under 0.20 g are shown in Figure 20. The peak stresses and stress amplitude in different locations under 0.20 g are shown in Table 8.

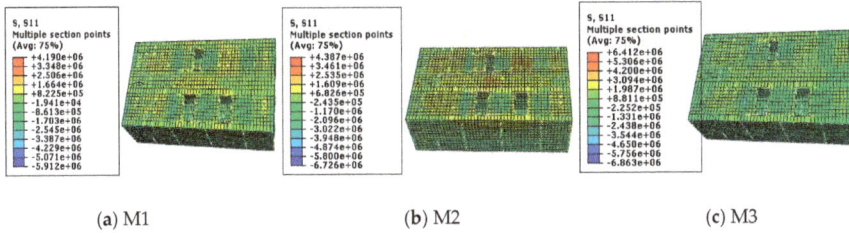

(a) M1   (b) M2   (c) M3

Figure 20. Stress nephograms of the underground substation.

Table 8. The peak stresses and stress amplitudes under Lanzhou motion.

Section	M1			M2		M3	
	Static (MPa)	Dynamic (MPa)	Amplification	Dynamic (MPa)	Amplification	Dynamic (MPa)	Amplification
−1 Floor	3.97	4.95	24.66	4.80	20.89	5.08	27.89
−2 Floor	3.61	4.22	16.90	4.13	14.40	5.28	46.23
−3 Floor	3.90	4.43	13.74	4.65	19.38	5.51	41.39
Side wall 1 *	1.99	2.27	14.09	2.10	5.44	2.23	11.98
Side wall 3	2.93	3.32	13.38	3.26	11.26	3.58	22.16
Side wall 4	5.43	6.56	20.80	6.11	12.59	6.51	19.97
Side wall 5	7.72	8.54	10.70	8.36	8.33	9.04	17.08
Side wall 6	4.15	4.70	13.14	4.65	12.36	5.41	30.23
Side wall A	4.14	4.71	13.88	4.40	6.48	4.73	14.49
Side wall B	4.09	4.14	1.09	4.10	0.25	4.22	3.07
Side wall C	3.73	5.19	39.08	4.79	28.28	5.14	37.69
Side wall D	2.48	2.60	4.70	2.59	4.43	2.81	13.19

* Side wall 1 means the number of the wall, i.e., located at the cross of axis 1. The rest have similar meanings.

With the increase in the elastic modulus of soil, stress amplitudes of layers and side walls do not simply decrease with the decrease of layer drift, but firstly decrease and then increase. The second-floor stress amplitude maximum value is 46.23% when the soil elastic modulus increases from two- to three-fold. The reason is that with the increase in the elastic modulus of soil, the deformation of soil reduces and the corresponding displacement of the structure reduces as well, which leads to a decrease in structure stress to some extent. However, some adverse factors, such as non-uniform stiffness, big holes in floors, large weight and the volume of electrical equipment, can lead to a nonlinear response in the internal forces.

4.2.3. Internal Force Responses

The peak internal forces of each section on surface-6 under different seismic motions are shown in Table 9 and the peak internal forces of different points on surface-6 are shown in Figure 21.

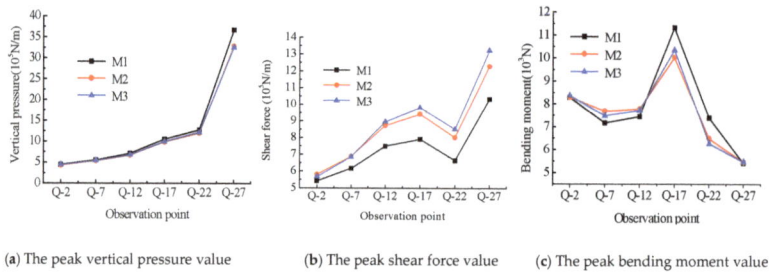

(a) The peak vertical pressure value   (b) The peak shear force value   (c) The peak bending moment value

Figure 21. The peak internal force of each section on surface-6 under El-Centro motion.

Table 9. The peak internal forces of each section on surface-6 under 0.20 g.

Section			Peak Vertical Pressure (10^5 N/m)			Peak Shear Force (10^5 N/m)			Peak Bending Moment (10^3 N·m/m)		
			M1	M2	M3	M1	M2	M3	M1	M2	M3
Floor 3	El-Centro motion	Top	4.43	4.28	4.40	5.42	5.67	5.81	8.28	8.28	8.36
		Bottom	5.57	5.32	5.39	6.17	6.80	6.85	7.16	7.67	7.87
	Lanzhou motion	Top	4.61	4.14	4.50	5.44	5.47	5.61	8.01	8.28	8.61
		Bottom	5.30	4.83	5.27	6.13	6.26	6.59	7.17	7.18	7.36
Floor 2	El-Centro motion	Top	7.14	6.69	6.81	7.50	8.72	8.94	7.45	7.77	7.96
		Bottom	10.51	9.80	9.92	7.92	9.40	9.79	9.32	10.01	10.33
	Lanzhou motion	Top	6.94	6.19	6.70	7.22	7.51	8.20	6.97	7.62	8.02
		Bottom	9.84	9.20	9.71	7.35	7.57	8.44	9.93	10.10	11.06
Floor 1	El-Centro motion	Top	12.77	11.91	12.11	6.64	8.00	8.51	5.39	6.47	6.75
		Bottom	36.70	32.79	32.48	10.33	12.27	13.22	5.21	5.46	5.67
	Lanzhou motion	Top	12.25	11.32	12.00	5.26	5.55	6.62	6.91	6.98	7.22
		Bottom	35.11	31.27	34.19	8.22	8.47	9.91	5.26	5.52	5.69

From Table 9 and Figure 21, it can be inferred that the shear forces of the underground substation gradually increase with the increase in the elastic modulus of the surrounding soil, while vertical pressure and bending moments change little under the same condition. The reason is that soil shear deformation reduced with the increase in the soil elastic modulus, and the corresponding displacements of the structure decreased as well.

The change of the soil's physical properties has a great effect on the seismic response of the underground substation. Therefore, the interaction between the underground substation and soil should be taken into account in the seismic design.

4.3. Seismic Response Influence of Structure–Electrical Equipment Interaction

Due to its heavy weight and huge size, electrical equipment fixed on the floors of the substation usually produced additional inertial force, which had an additional effect on the structure under seismic load and bending torsional load.

In previous seismic designs of underground substations, the interaction between structure and equipment was usually ignored. When the size and weight of electrical equipment is low, such a method is acceptable. Nowadays, the size and weight of electrical equipment is larger than before, and such an equivalent load method would result in some error. For this reason, Model 1 (M1) of the underground substation with the interaction between electrical equipment and structure and Model 2 (M2) without the interaction between them were established. According to a comparison and analysis of the internal force and seismic response in these two models, the seismic performance effect of electrical equipment on the underground substation can be determined.

4.3.1. Acceleration Responses

The acceleration amplification factors under 0.40 g, with the interaction between electrical equipment and structure, are shown in Table 10, in which the acceleration amplification factor is defined as the ratio between the peak acceleration of both the layer and input seismic motion. Similarly, the soil acceleration amplification factor is defined as the value at the same level of the structure.

Table 10. Acceleration amplification factors.

Acceleration Amplification Factor			Soil Surface	Roof	Floor 3	Floor 2	Floor 1	Equipment Top
El-Centro motion	Structure	X-direction	/	0.817	0.82	0.823	0.824	2.600
		Y-direction	/	0.498	0.582	0.661	0.738	1.194
	Soil	X-direction	0.752	0.810	0.985	1.064	1.219	/
		Y-direction	0.594	0.780	0.817	0.841	0.864	/
Lanzhou motion	Structure	X-direction	/	0.805	0.828	0.845	0.865	2.554
		Y-direction	/	0.502	0.515	0.545	0.599	1.096
	Soil	X-direction	0.812	1.287	1.327	1.319	1.457	/
		Y-direction	0.788	0.822	0.836	0.829	0.845	/
Taft motion	Structure	X-direction	/	0. 90	0.924	1.047	1.070	3.518
		Y-direction	/	0.69	0.759	0.807	0.954	1.494
	Soil	X-direction	1.267	1.342	1.551	1.583	1.772	/
		Y-direction	1.021	1.036	1.049	1.093	1.177	/

(1) Under the same direction and earthquake motion, acceleration amplification factors of soil and structure gradually increase; acceleration amplification factors in the horizontal X-direction are greater than those in the horizontal Y-direction. Meanwhile, the peak acceleration amplification factors of structure and soil are 1.070 and 1.772 (in the horizontal X-direction under Taft motion) respectively, which means that seismic responses are more severe in these cases than in other cases.

(2) Under different earthquake motions, acceleration amplification factors at the top of electrical equipment are different from each other. The peak acceleration amplification factor on the second floor under Taft motion is 3.52 (in the horizontal or the X-direction) which is much greater than 2.0 given in Code for the seismic design of electrical installations (2013, China). The acceleration amplification factors of electrical equipment in the X-direction are obviously greater than those on the same floor, which suggests that the acceleration response of the structure and electrical equipment is not synchronous. Therefore, the interaction between structure and equipment should be taken into account in the seismic design of an underground substation.

4.3.2. Displacement Responses

Time–history curves of displacement on the top floor in M1 and M2 under El-Centro motion and Lanzhou motion are shown in Figure 22. Layer drift and layer drift angle of each layer under 0.40 g are shown in Table 11.

Table 11. Layer drift (mm) and layer drift angle (rad).

	Section	El-Centro Motion		Lanzhou Motion	
		M1	M2	M1	M2
Floor 3	layer drift in the X-direction	3.110	2.737	2.949	2.589
	layer drift angle in the X-direction	1/1640	1/1863	1/1729	1/1970
	layer drift in the Y-direction	8.074	6.803	5.373	4.611
	layer drift angle in the Y-direction	1/632	1/750	1/949	1/1106
Floor 2	layer drift in the X-direction	2.984	2.623	2.808	2.463
	layer drift angle in the X-direction	1/1609	1/1830	1/1709	1/1949
	layer drift in the Y-direction	7.660	6.454	5.185	4.510
	layer drift angle in the Y-direction	1/627	1/744	1/926	1/1064
Floor 1	layer drift in the X-direction	3.135	2.754	2.952	2.610
	layer drift angle in the X-direction	1/1531	1/1743	1/1626	1/1839
	layer drift in the Y-direction	8.132	6.796	5.402	4.621
	layer drift angle in the Y-direction	1/590	1/706	1/889	1/1039

(a) Displacement time–history curves under El-Centro motion

(b) Displacement time -history curves under Lanzhou motion

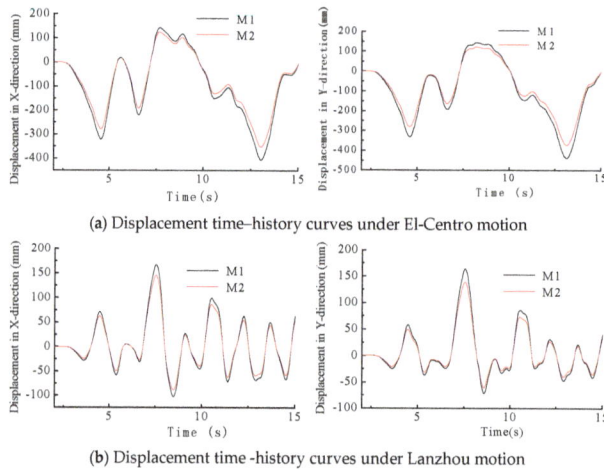

Figure 22. Time–history curves of displacement on the top floor.

From Figure 22 and Table 11, it can be seen that layer drift and layer drift angles of M1 are larger than those of M2 under 0.40 g. The reason is mainly that electrical equipment produces additional inertial force and extra relative displacement, which leads to structure displacement increasing at the same time. Therefore, the interaction of structure–equipment should be involved in seismic design. Only when the effect of electrical equipment is taken into account is the seismic design of an underground substation safe.

4.3.3. Internal Force Responses

Shear force and bending moment time–history curves in the X-direction at the bottom of side wall 4 under 0.40 g are shown in Figure 23. The peak internal forces of the side wall under 0.40 g are shown in Table 12.

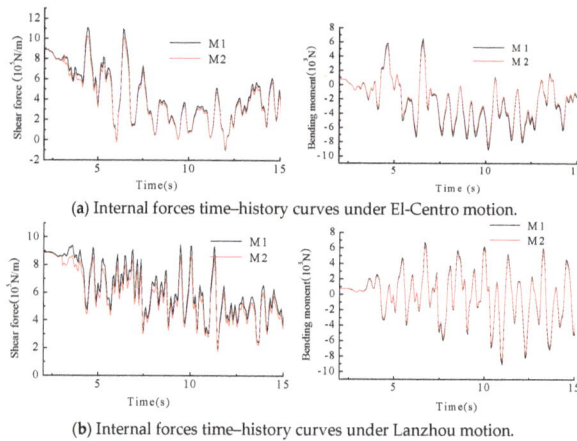

(a) Internal forces time–history curves under El-Centro motion.

(b) Internal forces time–history curves under Lanzhou motion.

Figure 23. Shear force and bending moment time–history curves in the X-direction at the bottom of side wall 4.

Table 12. Peak internal forces of the side wall.

Section		Shear Force (10^5 N/m)			Bending Moment (10^3 N·m/m)		
		M1	M2	Increase (%)	M1	M2	Increase (%)
Side wall 1	El-Centro motion	6.53	5.80	12.59	135	122	10.66
	Lanzhou motion	6.01	5.49	9.47	130	117	11.11
Side wall 3	El-Centro motion	10.98	9.79	12.16	947	850	11.41
	Lanzhou motion	9.76	8.68	12.44	893	812	9.98
Side wall 4	El-Centro motion	11.08	9.42	17.62	650	542	19.93
	Lanzhou motion	9.43	8.10	16.42	818	683	19.77
Side wall 5	El-Centro motion	11.70	10.57	10.69	1234	1132	9.01
	Lanzhou motion	10.02	8.83	13.48	566	517	9.48
Side wall 6	El-Centro motion	3.76	3.16	18.99	126	109	15.60
	Lanzhou motion	3.89	3.31	17.52	99	87	13.79

From Figure 23 and Table 12, it is shown that the difference of internal force is obvious in M1 and M2, in which the peak shear amplification is 18.99% from M2 to M1 and the peak bending moment difference increased by 19.93%. Such data reaffirm that the interaction between electrical equipment and structure could not be ignored.

4.4. Influence of Burial Depth

Burial depth is one of the important factors in the seismic response of the underground structure. The same underground structures may have different seismic responses in different burial depths. To study the influence of different burial depths on the seismic response of an underground substation in the horizontal direction, five conditions including different burial depth at 1.2 m (M1), 3 m (M2), 5 m (M3), 7 m (M4) and 9 m (M5) were chosen for the samples.

4.4.1. Displacement Responses

Layer drifts and drift angles under 0.20 g are shown in Table 13 under El-Centro motion.

Table 13. Layer drift (mm) and drift angle (rad) under El-Centro motion.

Section		M1	M2	M3	M4	M5
Floor 3	layer drift in the X-direction	1.76	2.25	2.42	3.06	2.97
	layer drift angle in the X-direction	1/2898	1/2267	1/2107	1/1667	1/1717
	layer drift in the Y-direction	4.03	2.88	3.82	4.01	3.91
	layer drift angle in the Y-direction	1/1266	1/1770	1/1335	1/1271	1/1304
Floor 2	layer drift in the X-direction	1.65	2.10	2.23	2.69	1.81
	layer drift angle in the X-direction	1/2909	1/2286	1/2152	1/1784	1/2652
	layer drift in the Y-direction	3.81	2.74	3.60	3.76	3.62
	layer drift angle in the Y-direction	1/1260	1/1752	1/1333	1/1277	1/1326
Floor 1	layer drift in the X-direction	1.79	2.25	2.36	5.93	2.83
	layer drift angle in the X-direction	1/2682	1/2133	1/2034	1/809	1/1696
	layer drift in the Y-direction	3.83	2.74	3.59	3.76	3.73
	layer drift angle in the Y-direction	1/1253	1/1752	1/1337	1/1277	1/1287

(1) With the increase in buried depth, the relative horizontal displacements in the X-direction first increase and then decrease, and then reach the maximum values when the buried depth is 7.0 m. Simultaneously, the relative horizontal displacements in the Y-direction first decrease and then increase and reach the maximum values when the buried depth is 1.2 m. The layer drift of each floor has the same trend and the maximum layer drift angle value in the X-direction reaches 1/809 (the third underground floor when the buried depth is 7.0 m), which is close to the limit value of 1/800 stated in Code for seismic design of buildings (2010, China).

(2) It is usually believed that the deeper the structures are buried, the safer they are under earthquake motions. However, the rule is unsuitable for some special industrial structures such

as an underground substation. The reasons are that equipment of high quality and stiffness with complex shapes makes displacements of structure complicated, and overlaying soil reduces the overall displacement of the structure with the increase in buried depth. The layer drift is critical in the Y-direction when the buried depth is shallow, while it is critical in the X-direction with the increase in buried depth. In order to prevent the danger caused by excessive displacement, buried depths which can cause a large seismic response should be avoided in underground substation seismic design.

4.4.2. Internal Force Responses

It has been previously demonstrated that the side wall is the weak area in the structure; therefore, shear force of side wall 3 is chosen to be analyzed and the results are shown in Table 14, in which the static values are defined as static computational results under the effect of gravity, and dynamic values are defined as dynamical computational results under earthquake motions.

Table 14. Shear forces of internal side wall 3 in different buried depths.

		Top of Floor 3	Bottom of Floor 3	Top of Floor 2	Bottom of Floor 2	Top of Floor 1	Bottom of Floor 1
	Static ((10^5 N/m)	3.53	4.20	5.59	8.20	11.18	16.88
M1	Dynamic (10^5 N/m)	4.57	5.46	7.29	10.58	14.41	21.56
	Amplification (%)	29.51	30.02	30.43	28.94	28.85	27.71
	Static (10^5 N/m)	6.19	6.42	7.72	10.42	13.61	20.15
M2	Dynamic (10^5 N/m)	7.74	8.12	9.77	13.18	17.16	25.41
	Amplification (%)	24.88	26.46	26.56	26.48	26.11	26.10
	Static (10^5 N/m)	8.74	9.03	9.94	12.70	16.06	23.43
M3	Dynamic (10^5 N/m)	11.06	11.26	12.61	16.17	20.40	29.78
	Amplification (%)	26.62	24.68	26.86	27.36	26.98	27.09
	Static (10^5 N/m)	11.39	12.46	12.49	15.40	18.98	28.43
M4	Dynamic (10^5 N/m)	14.24	15.00	15.74	19.68	24.29	35.85
	Amplification (%)	25.06	20.35	25.96	27.77	27.99	26.11
	Static (10^5 N/m)	13.54	14.96	14.53	17.41	21.09	30.13
M5	Dynamic (10^5 N/m)	17.22	18.50	18.56	22.37	26.91	38.52
	Amplification (%)	27.18	23.69	27.70	28.46	27.61	27.85

Comparing shear force amplifications at the top and bottom of side walls in different buried depths, it can be concluded that dynamic load has a greater influence on the top of side walls when the buried depth is 1.2 m when the shear force amplifications of the top of the side wall are larger than those of bottom. With the increase of buried depths, dynamic loads have a greater effect on the base of side walls, and the shear force amplifications on the top of the side wall are less than those of the bottom, which causes shear forces within the bottom of the wall to reach the maximum values under static and dynamic loads. Thus, the bottom of the side wall is the weak part of the underground substation. Therefore, it should be given more attention during the seismic design of such a system.

5. Conclusions

Three-dimension finite element models involving the interaction of soil–structure–equipment are established, and dynamic numerical simulation analysis of such models is performed. By changing some parameters such as ground motion input motions, properties of surrounding soils, burial depth, etc., the seismic performance of underground substations is proposed, and simulated results, such as acceleration amplification coefficient, displacement, stress and internal forces, are obtained from research. The findings are summarized as follows:

(1) The coupling boundary is selected as the boundary condition of soil–structure, which is feasible in a seismic response of an underground substation.

(2) The seismic response of an underground substation is more sensitive to vertical earthquake motion than the seismic response of normal structures. Therefore, vertical earthquake motions should be taken into account in the seismic design of such underground structures.

(3) Burial depth and elastic modulus are the main factors for the seismic performance of the underground electrical substation. With the increase in burial depth, layer drifts of an underground substation first increase and then decrease. With the increase in the elastic modulus of soil, the constraint of soil on structure increases and the deformation of structure decreases.

(4) The acceleration amplification factors of electrical equipment are obviously greater than those of the same layer of the structure, and the peak drift of equipment is not synchronous with structure. Therefore, the interaction between structure and equipment should be taken into account in the seismic design of underground substations and some seismic measures should be proposed to control the dynamic response of electrical equipment.

(5) The bottom of the side walls is the weak part of an underground substation. Therefore, it is necessary to increase the stiffness of side walls and strengthen the connection between the bottom floor and side walls.

Acknowledgments: The authors wish to express their gratitude towards the Major Program of the National Natural Science Foundation (No. 51590914) (in China), the National Natural Science Foundation Project (No. 51578450, No. 51378225) (in China).

Author Contributions: Bo Wen and Lu Zhang conceived and designed the models; Lu Zhang analyzed the data; Ditao Niu contributed analysis tools; Bo Wen wrote the paper; and Muhua Zhang searched some references.

Conflicts of Interest: The authors declare no conflict of interest. And the founding sponsors had no role in the design of the study; in the collection, analyses, or interpretation of data; in the writing of the manuscript, and in the decision to publish the results.

References

1. Suarez, L.E.; Singh, M.P. Floor Spectra with Equipment Structure Equipment Interaction Effects. *J. Eng. Mech.* **1989**, *115*, 247–264. [CrossRef]
2. Pires, J.A.; Ang, A.H.-S.; Villaverde, R. Seismic reliability of electrical power transmission systems. *Nucl. Eng. Des.* **1996**, *160*, 427–439. [CrossRef]
3. Brzan-Zurita, E.; Bielak, J.; Digioia, A.M. Seismic Design of Substation Structures. In Proceedings of the Electrical Transmission and Substation Structures Conference ASCE, Fort Worth, TX, USA, 8–12 November 2009; pp. 1–12.
4. Kempner, L.; Reston, V.A., Jr. (Eds.) *Substation Design Guide*; ASCE: Oakland, CA, USA, 2008.
5. Knight, B.T.; Kempner, L. Vulnerabilities and Retrofit of High-Voltage Electrical Substation Facilities. In Proceedings of the Technical Council on Lifeline Earthquake Engineering Conference, Oakland, CA, USA, 28 June–1 July 2009; pp. 232–243.
6. International Atomic Energy Agency (IAEA). *Seismic Evaluation of Existing Nuclear Installations*; IAEA: Vienna, Austria, 2008.
7. Wen, B.; Niu, D.T. Seismic vulnerability analysis for the main building of the large substation. *China Civ. Eng. J.* **2013**, *46*, 19–23. (In Chinese)
8. Wen, B.; Taciroglu, E.; Niu, D.T. Shake table testing and numerical analysis of transformer substations including main plant and electrical equipment interaction. *Adv. Struct. Eng.* **2015**, *18*, 1959–1980. [CrossRef]
9. Yuan, H.F.; Walker, R.E. The Investigation of a Simple Soil-Structure Interaction Model. In *Dynamic Waves in Civil Engineering*; Howells, D.A., Haigh, I.P., Taylor, C., Eds.; Wiley: New York, NY, UAS, 1970; pp. 247–266.
10. Rodriguez, B. Numerical Analysis of an Experimental Tunnel. Proceeding of the 11th International Conference on Soil Mechanics and Foundation Engineering San Francisco, San Francisco, CA, USA, 12–16 August 1985; pp. 789–792.
11. Penzien, J. Seismically Induced Racking of Tunnel Linings. *Earthq. Eng. Struct. Dyn.* **2000**, *29*, 683–691. [CrossRef]
12. Huo, H. Seismic Design and Analysis of Rectangular Underground Structures. Tunneling and Underground Space Tecnology United Kingdom. Ph.D. Thesis, Purdue University, West Lafayette, IN, USA, 2000.

13. Gong, B.N.; Zhao, D.P. Experimental Study on Dynamic Interaction of Underground Structure and Soil. *Undergr. Space* **2002**, *22*, 320–324. (In Chinese)
14. Zhuang, H.Y. *Study on Dynamic Contact Properties of Soil-Subway Underground Structure Interaction System*; Nanjing University of Technology: Nanjing, China, 2006. (In Chinese)
15. Chen, G.X.; Zhuang, H.Y.; Du, X.L. Analysis of large-scale shaking table test of dynamic soil-subway station interaction. *Earthq. Eng. Eng. Vib.* **2007**, *27*, 171–176. (In Chinese)
16. Li, L.Y.; Du, X.L.; Li, L. Review on Mechanical Performanical of Soil-Structure Contact. *Adv. Mech.* **2009**, *39*, 588–597. (In Chinese)
17. Tao, L.J.; Wang, W.P.; Zhang, B. Difference law study of seismic design methods for subway structures. *China Civ. Eng. J.* **2012**, *45*, 170–176. (In Chinese)
18. Geng, P.; He, C.; Tang, J.; Yan, Q.; Xu, Y. Appropriate Seismic Analysis Method of Subway Shield Tunnels in Soft Ground. *Int. Efforts Lifeline Earthq. Eng.* **2013**, 283–290. [CrossRef]
19. Debiasi, E.; Gajo, A.; Zonta, D. On the Seismic Response of Shallow-Buried Rectangular Structures. *Tunn. Undergr. Space Technol.* **2013**, *38*, 99–113. [CrossRef]
20. Liu, J.B.; Wang, W.H.; Dasgupta, G. Pushover analysis of underground structures: Method and Application. *Sci. China Technol. Sci.* **2014**, *57*, 423–437. (In Chinese) [CrossRef]
21. Kang, G.H.; Tobita, T.; Lai, S. Seismic simulation of liquefaction-induced uplift behavior of a hollow cylinder structure buried in shallow ground. *Soil Dyn. Earthq. Eng.* **2014**, *64*, 85–94. [CrossRef]
22. Pitilakis, K.; Tsinidis, G.; Leanza, A.; Maugeri, M. Seismic behavior of circular tunnels accounting for above ground structures interaction effects. *Soil Dyn. Earthq. Eng.* **2014**, *67*, 1–15. [CrossRef]
23. Clough, R.; Penzien, J. *Dynamics of Structures*; McGraw-Hill: New York, NY, USA, 1993.
24. Zhang, F.; Cheng, C.S. A modified Newton method for radial distribution system power flow analysis. *IEEE Trans. Power Syst.* **2002**, *12*, 389–397. [CrossRef]
25. *35kV–220kV Urban Underground Substation Design Rules*; China Electric Power Press: Beijing, China, 2005. (In Chinese)
26. *Chinese Loading Code for Design of Building Structures*; GB50009-2012; Architecture & Building Press: Beijing, China, 2012. (In Chinese)
27. *Chinese Code for Seismic Design of Buildings*; GB50011-2010; Architecture & Building Press: Beijing, China, 2010. (In Chinese)
28. *Chinese Code for Seismic Design of Electrical Installations*; GB50260-2013; Architecture & Building Press: Beijing, China, 2013. (In Chinese)
29. Hibbitt, K.; Sorensen, I. *ABAQUS User Subroutines Reference Manual*; Elsevier Science Publishers: Philadelphia, PA, USA, 2006.
30. Liu, J.L.; Luan, M.T.; Xu, C.S. Study on parametric characters of Drucker-prager criterion. *Chin. J. Rock Mech. Eng.* **2006**, *25*, 4009–4015. (In Chinese)
31. Liu, J.B.; Gu, Y.; Du, Y.X. Consistent viscous-spring artificial boundaries and viscous-spring boundary elements. *Chin. J. Geotech. Eng.* **2006**, *28*, 1070–1075. (In Chinese)
32. Huang, S.; Chen, W.Z.; Wu, G.J. Study of method of earthquake input in a seismic analysis for underground engineering. *Chin. J. Rock Mech. Eng.* **2010**, *29*, 1254–1262.
33. Lou, M.L.; Chen, H.J. *Research on the Effects of Lateral Boundary of Pile Seismic Response*; Tongji University Press: Shanghai, China, 1999. (In Chinese)
34. Lou, M.L.; Wang, W.J.; Zhu, T. Soil lateral boundary effect in shaking table model test of soil-structure system. *Earthq. Eng. Eng. Vib.* **2000**, *20*, 30–36. (In Chinese)
35. Chen, G.X.; Chen, S.; Zuo, X. Shaking-table tests and numerical simulations on a subway structure in soft soil. *Soil Dyn. Earthq. Eng.* **2015**, *76*, 13–28.
36. Novak, M.; Bereduge, Y.O. Vertical Vibration of Embedded Footings. *J. Soil Mech. Found. Div ASCE* **1972**, *98*, 1291–1310.
37. Idriss, I.M.; Sun, J. User's Manual for SHAKE91. *Center Geotech. Model.* **1992**, *388*, 279–360.

38. Liang, X.W.; Qian, L.; Tan, L.N. The Concrete Damage Plastic Constitutive Relation Research Based on ABAQUS. In Proceedings of the 8th Session of the National Earthquake Engineering, Chongqing, China, 24–26 December 2010; Volume 32, pp. 646–648. (In Chinese)
39. Meng, W.Y.; Wang, J.F.; Zhang, R. Comparison Analysis of Constitutive Models of Reinforced Concrete Structures Based on ABAQUS. *J. North China Inst. Water Conserv. Hydroelectr. Power* **2012**, *33*, 40–42. (In Chinese)

applied
sciences

MDPI

Article

Research on the Rational Yield Ratio of Isolation System and Its Application to the Design of Seismically Isolated Reinforced Concrete Frame-Core Tube Tall Buildings

Aiqun Li [1,2,3], Cantian Yang [1,2], Linlin Xie [1,2,*], Lide Liu [1,2] and Demin Zeng [1,2]

1   Beijing Advanced Innovation Center for Future Urban Design, Beijing University of Civil Engineering and Architecture, Beijing 100044, China; liaiqun@bucea.edu.cn (A.L.); yangcantian@outlook.com (C.Y.); lukixwr@sina.com (L.L.); zengdemin@vip.163.com (D.Z.)
2   School of Civil and Transportation Engineering, Beijing University of Civil Engineering and Architecture, Beijing 100044, China
3   School of Civil Engineering, Southeast University, Nanjing 210096, China
*   Correspondence: xielinlin@bucea.edu.cn; Tel.: +86-10-6120-9367

Received: 10 October 2017; Accepted: 15 November 2017; Published: 19 November 2017

Featured Application: A high-efficiency design method based on the rational yield ratio of isolation system is proposed and applied to the design of the seismically isolated RC frame-core tube tall buildings. The research outcome of this paper can guide the design of such buildings located in high seismic region and significantly improve the design efficiency.

Abstract: Resilience-based seismic design of reinforced concrete (RC) tall buildings has become an important trend in earthquake engineering. Seismic isolation technology is an effective and important method to improve the resiliency of RC frame-core tube tall buildings located in high seismic regions. However, the traditional design method for this type of building does not focus on the key design parameter, namely, the yield ratio of the isolation system and has therefore been proved to be highly inefficient. To address these issues, the rational yield ratio of isolation system for such buildings is investigated based on 28 carefully designed cases, considering the influences of total heights, yield ratios and seismically isolated schemes. The rational range of the yield ratio is recommended to be 2-3%. Based on this, a high-efficiency design method is proposed for seismically isolated RC frame-core tube tall buildings. Subsequently, a seismically isolated RC frame-core tube tall building with a height of 84.1 m is designed using the proposed design method. The rationality, reliability and efficiency of the proposed method are validated. The research outcome can serve as a reference for further development of the seismic design method for seismically isolated RC frame-core tube tall buildings.

Keywords: seismically isolation; RC frame-core tube tall building; rational yield ratio of isolation system; high-efficiency design method

1. Introduction

In recent years, studies on the resilience-based seismic design of next-generation cities have attracted much attention in earthquake engineering [1–4]. To achieve a seismically resilient city, it is essential to improve the seismic resilience of important buildings that can highly affect the normal operation of the city, such as the government buildings, hospitals and telecommunications buildings, which are usually constructed using reinforced concrete (RC) tall structure. For RC tall buildings located in high seismic regions, it is difficult to achieve a seismic resilient structure through traditional seismic design [5]. It is well acknowledged that the seismic isolation technology can significantly reduce or even eliminate the damage of structural and nonstructural components [6–16], thus enabling the buildings to achieve a resilient performance after an earthquake. Hence, the seismic isolation technology is considered an effective and important method to realize resilient RC tall buildings located in high seismic regions. The seismic design of one type of typical seismically isolated RC tall buildings, namely, the RC frame-core tube tall building, is studied here.

In comparison with the seismic design of seismically isolated multi-story buildings, it is somewhat difficult to design seismically isolated tall buildings, which satisfy all the requirements specified in the Chinese Code for seismic design of buildings (GB50011-2010) [17], including the efficiency of isolation system, the maximum bearing displacement (MBD) and the maximum tensile and compressive stresses of isolators. This is due to the characteristics of the RC frame-core tall buildings, including long fundamental period, large weight and significant overturning effect [7,9,18]. Various problems may be encountered during the design of such isolated tall building, especially in the design of the isolation system. Specifically, some of the main problems are as follows: (1) Numerous parameters are required to be determined. The selection of the isolators, including the number, size and type (which includes the lead rubber bearing (LRB) and the natural rubber bearing (NRB)) of the isolators, are extremely complicated. In addition, nonlinear fluid viscous dampers (NFVDs) may be required to control the large MBD, which is usually observed in the isolation system. (2) Design experience and comprehensive investigations on such buildings are less reported than those on the multi-story buildings. It is worth mentioning here that a step-by-step iterative design method based on a refined finite element (FE) model is usually adopted for such buildings, leading to an overwhelming workload to create such model and implement the time history analysis [9]. The traditional design method of the isolated tall buildings is conclusively considered undesirable and inefficient for the RC frame-core tube tall buildings.

In light of the issues discussed above, this study aims to (1) identify and propose a rational range of the critical design parameter for the isolation system of the RC frame-core tube buildings, (2) develop a high-efficiency design method based on such range and (3) validate the rationality and reliability of this method. The yield ratio of the isolation system (referred to as the "yield ratio" hereafter), Q_y/W, is defined as the ratio of the total yield force of the LRBs (Q_y) to the total seismic weight of the structure (W) [19–23]. Previous research has indicated that the yield ratio is a critical design parameter, which significantly affects the efficiency of the isolation system and the seismic performance of the entire building [19–23]. It is also notable that yield ratio is directly related to the quantity of the LRB isolators. Hence, a rational range of yield ratio can be theoretically used to guide the design of an isolation system. Various investigations on the rational value of the yield ratio have been conducted for multi-story buildings. Pourzeynali et al. [24] studied the optimal values of the parameters of the base isolation system. Shen et al. [25] investigated the influence of the yield ratio on the seismic responses of a multi-story isolated structure and found that there existed a rational range for the yield ratio. Wang et al. [26], Li [21], Providakis et al. [20] and Mollaioli et al. [22] also investigated the rational range of the yield ratio.

These investigations indicate that the rational range of the yield ratio is highly related to the seismic responses of structures. It is notable that the total height and the seismically isolated scheme also have a certain extent of influence on the seismic response of such buildings and therefore, affect the rational range of the yield ratio. However, investigations on the rational range of the yield ratio with consideration of such factors, as well as the high-efficiency design method based on this, have been rarely reported for seismically isolated RC frame-core tube tall buildings.

As described above, based on two real engineering practices (seismically isolated RC frame-core tube buildings) with different heights located in high seismic regions, 28 study cases with different yield ratios and seismically isolated schemes are designed. Subsequently, the influence of the yield ratio on the efficiency of the isolation system and the MBD are investigated. A rational range of the yield ratio is recommended and is used to propose a high-efficiency design method for the isolated tall building. Subsequently, a seismically isolated RC frame-core tube building with a height of 84.1 m is designed using the proposed design method. The rationality, reliability and efficiency of the proposed method are validated. The research outcome can serve as a reference for further development of seismic design method for seismically isolated RC frame-core tube tall buildings.

2. Rational Range of the Yield Ratio of Isolation System

As mentioned above, the yield ratio of isolation system is a critical design parameter for the seismically isolated RC frame-core tube tall buildings. To obtain a rational value for this ratio, a number of rational and comprehensive study cases are required. Specifically, (1) the cases based on real engineering practices, which are located in high seismic region with different heights, are preferred; (2) different seismically isolated schemes, which are determined according to real architectural design requirements, should be considered; (3) a number of yield ratios of isolation system, covering a sufficiently large range, are required for such investigation. To achieve these, two carefully designed real engineering practices with different heights are selected as the prototype buildings. Based on the prototype buildings, the study cases accounting for different seismically isolated schemes and yield ratios of isolation system are designed to investigate the rational yield ratio of isolation system.

2.1. Overview of the Prototype Buildings

The selected two real engineering practices are seismically isolated RC frame-core tube tall buildings located in a near-fault region with a seismic intensity of 8.5 degree. According to GB50011-2010 [17], the 8.5 degree means that the correspond peak ground acceleration (PGA) values of the service level earthquake (SLE) (i.e., 63% probability of exceedance in 50 years), the design basis earthquake (DBE) (i.e., 10% probability of exceedance in 50 years) and the maximum considered earthquake (MCE) (i.e., 2% probability of exceedance in 50 years) are 110 cm/s^2, 300 cm/s^2 and 510 cm/s^2, respectively. The total heights of these two buildings above the ground are 79.2 m with 22 stories (named as C1) and 65.8 m with 17 stories (named as C4), respectively. The height-to-width ratios are 2.3 and 1.91 for C1 and C4, respectively. The three-dimensional views and the plan views of the prototype buildings are presented in Figures 1 and 2, respectively. The site condition belongs to the Site Class III and the second group in the GB50011-2010 [17]. The characteristic period of the site is 0.55 s. The closet distance from these two buildings to the rupture plane (i.e., R_{rup}) is 7.5 km. When R_{rup} is lower than 10 km and higher than 5 km, a near field influence coefficient with a value not less than 1.25 is introduced to consider the near field effect according to the GB50011-2010 [17]. Because the seismic design load of these buildings is extremely large and the occupancy importance of these buildings is very high, a special advisory committee (referred to as "the committee" hereafter) was established to guide and monitor the seismically isolated design. The near field influence coefficient was recommended to be 1.25 through a detailed discussion by the committee, which means the PGA values of the SLE, DBE and MCE increase to 137.5 cm/s^2, 375 cm/s^2 and 637.5 cm/s^2, respectively.

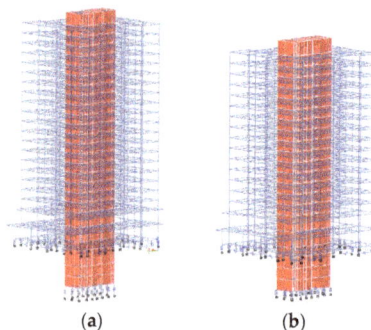

(a)   (b)

Figure 1. Three-dimensional view of prototype buildings: (**a**) C1 and (**b**) C4.

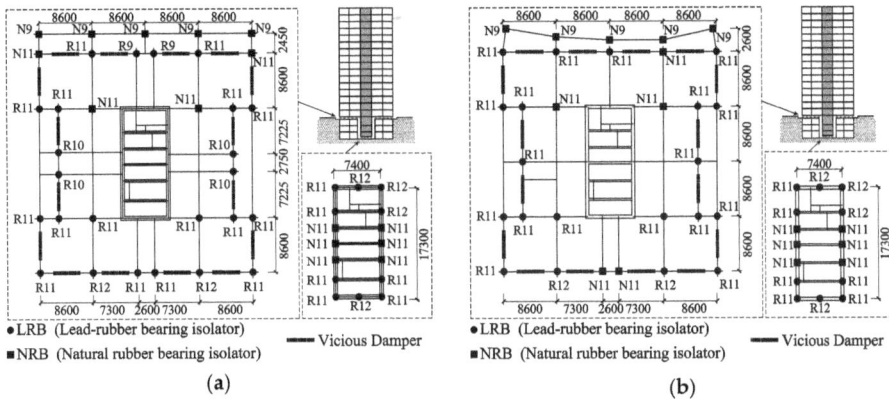

Figure 2. Isolation system of prototype reinforced concrete (RC) frame-core tube tall buildings: (a) C1 and (b) C4.

It is notable that the basements of C1 and C4 are designed as 4 stories with a total height of 16.05 m and 14.15 m, respectively. The conventional base-isolated (CBI) scheme, which isolates the superstructure in the elevation of ±0 m, cannot satisfy the architectural design requirements. Meanwhile, the base-isolated scheme, which isolates the entire structure at the bottom of the basement, leads to a total height of approximately 95 m, thus leading to a certain extent of tensile stress in the isolators. Hence, a partially basement-isolated (PBI) scheme as shown in Figure 2 is recommended by the committee and adopted for these two buildings. Specifically, the columns are isolated in the elevation of ±0 m, while the core tubes are isolated at the bottom of the basement.

Because of the limit width of the isolation gap, which is recommended to be 600 mm by the committee, 16 NFVDs are adopted to control the MBD under the MCE. Specifically, by adopting the NFVDs, the MBD under the MCE are effectively reduced from 932 to 488 mm and 656 to 488 mm for C1 and C4, respectively. It is worth mentioning here that the authors have designed 29 different RC tall buildings located in this region and 16 NFVDs with approximately identical parameters are adopted for these buildings. The damping exponent and damping coefficient of the NFVDs are approximately 0.3 and 210 kN·s/mm. The maximum damping forces of the NFVDs in these buildings are all almost 2000 kN. Hence, the variation in numbers and parameters of NFVDs are herein not considered. The layouts of the LRB isolators, the NRB isolators and the NFVDs are schematically presented in Figure 2. The characteristic parameters of all isolators are presented in Table 1.

Table 1. Characteristic parameters of the isolators.

Type	NRB 900	NRB 1100	LRB 900	LRB 1000	LRB 1100	LRB 1200
Notation	N9	N11	R9	R10	R11	R12
Effective diameter (mm)	900	1100	900	1000	1100	1200
Total rubber thickness (mm)	176	216	176	197	216	235
Vertical stiffness (kN/m)	3,630,000	4,519,000	4,168,000	4,639,000	5,550,000	5,940,000
Equivalent stiffness at 100% shear strain (kN/m)	1110	1358	2070	2300	2450	2600
Post-yield stiffness (kN/m)	/	/	1070	1190	1310	1470
Horizontal yield force (kN)	/	/	238	294	355	410
Rubber shear modulus (N/mm^2)	0.32	0.32	0.32	0.32	0.32	0.32

The critical design parameters and indexes of the prototype buildings are listed in Table 2. The fundamental period of the corresponding fixed-base structure is denoted as T_f. The isolation period, T_{is}, is calculated using the equivalent stiffness of the isolators when the horizontal shear strain of isolator is 100% according to the GB50011-2010 [17].

Table 2. Critical design parameters and indexes of prototype structures.

Indexes	C1	C4
T_f (s)	1.59	1.52
T_{is} (s)	4.44	4.08
Q_y/W (%)	2.7	3.2
β	0.36	0.36
MBD (mm)	488	488
σ_{max}^g (MPa)	12.38	13.17
σ_{max}^p (MPa)	29.07	26.43
σ_{max}^t (MPa)	0.00	0.09

The horizontal seismic absorbing coefficient, β, which indicates the efficiency of the isolation system, is one of the most important indexes in the design of seismically isolated structures. It is defined as the maximum ratio of the shear force of the isolated structure to that of a fixed-base structure in each story subjected to the DBE [17]. For seismically isolated building with NFVDs, β is required to be no more than 0.38 according to the GB50011-2010 [17]. The horizontal seismic absorbing coefficient of C1 and C4 are both 0.36, which satisfy this specification.

The MBD under the MCE is another important design index that determines the width of the isolation gap (i.e., D_g). Specifically, MBD = max$\{\eta_i \cdot u_c\}$, where η_i is the torsion influence coefficient (i.e., a scale factor to consider the effects of torsion) of ith isolator with a value of 1.15 for the prototype buildings and u_c is the maximum displacement of the mass center of the isolation system under the MCE [17]. For each isolator, MBD is required to be no more than the smaller of the two values of 0.55 times the effective diameter of the isolator and 3.0 times the total rubber thickness. The MBD of C1 and C4 are both 488 mm as shown in Table 2, which strictly satisfy the specification of the GB50011-2010 [17]. In addition, D_g is required to be no less than 1.2 times the MBD. As for the prototype buildings, D_g is 600 mm, which is larger than 1.2 times the MBD (i.e., 585.6 mm).

The maximum compressive and tensile stresses of the isolators are also important design indexes for the seismically isolated structures. According to the GB50011-2010 [17], the compressive stress of isolators under gravity load (i.e., σ_{max}^g) should be no more than 15 MPa. Furthermore, the compressive and tensile stresses of isolators under the MCE (i.e., σ_{max}^p and σ_{max}^t) should be no more than 30 MPa and 1 MPa, respectively. The typical results presented in Table 2 indicate that the design of the prototype buildings can meet the abovementioned requirements.

2.2. Design of the Study Cases

Through the introduction of the prototype buildings, it can be seen that two different heights and one seismically isolated scheme (i.e., PBI scheme) have been taken into consideration. As mentioned above, different schemes of the isolation system, which are usually determined by the real architectural design requirements, may be adopted in the design of seismically isolated RC frame-core tube tall buildings. Hence, in addition to the PBI scheme, the conventional base-isolated scheme should also be taken into account. The cases using the CBI scheme are herein designed based on the structures of C1 and C4 above the ground. Subsequently, different yield ratios are considered. 28 study cases with different yield ratios and seismically isolated schemes were redesigned based on the prototype buildings. The upper and lower limits of the yield ratios are approximately 1.5% and 3.5%, respectively. Different yield ratios are achieved through the adjustment of the type and quantity of the isolators. Detailed information (consisting of the height, seismically isolated scheme and yield ratio) and the corresponding T_{is} of these cases are presented in Table 3.

Table 3. Detailed information and corresponding T_{is} of the study cases.

Prototype Buildings	Seismically Isolated Schemes	Yield Ratio/%	T_{is}/s
C1 (79.2 m)	PBI scheme	1.4	4.95
		1.8	4.70
		2.0	4.63
		2.2	4.58
		2.7	4.44
		3.0	4.36
		3.4	4.27
	CBI scheme	1.5	4.65
		1.8	4.55
		2.0	4.48
		2.2	4.43
		2.7	4.30
		3.0	4.23
		3.5	4.15
C4 (65.8 m)	PBI scheme	1.4	4.51
		1.8	4.40
		2.0	4.35
		2.2	4.30
		2.6	4.21
		3.0	4.12
		3.5	4.00
	CBI scheme	1.5	4.35
		1.8	4.27
		2.0	4.22
		2.2	4.17
		2.5	4.11
		3.0	3.99
		3.5	3.90

The single-degree-of-freedom (SDOF) model and multi-degree-of-freedom (MDOF) shear model are widely used to guide the preliminary design and conceptual research of seismically base-isolated structures exhibiting significant shear deformation mode [9,22,24,27–29]. However, the RC frame-core tube buildings exhibit a flexural-shear deformation mode [30]. A reliable numerical model that can consider such deformation characteristic is required. The refined finite element (FE) model, which is widely acknowledged to be capable of reflecting such deformation characteristic [31–34], is adopted here. The general-purpose commercial software ETABS [35] is used to establish the refined FE model and conduct the seismically isolated design as well as the seismic performance assessment of the abovementioned 28 cases. It is notable that the superstructure of the isolated buildings usually experiences minor damage as expected; therefore, an elastic model for the superstructure is usually adopted [36,37]. Because of this, the beams and columns of the superstructure are modeled with elastic frame elements. The shear walls are simulated using the elastic shell element. In contrast, as the plastic response of the seismically isolated structures is mostly concentrated in the isolation system, a plastic model is usually adopted for the isolation system. Hence, the Rubber Isolator element and the Gap element are herein adopted for the isolators, while the Damper-Exponential element is used for the NFVDs. The abovementioned modeling methods are widely adopted and have been proved to be reliable [20,38].

For the ground motions, two natural ground motion records and one artificial ground motion are used to design the study cases and investigate the rational yield ratio according to GB50011-2010 [17]. The response spectrum of each ground motion is compared with the design response spectrum as shown in Figure 3. The ranges of interest for the period are 1.31–1.59 s for the fixed-base structures and 3.90–4.95 s for the seismically isolated structures. It can be found that the mean response spectrum agrees well with the design response spectrum at these period ranges. The ground motions are scaled to 375 cm/s^2 and 637.5 cm/s^2 to conduct time history analyses (THAs) under DBE and MCE, respectively. The indexes, including σ^g_{max}, σ^p_{max} and σ^t_{max}, all meet the requirements specified in the GB50011-2010 [17]. The β and the MBD under the MCE are critical indexes to evaluate a rational yield ratio. Hence, these two critical indexes of 28 study cases will be presented and discussed in detail in the following section.

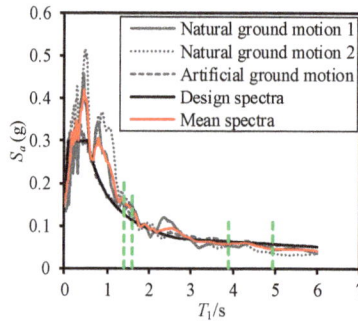

Figure 3. Response spectra of design ground motions.

2.3. Rational Value of the Yield Ratio

To identify the rational yield ratio for the RC frame-core tube tall buildings located in high seismic regions, the following two principles are followed: (1) the horizontal seismic absorbing coefficient, β, is required to be no more than 0.38 as specified in the GB50011-2010 [17]; (2) the MBD under the MCE should not exceed 540 mm. In consideration of the difficulties in design of such seismically isolated tall buildings, the committee recommended that the limit width of the isolation gap could increase to 650 mm, which meant that the MBD could raise to approximate 540 mm (i.e., $650/1.2 \approx 540$ mm).

The values of β and MBD for the 28 study cases with different yield ratios and different seismically isolated schemes are calculated and presented in Figures 4–7. The following conclusions can be drawn:

(1) The recommended upper limit of the yield ratio is 3% and the upper limit is determined by the β. When the yield ratio is no more than 3%, the value of β of all 28 study cases do not exceed 0.38, which means that the expected efficiency of the isolation system is successfully achieved and 3% can be regarded as the upper limit. In addition, β decreases with the decrease in yield ratio, indicating that a lower yield ratio can achieve a better isolation efficiency. However, the value of β becomes steady when the yield ratio decreases to 2%, which means the continuous decrease of the yield ratio below 2% cannot lead to better isolation efficiency.

(2) The upper limit of the yield ratio can increase to a certain level when the building height is smaller than 80 m or the CBI scheme is adopted. The upper limit of the yield ratios of C1 with the height of 79.2 m (3% for the PBI scheme) is lower than that of C4 with the height of 65.8 m (3.5% for the PBI scheme), indicating a higher upper limit of the yield ratio can be adopted for a building with a height lower than approximately 80 m. In addition, when the yield ratios of the PBI scheme and the CBI scheme both reached 3% for C1, the value of β for the CBI scheme is lower than that for the PBI scheme, indicating that a larger value of the yield ratio can be adopted for the CBI scheme.

(3) The recommended lower limit of the yield ratio is 2%. When the yield ratio is no less than 2%, the MBD under the MCE are all lower than 540 mm, which satisfy the requirements specified by the limit value of the isolation gap. The MBD generally decreases with the increase of the yield ratio, indicating that a higher yield ratio can better control the MBD, thereby satisfying the demand of the isolation gap. In addition, with the increase of the height of the building, the lower limit of the yield ratio can adopt a smaller value (e.g., 1.8% for C1 with the PBI scheme). However, the isolation efficiency becomes steady when the yield ratio decreases to 2% as mentioned above, while

a considerable increase of the MBD is observed. Hence, the lower limit of the yield ratio with a value of 2% is recommended.

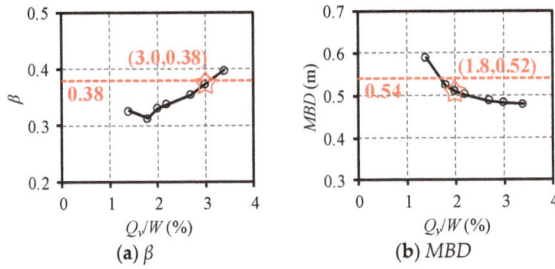

Figure 4. C1 with partially basement-isolated (PBI) scheme.

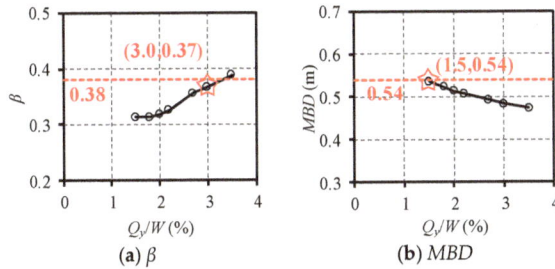

Figure 5. C1 with conventional base-isolated (CBI) scheme.

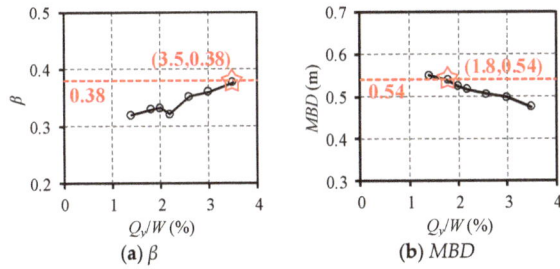

Figure 6. C4 with PBI scheme.

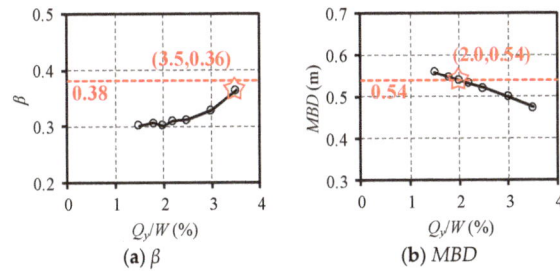

Figure 7. C4 with CBI scheme.

Based on the above discussion, it can be concluded that the upper and lower limits of the yield ratio are determined according to the requirements of the horizontal seismic absorbing coefficient and the width of the isolation gap, respectively. A yield ratio in the range of 2–3% is considered rational for the seismically isolated RC frame-core tube tall buildings.

3. Design Method for Seismically Isolated RC Frame-Core Tube Tall Building

3.1. Traditional Design Method

For the traditional design method of seismically isolated tall buildings, the seismic performance targets, including the limits of the β and the MBD under the MCE, are determined first. According to the GB50011-2010 [17], β is required to be no more than 0.40 and 0.38 for the structures without and with the NFVDs in the isolation system, respectively. This indicates that the seismic forces of the superstructure are similar to that of a fixed-base structure with one-degree lower seismic intensity. Due to this fact, the seismically isolated structures are designed following a step-by-step iterative design method as shown in Figure 8.

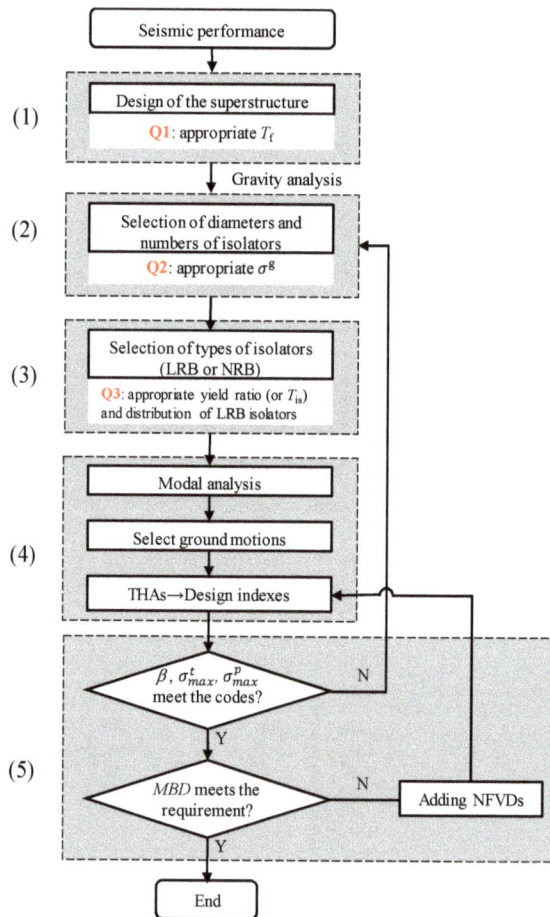

Figure 8. Traditional design method for seismically isolated tall building. (Q1~Q3 denote the critical problems).

The design method can be divided into 5 steps:

(1) Design of the superstructure.

The superstructure is first designed as a fixed-base structure with a one-degree lower seismic intensity. An appropriate superstructure is the basis for the seismic design of the isolated building. The T_f is the most critical index to be determined in this step, because it is highly associated with the seismic force of the structure and the seismic absorbing coefficient.

(2) Selection of the isolator numbers and sizes.

The numbers and sizes of isolators are determined according to the appropriate compressive stress of each isolator under gravity load (referred to as σ^g hereafter). It is notable that the requirements of σ^g for different isolators, including the isolators under the columns, the corner of core tube, the outer wall and the inner wall, are different. Due to the significant overturning effect of the RC frame-core tube tall buildings, the isolators under the columns and the corner of the core tube may possibly be under tension, a higher σ^g is usually required for these isolators. However, an appropriate σ^g for different isolators of RC frame-core tube buildings has rarely been investigated and reported, which is critical for the selection of numbers and sizes of isolators.

(3) Selection of the isolator types.

Determining the type of the isolators (LRB or NRB) is a key step in the design of such buildings. This can significantly affect the seismic performance of the isolation system and the entire structure. Specifically, an appropriate yield ratio is required first, as mentioned above. In addition, a rational distribution of the LRB isolators should be carefully designed to avoid torsion effect. However, a comprehensive investigation on the appropriate yield ratio and the distribution of the LRB isolators has rarely been proposed for the RC frame-core tube seismically isolated tall buildings.

(4) Calculation of the design indexes.

After the preliminary isolation system is established, critical design indexes, including β, MBD under the MCE, σ_{max}^p and σ_{max}^t, are assessed to validate the feasibility of the isolation system. The T_{is} is initially obtained through modal analysis. Then, a set of appropriate ground motions are selected according to T_f and T_{is}. The THAs are conducted for these two structures and critical design indexes are calculated.

(5) Validation and/ or modification of isolation system.

Three design indexes, including β, σ_{max}^p and σ_{max}^t, are evaluated. If one of these indexes cannot meet the requirements, the sizes, numbers or types of the isolators will be constantly modified (i.e., iterate from step 2) until these indexes can satisfy the requirements. Subsequently, the MBD under the MCE is examined. The introduction of the NFVDs in the isolation system is recommended, if the MBD under the MCE exceeds the limit value specified by the allowable width of the isolation gap.

Numerous iterations are usually required to achieve a satisfactory isolation system for the traditional design method, leading to extremely low efficiency. This is attributed to the following three critical problems:

(1) An appropriate T_f is currently not available for RC frame-core tube buildings.

(2) The appropriate ranges of σ^g for the isolators at different locations, especially those under the columns and the corner of the core tube, are not proposed yet.

(3) A rational range of the yield ratio and associated distribution of LRB isolators have not been recommended.

Based on the above, it can be concluded that a high-efficiency design method with emphasis on the solutions of abovementioned three critical problems is required.

3.2. Proposed High-Efficiency Design Method

According to the above requirements, solutions for the three critical problems are firstly proposed here.

(1) For the appropriate T_f, a statistical regression analysis is conducted based on the seismically isolated tall buildings designed by the authors, as shown in Figure 9. An equation, which describes

the relationship between the height and T_f, is obtained as shown in Equation (1). Here, H represents the total height of the structure.

$$T_f = 0.193H^{0.5} \tag{1}$$

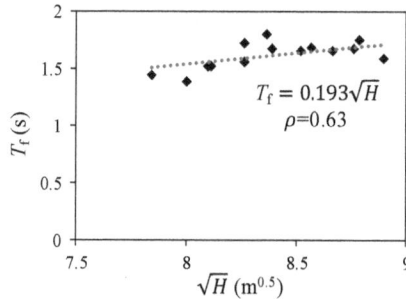

Figure 9. Relationship between the height and fundamental period of fixed-base structure.

(2) A certain extent of tensile stress may be observed for the isolators under the columns due to the overturning effect, a relatively large σ^g is therefore usually adopted for these isolators. However, a σ^g with excessive value is also not recommended because it may lead to a large σ^c_{max}, which may exceed the specified limit value. Based on the design experience, the σ^g is recommended to be approximately 10–12 MPa for the isolators under the columns. It is notable that the stress of the isolators under the corner of the core tube when subjected to MCE is lower than this of the isolators under the columns. Hence, a lower σ^g can be adopted for such isolators and a σ^g with the value ranged from 8 to 10 MPa is recommended for the isolators under the corner of the core tube, which is also based on the design experience.

(3) The rational range of the yield ratio has been investigated in Section 2 and has been recommended in the range of 2–3%. To avoid the torsion effect, the LRB isolators are recommended for the columns and the corner of the core tube in the RC frame-core tube tall buildings. If these LRB isolators can lead to an acceptable yield ratio, the requirements of β, σ^p_{max} and σ^t_{max} are considered easily satisfied.

Based on the above discussion, a new high-efficiency design method, as shown in Figure 10, is proposed for the seismically isolated RC frame-core tube tall buildings. The entire design procedure is presented as follows:

(1) Design of the superstructure.

According to the height of the building, an appropriate value of T_f is calculated using Equation (1). Then, the superstructure is designed as a fixed-base structure with one-degree lower seismic intensity and with a T_f value approximate to the calculated value.

(2) Selection of isolator numbers and sizes.

Gravity analysis is conducted to calculate the axial forces under the columns and the shear walls. Then, according to the recommended compressive stress for different isolators, the demand for the area of the corresponding isolators can be calculated and appropriate numbers and sizes of the isolators can be selected.

(3) Selection of the isolator types.

LRB isolators are adopted for the isolators under the columns and the corner of the core tube. Then, the yield ratio is calculated. If the calculated yield ratio is lower than 2%, more LRB isolators are recommended to be adopted under the external wall. If the calculated yield ratio is higher than 3%, the LRB isolators under the corner of the core tube can be reduced first, followed by those under the columns.

(4) Calculation of the design indexes.

Modal analysis is conducted firstly to calculate the T_{is} and appropriate ground motions are selected. Based on these, THAs are performed to calculate the critical design indexes.

(5) Validation and/or modification of isolation system.

Because of the rational yield ratio adopted, β, σ_{max}^p and σ_{max}^t can usually satisfy the requirements. If not, only limit modifications of the isolators are required. Subsequently, the MBD under the MCE is examined. If the MBD exceeds the limit value, the NFVDs are recommended to be introduced in the isolation system. It is notable that the quantity and parameters of the NFVDs can be determined with a few number of iterations.

The application of this high-efficiency design method will be explained in detail in the following section.

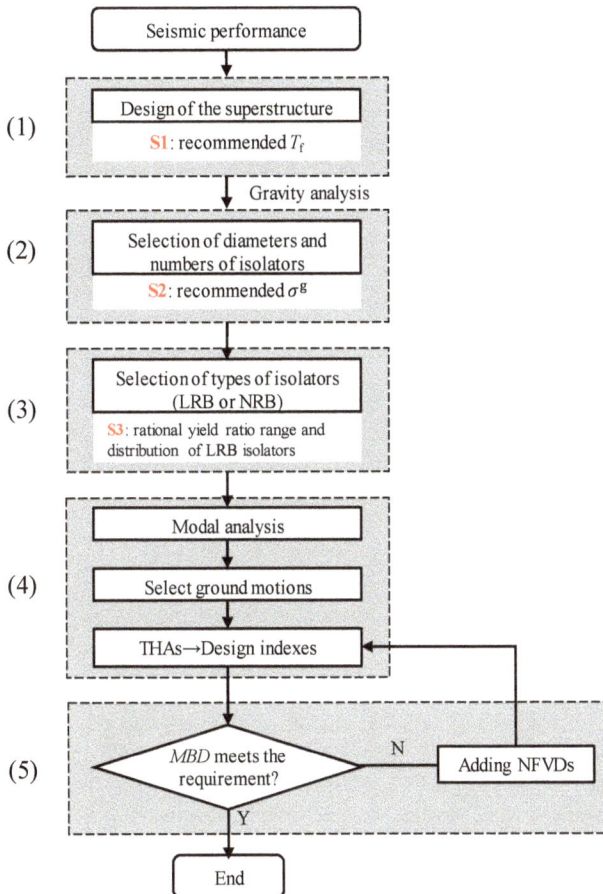

Figure 10. Proposed high-efficiency design method for seismically isolated tall building based on rational range of yield ratios. (S1~S3 denote the solutions to the critical problems).

4. Application of the Design Method

To validate the reliability and applicability of the proposed design method, a seismically isolated RC frame-core tube tall building is designed using this method. The seismic intensity and the site condition of this building are identical to those of the prototype buildings in Section 2. The near field influence coefficient with a value of 1.25 is also adopted. The PGA values of the SLE, DBE and MCE are also 137.5 cm/s², 375 cm/s² and 637.5 cm/s², respectively. The total height of the building is 84.1 m with 23 stories above the ground and the height-to-width ratio is 2.06. The three-dimensional view of the entire building and one typical floor are presented in Figures 11 and 12, respectively.

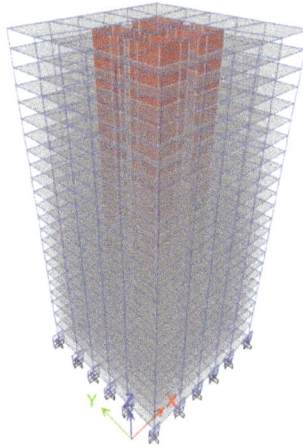

Figure 11. Three-dimensional view of the isolated RC frame-core tube building.

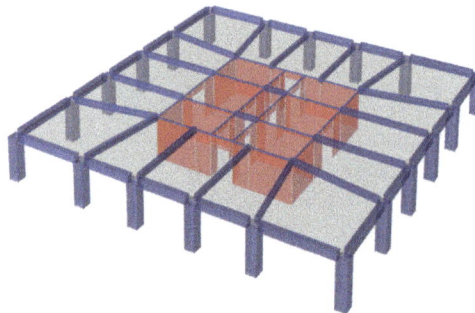

Figure 12. Three-dimensional view of one typical floor.

According to the GB50011-2010 [17], β is required to be no more than 0.40 and 0.38 for the structures without and with NFVDs in the isolation system, respectively. This indicates that the seismic forces of the superstructure are similar to those of a fixed-base structure with one-degree lower seismic intensity (i.e., 7.5 degree). The D_g value of this building is 600 mm, leading to an upper limit value of the MBD under the MCE of 500 mm, which is identical to that of the above prototype buildings.

4.1. Design of the Superstructure

The superstructure of the isolated building is designed as a fixed-base structure with a seismic intensity of 7.5 degree, along with a near field coefficient of 1.25. The corresponding PGA values of the SLE, DBE and MCE are also 68.75 cm/s², 187.5 cm/s² and 387.5 cm/s², respectively. The seismic design of the superstructure is conducted using the structural design software PKPM [39] under the SLE.

The appropriate T_f is calculated using Equation (1) with a value of 1.77 s. Hence, the superstructure is designed with a T_f of 1.798 s and 1.697 s for the Y-direction and the X-direction, respectively, which can approximately satisfy the requirement. The section size and the concrete strength of the superstructure are listed in Table 4. The total gravity load of the entire structure is 617338 kN. The distribution of the inter-story drift ratio of the superstructure under response spectrum analysis is presented in Figure 13. The corresponding maximum inter-story drift ratios of the X and Y directions are 0.066% and 0.072%, respectively, which are both lower than the limit value (i.e., 0.125%) specified in the GB50011-2010 [17].

Table 4. Concrete strength and section size of components.

Component	Story	Concrete	Section/mm	
Column	1–8	C50	1200×1200	
	9–10	C50	1000×1000	
	11–15	C40	1000×1000	
	16–18	C40	800×800	
	19–27	C30	800×800	
Frame beam	1–15	C30	600×900	
	16–27	C30	500×900	
Floor beam	1–27	C30	350×750	
			External	Internal
Shear wall	1–10	C50	400	300
	11–15	C40	400	200
	16–18	C40	300	200
	19–27	C30	300	200
			External	Internal
Height of coupling beam	1–3	Same as shear wall	1000	1000
	4–27		600	1000

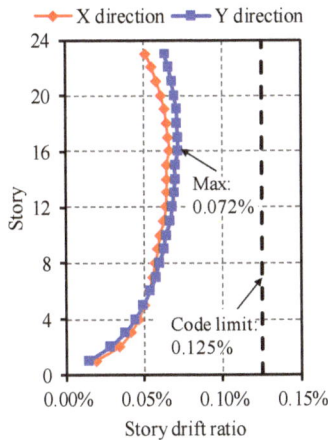

Figure 13. Distribution of inter-story drift ratio of superstructure.

It is worth mentioning here that the design of the seismically isolated building is mostly conducted using ETABS. Therefore, a reliable refined FE model should be built using ETABS. To validate the reliability of this model, typical analytical results obtained from ETABS are compared with those from PKPM. After conducting the modal analysis and the response spectrum analysis under the SLE, the fundamental periods and distribution of the inter-story drift ratios predicted by ETABS and PKPM are found to agree well with each other, with only 3% of difference. This value can be considered acceptable and therefore, validates the reliability of the refined FE model.

4.2. Selection of the Isolator Numbers and Sizes

Gravity analysis is conducted to obtain the axial force of the columns and shear walls at the ground story, in order to determine the sizes and numbers of the isolators under these columns and shear walls. As mentioned above, the σ^g of isolators under the columns and the corner of the core tube are recommended to be approximately 10–12 MPa and 8–10 MPa, respectively. To achieve these, the sizes and the diameters of the isolators under the columns and the shear walls are determined as follows:

(1) The area of the isolators under the columns except those under the corner columns is required to be no less than 1,452,250 mm^2. Hence, two isolators with diameter of 1000 mm are adopted, with the corresponding value of σ^g found to be approximately 11 MPa. The identical numbers and diameters of the isolators under the corner columns are adopted here for convenience, although a relatively lower σ^g with a value of 9.1 MPa is achieved.

(2) The area of the isolators under the corner of the core tube is required to be no less than 1,491,500 mm^2. Two isolators with diameter of 1000 mm are also used, leading to the value of 9.5 MPa for σ^g.

(3) For the external and inner shear walls, the isolators with a diameter of 1000 mm and 900 mm are adopted, respectively. This difference in diameters is attributed to the fact that the axial force of the external shear walls is larger than that of the inner shear walls.

Based on the above discussion, the sizes and numbers of isolators are determined and the process is schematically illustrated in Figure 14. In addition, the σ^g value of each isolator is also presented in Figure 14.

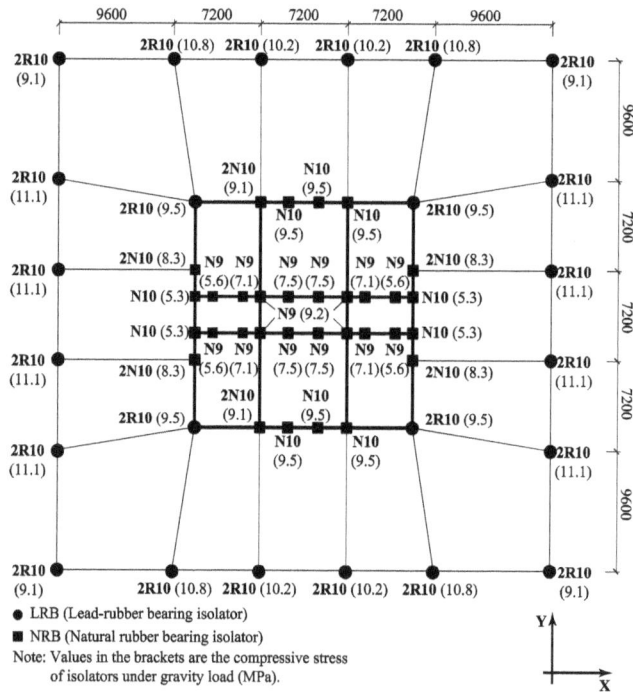

Figure 14. Compressive stress of isolators under gravity load.

4.3. Selection of the Isolator Types (LRB or NRB)

The torsion effect of the seismically isolated tall buildings should be controlled strictly. To avoid the torsion, the rigid center of the isolation system is required to be approximately identical to the mass center of the superstructure, which is specified in the technical specification for seismic-isolation with laminated rubber bearing isolators (CECS126:2001) [40]. To achieve this, LRB isolators are usually adopted for the columns and the corner of the core tube in RC frame-core tube tall buildings. In addition, an index named eccentricity of the isolation system is introduced to quantitatively evaluate the potential torsion effect.

Based on the above, LRB isolators are adopted for the isolators under the columns and the corner of the core tube, while NRB isolators are adopted for other isolators as shown in Figure 14. The yield ratio calculated has a value of 2.3%, which lies in the proposed rational range of the yield ratio (i.e., 2–3%), indicating that there is no need to add other LRB isolators. Then, the eccentricity of isolation system is calculated to have the values of 0.0043% and 0.0007% in the X and Y directions, respectively, which are both lower than the limit of 3%.

4.4. Calculation of the Design Indexes

The fundamental periods of the isolated structure are 4.294 s and 4.242 s for the Y direction and the X direction, respectively. Based on the T_f and T_{is} values, three ground motions, including two natural ground motion records and one artificial ground motion, are selected according to GB50011-2010 [17]. The response spectrum of each ground motion is compared to the design response spectrum as shown in Figure 15. It can be found that the mean response spectrum agrees well with the design response spectrum at the periods of T_f and T_{is}. Subsequently, THAs are conducted for these two structures. The critical design indexes, including β under DBE, MBD under the MCE, σ_{max}^t and σ_{max}^p are calculated. The critical design indexes and parameters are presented in Table 5.

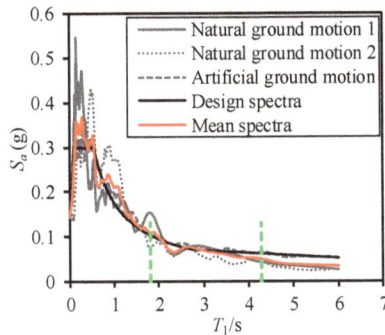

Figure 15. Response spectra of the selected ground motions.

Table 5. Critical design parameters and indexes of the isolated tall building.

Index	Value
T_f (s)	1.798
T_{is} (s)	4.294
Q_y/W (%)	2.3
β	$0.40 \leq 0.40$
MBD (mm)	$788 > 500$
σ_{max}^g (MPa)	$11.1 < 15$
σ_{max}^p (MPa)	$29.95 < 30$
σ_{max}^t (MPa)	$0.00 < 1$

4.5. Validation and Modification of the Isolation System

It can be found from Table 5 that the efficiency of the isolation system (i.e., β) and the stresses of the isolators under the gravity and seismic loads all satisfy the requirements specified in the GB50011-2010 [17], which is consistent with the expected results of the proposed design method. However, the MBD under the MCE with a value of 788 mm exceeds the limit value of 500 mm. As expected and recommended by the proposed method, the NFVDs are introduced for further control.

From the design experience of the 29 seismically isolated tall buildings conducted by the authors, 16 NFVDs are adopted in the isolation system as shown in Figure 16. The parameters of the NFVDs are determined by a few number of trials. When the damping exponent and the damping coefficient reach 0.3 and 220 kN·s/mm, respectively, the MBD under the MCE decreases to 493 mm, which is lower than 500 mm. The maximum damping forces of these NFVDs are approximately 2000 kN. The corresponding critical design parameters and indexes are presented in Table 6. It is obvious that the seismically isolated structure achieves the expected seismic performance.

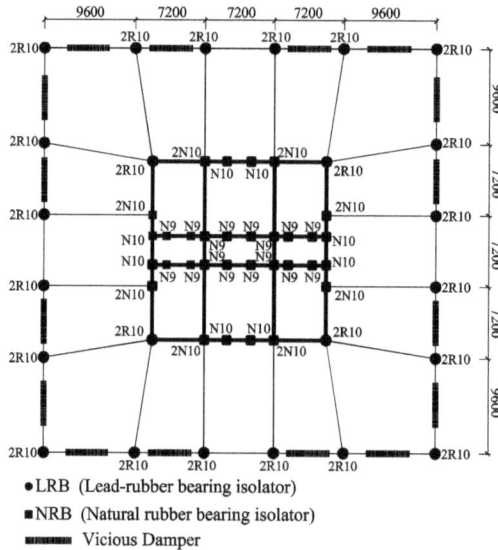

Figure 16. Isolation system with nonlinear fluid viscous dampers (NFVDs) of the RC frame-core tube building.

Table 6. Critical design parameters and indexes of the isolated tall building with NFVDs.

Index	Value
T_f (s)	1.798
T_{is} (s)	4.294
Q_y/W (%)	2.3
β	$0.37 < 0.38$
MBD (mm)	$492 < 500$
σ_{max}^g (MPa)	$11.1 < 15$
σ_{max}^p (MPa)	$27.5 < 30$
σ_{max}^t (MPa)	$0.00 < 1$

The above discussion demonstrates that the numbers, sizes and types of isolators can be directly determined with the proposed method. Only limited numbers of iteration are conducted to determine the quantity and the properties of the NFVDs. The efficiency can be substantially increased with the proposed design method.

5. Conclusions

The rational yield ratios of the isolation system for the RC frame-core tube buildings are investigated using 28 carefully designed study cases, which are designed based on two real engineering practices and considers the influences of total heights, yield ratios and seismically isolated schemes. Based on this, a high-efficiency design method is proposed for such buildings. Subsequently, a seismically isolated RC frame-core tube tall building with a height of 84.1 m is designed using the proposed design method. The following conclusions can be drawn:

(1) The upper and the lower limits of the yield ratio are determined according to the requirements of the horizontal seismic absorbing coefficient and the width of the isolation gap, respectively.

(2) A yield ratio in the range of 2–3% is considered rational and recommended for the seismically isolated RC frame-core tube tall buildings.

(3) A typical seismically isolated RC frame-core tube tall building is successfully designed using the proposed method, thereby validating the rationality, reliability and efficiency of the proposed method.

This paper focuses on the seismically isolated RC frame-core tube tall buildings with moderate height-to-width ratios, in which significantly overturning effect is not observed. Further investigations on such building with significantly overturning effect are required to be conducted. The research outcome can serve as a reference for further development of the seismic design method for seismically isolated RC frame-core tube tall buildings.

Acknowledgments: The authors are grateful for the financial support received from the National Key Technology Research and Development Program of China (No. 2017YFC0703602), the Beijing Advanced Innovation Center for Future Urban Design (No. UDC2016030200) and the Research project of Beijing University of Civil Engineering and Architecture (KYJJ2017005).

Author Contributions: Aiqun Li and Linlin Xie conceived the concept, Linlin Xie guided the students to complete the specific research and proposed the high-efficiency design method, Cantian Yang and Lide Liu investigated the rational yield ratio and applied the design method, Demin Zeng reviewed the study cases, Cantian Yang wrote the paper.

Conflicts of Interest: The authors declare no conflict of interest.

References

1. Formisano, A.; Chieffo, N.; Marius, M. Fragility and resilience curves of an urbanised sector in the district of Naples. *Int. J. Constr. Res. Civ. Eng.* **2017**, *3*, 58–67. [CrossRef]

2. Bozza, A.; Asprone, D.; Fabbrocino, F. Urban resilience: A civil engineering perspective. *Sustainability* **2017**, *9*. [CrossRef]

3. Tipler, J.; Deierlein, G.; Almufti, I. *Seismic Resilience of tall Buildings-Benchmarking Performance and Quantifying Improvements*; Stanford University: Stanford, CA, USA, 2014.

4. Pant, D.; Montgomery, M.; Christopoulos, C.; Xu, B.; Poon, D. Viscoelastic coupling dampers for the enhanced seismic resilience of a megatall building. In Proceedings of the 16th World Conference on Earthquake Engineering, Santiago, Chile, 9–13 January 2017.

5. Tian, Y.; Lu, X.; Lu, X.Z.; Li, M.K.; Guan, H. Quantifying the seismic resilience of two tall buildings designed using Chinese and US codes. *Earthq. Struct.* **2016**, *11*, 925–942. [CrossRef]

6. Heaton, T.; Hall, J.; Wald, D.; Halling, M. Response of highrise and base-isolated buildings to a hypothetical M(w)7.0 blind thrust earthquake. *Science* **1995**, *267*, 206–211. [CrossRef] [PubMed]

7. Pan, P.; Zamfirescu, D.; Nakashima, M.; Nakayasu, N.; Kashiwa, H. Base-isolation design practice in Japan: Introduction to the post-Kobe approach. *J. Earthq. Eng.* **2005**, *9*, 147–171. [CrossRef]

8. Zhu, H.; Zhou, F.; Yuan, Y. Development and analysis of the research on base isolated structures. *Eng. Mech.* **2014**, *3*. [CrossRef]

9.   Higashino, M.; Okamoto, S. *Response Control and Seismic Isolation of Buildings*; Routledge: Abingdon, UK, 2006; ISBN 978-0-415-36623-6.
10.   Zhou, Y.; Chen, P. Shaking table tests and numerical studies on the effect of viscous dampers on an isolated RC building by friction pendulum bearings. *Soil Dyn. Earthq. Eng.* **2017**, *100*, 330–344. [CrossRef]
11.   Pan, P.; Shen, S.; Shen, Z.; Gong, R. Experimental investigation on the effectiveness of laminated rubber bearings to isolate metro generated vibration. *Measurement* **2017**. [CrossRef]
12.   Ye, K.; Li, L.; Zhu, H. A modified Kelvin impact model for pounding simulation of base-isolated building with adjacent structures. *Earthq. Eng. Eng. Vib.* **2009**, *8*, 433–446. [CrossRef]
13.   Ubertini, F.; Comodini, F.; Fulco, A.; Mezzi, M. A Simplified parametric study on occupant comfort conditions in base isolated buildings under wind loading. *Adv. Civ. Eng.* **2017**. [CrossRef]
14.   Sorace, S.; Terenzi, G. Analysis and demonstrative application of a base isolation/supplemental damping technology. *Earthq. Spectra* **2008**, *24*, 775–793. [CrossRef]
15.   Cancellara, D.; De Angelis, F. Assessment and dynamic nonlinear analysis of different base isolation systems for a multi-storey RC building irregular in plan. *Comput. Struct.* **2017**, *180*, 74–88. [CrossRef]
16.   Becker, T.; Bao, Y.; Mahin, S. Extreme behavior in a triple friction pendulum isolated frame. *Earthq. Eng. Struct. Dyn.* **2017**. [CrossRef]
17.   China Ministry of Construction. *Code for Design of Concrete Structures (GB 50010-2010)*; China Ministry of Construction: Beijing, China, 2010.
18.   Hu, K.; Zhou, Y.; Jiang, L.; Chen, P.; Qu, G. A mechanical tension-resistant device for lead rubber bearings. *Eng. Struct.* **2017**, *152*, 238–250. [CrossRef]
19.   Park, J.-G.; Otsuka, H. Optimal yield level of bilinear seismic isolation devices. *Earthq. Eng. Struct. Dyn.* **1999**, *28*, 941–955. [CrossRef]
20.   Providakis, C.P. Effect of LRB isolators and supplemental viscous dampers on seismic isolated buildings under near-fault excitations. *Eng. Struct.* **2008**, *30*, 1187–1198. [CrossRef]
21.   Li, B. Computational Analysis and Research on Relevant Parameters of Isolation Layer and Seismic Response Based on Base-Isolated Benchmark Building. Master's Thesis, Guangzhou University, Guangzhou, China, 2013.
22.   Mollaioli, F.; Lucchini, A.; Cheng, Y.; Monti, G. Intensity measures for the seismic response prediction of base-isolated buildings. *Bull. Earthq. Eng.* **2013**, *11*, 1841–1866. [CrossRef]
23.   Avşar, Ö.; Özdemir, G. Response of seismic-isolated bridges in relation to intensity measures of ordinary and pulselike ground motions. *J. Bridge Eng.* **2011**, *18*, 250–260. [CrossRef]
24.   Pourzeynali, S.; Zarif, M. Multi-objective optimization of seismically isolated high-rise building structures using genetic algorithms. *J. Sound Vib.* **2008**, *311*, 1141–1160. [CrossRef]
25.   Shen, S.; Liu, W.; Du, D.; Wang, S. Effects of nonlinear seismic response of isolated structures on spectrum characteristics of ground motions. *J. Nanjing Univ. Technol. Nat. Sci. Ed.* **2013**, *35*, 1–5. [CrossRef]
26.   Wang, S.; Zhao, X.; Miao, Q.; Liu, W.; Du, D. Parameter optimization of isolator in structure with added floors and related shaking tabsle tests. *J. Vib. Eng.* **2013**, *26*, 722–731. [CrossRef]
27.   Ma, C.F.; Zhang, Y.H.; Tan, P.; Zhou, F.L. Seismic response of base-isolated high-rise buildings under fully nonstationary excitation. *Shock Vib.* **2014**, *2014*. [CrossRef]
28.   Takewaki, I. Robustness of base-isolated high-rise buildings under code-specified ground motions. *Struct. Des. Tall Spec. Build.* **2008**, *17*, 257–271. [CrossRef]
29.   Kaplan, H.; Seireg, A. Optimal design of a base isolated system for a high-rise steel structure. *Earthq. Eng. Struct. Dyn.* **2001**, *30*, 287–302. [CrossRef]
30.   Xiong, C.; Lu, X.Z.; Guan, H.; Xu, Z. A nonlinear computational model for regional seismic simulation of tall buildings. *B. Earthq. Eng.* **2016**, *14*, 1047–1069. [CrossRef]
31.   Lu, X.; Xie, L.; Guan, H.; Huang, Y.; Lu, X. A shear wall element for nonlinear seismic analysis of super-tall buildings using OpenSees. *Finite Elem. Anal. Des.* **2015**, *98*, 14–25. [CrossRef]
32.   Lu, X.; Lu, X.Z.; Zhang, W.K.; Ye, L.P. Collapse simulation of a super high-rise building subjected to extremely strong earthquakes. *Sci. China Technol. Sci.* **2011**, *54*, 2549–2560. [CrossRef]
33.   Pan, P.; Li, W.; Nie, X.; Deng, K.; Sun, J. Seismic performance of a reinforced concrete frame equipped with a double-stage yield buckling restrained brace. *Struct. Des. Tall Spec. Build.* **2017**, *26*. [CrossRef]
34.   Su, N.; Lu, X.; Zhou, Y.; Yang, T.Y. Estimating the peak structural response of high-rise structures using spectral value-based intensity measures. *Struct. Des. Tall Spec. Build.* **2017**, *26*. [CrossRef]

35. Computers and Structures Inc. *ETABS (Version 2016). Windows*; Computers and Structures Inc.: Berkeley, CA, USA, 2016.

36. Faal, H.N.; Poursha, M. Applicability of the N2, extended N2 and modal pushover analysis methods for the seismic evaluation of base-isolated building frames with lead rubber bearings (LRBs). *Soil Dyn. Earthq. Eng.* **2017**, *98*, 84–100. [CrossRef]

37. Cardone, D.; Gesualdi, G. Seismic rehabilitation of existing reinforced concrete buildings with seismic isolation: A case study. *Earthq. Spectra* **2013**, *30*, 1619–1642. [CrossRef]

38. Chen, J.; Ding, J. Aseismic design of tall building of concrete filled square steel tube with energy dissipation braces. *J. Tongji Univ. Nat. Sci.* **2006**, *34*, 1431–1435.

39. China Academy of Building Research. *PKPM (Version 2010). Windows*; China Academy of Building Research: Beijing, China, 2010.

40. Guangzhou University; China Academy of Building Research. *Technical Specification for Seismic Isolation with Laminated Rubber Bearing Isolators (CECS126:2001)*; CECS: Beijing, China, 2001.

applied
sciences

MDPI

Article

Base Pounding Model and Response Analysis of Base-Isolated Structures under Earthquake Excitation

Chengqing Liu [1], Wei Yang [1], Zhengxi Yan [1], Zheng Lu [2,]* and Nan Luo [1]

[1] School of Civil Engineering, Southwest Jiaotong University, Chengdu 610031, China;
    lcqjd@swjtu.edu.cn (C.L.); XNJDL@my.swjtu.edu.cn (W.Y.); yenchenghsi@gmail.com (Z.Y.);
    NanLuo@swjtu.edu.cn (N.L.)
[2] State Key Laboratory of Disaster Reduction in Civil Engineering, Tongji University, Shanghai 200092, China
*   Correspondence: luzheng111@tongji.edu.cn; Tel.: +86-021-6598-6186

Received: 31 October 2017; Accepted: 27 November 2017; Published: 29 November 2017

Abstract: In order to study the base pounding effects of base-isolated structure under earthquake excitations, a base pounding theoretical model with a linear spring-gap element is proposed. A finite element analysis program is used in numerical simulation of seismic response of based-isolated structure when considering base pounding. The effects of the structure pounding against adjacent structures are studied, and the seismic response of a base-isolated structure with lead-rubber bearing and a base-isolated structure with friction pendulum isolation bearing are analyzed. The results indicate that: the model offers much flexibility to analyze base pounding effects. There is a most clearance unfavorable width between adjacent structures. The structural response increases with pounding. Significant amplification of the story shear-force, velocity, and acceleration were observed. Increasing the number of stories in a building leads to an initial increase in impact force, followed by a decrease in such force. As a result, it is necessary to consider base pounding in the seismic design of base-isolated structures.

Keywords: base isolation; isolated structure; base pounding model; time-history analysis; seismic response; impact response

1. Introduction

As one of the most destructive natural disasters, an earthquake can cause heavy casualties, and great damage to buildings, bridges, and roads. One of the most devastating earthquakes in recent years is the 2008 Sichuan earthquake, which killed more than 69,000 people, left more than 18,000 missing, and caused a direct economic loss of 845.1 billion yuan. Earthquakes can cause great damage. Therefore, the study and application of seismic engineering are of great significance. With the development of science and technology, many meaningful anti-seismic methods, including energy dissipation, vibration control, and based isolation were developed [1–5]. Since the base isolated system was first applied in the 1970s, a lot of relevant research has been conducted [6–10]. Energy dissipation devices [11–13], which can dissipate seismic energy and efficiently reduce structural damages, are set between the foundation and the superstructure. Lead-rubber bearing and friction pendulum isolation bearing are usually used as energy dissipation device for base isolation. However base-isolated structures usually experience large horizontal displacements during strong earthquakes due to their weak horizontal stiffness. Hence, there is a great possibility of the structure pounding against adjacent structures [14,15].

Studies on base pounding effects during a strong earthquake are rare. The earliest studies of the width of clearance and foundation stiffness effects were was performed by Tsai [16] and Malhotra [17]. Other teams [18–26] conducted extensive research on response of the structures pounding against adjacent structures, and on how to reduce seismic energy through theoretical studies and numerical simulations. Mavronicola and colleagues [27] used a smooth bilinear (Bouc-Wen) model to simulate the seismic isolation system, while the Kelvin-Voigt [28] impact model and other models were adopted in structural response analysis under strong excitations. The accuracy and flexibility of these impact models were discussed. A typical four-story fixed-base RC building that was subjected to seismic pounding was analyzed in Pant and Wijeyewichrema [29]. Three-dimensional finite element analyses were conducted considering material and geometric nonlinearities. Fan et al. [30] considered pounding responses with different system parameters, such as impact model, size of gap, and natural vibration period. Many factors were considered in Ye's study [31], including superstructure's stiffness, impaction stiffness, the mechanical properties of the bearing, and the different width of clearance.

On the basis of previous research work, a new base pounding theoretical model with linear spring-gap element is proposed. Assuming that the superstructure is linear-elastic, the colliding unit presented in Figure 1 adopts the linear spring with gap, and the collision analysis of the base isolation structure under strong earthquakes is conducted. Seismic response analysis of base-isolated structure considering base pounding by this model is discussed in this paper. In order to compare the difference in response between the base-isolated structure with lead-rubber bearing and the base-isolated structure with friction pendulum isolation bearing, two types of finite element models are used in analysis. Finite element models with different gap have were used to determine the maximum node acceleration in top story and the most unfavorable width of clearance between adjacent structures. The values of impact force, story shear-force, displacement, velocity and acceleration are obtained. Finally, such values are compared to previous research to verify its rationality.

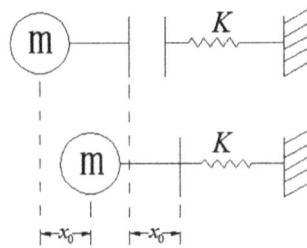

Figure 1. Linear spring-gap element.

2. Models and Equations of Motion

2.1. Base Pounding Model

There are two methods to investigate impact behavior, the classical dynamics method and the contact element method. The classical one cannot reflect the change of impact force, deformation and collision duration and other elements. Furthermore, it is difficult to implement in finite element analysis. Therefore, it has limited scope of use [32–34]. The contact element method is easy to implement in software with high precision. Consequently, the contact element method is adopted in this paper. Research conducted by Fan et al. [30] shows that linear viscoelastic model can provide enough accuracy in engineering. Thus, a linear spring-gap element was used in this base pounding theoretical model. Figure 1 presents the linear spring-gap element. Figure 2 and Equation (1) present its force-displacement relation.

$$f_p = \begin{cases} 0 & |x_0| < x_{gap} \\ k(|x_b| - x_{gap}) & |x_0| \geq x_{gap} \end{cases}, \tag{1}$$

where f_p is the impact force, k is the stiffness of linear spring-gap element, x_0 is the relative displacement, and x_{gap} is the initial width of clearance.

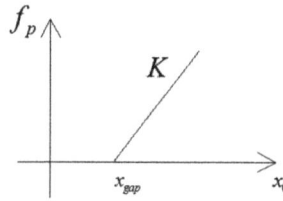

Figure 2. Force-displacement curve.

2.2. Equations of Motion

Assuming that the stiffness of the floor slabs in-plane is infinite and the masses of the floor slabs are lumped at the floor levels, base pounding models were built with linear spring-gap elements. Figures 3 and 4 present the models of a base-isolated structure with lead-rubber bearing and with friction pendulum isolation bearing, respectively. The equations of motion of the superstructure are expressed in Equation (2).

$$\begin{cases} m_1 \ddot{x}_1 + c_1 \dot{x}_1 - c_1 \dot{x}_0 + f_1 - f_2 = -m_1 \ddot{x}_g \\ m_i \ddot{x}_i + c_i \dot{x}_i + f_i - f_n = -m_i \ddot{x}_g \\ m_n \ddot{x}_n + c_n \dot{x}_n + f_n = -m_n \ddot{x}_g \end{cases}, \qquad (2)$$

where $\dot{x}_i, \ddot{x}_i (i = 1, 2 \cdots n)$ are the relative velocities and accelerations of floor i, respectively, $\ddot{x}_g$ is the earthquake ground motion acceleration, $m_i (i = 1, 2 \cdots n)$ and $c_i (i = 1, 2 \cdots n)$ are the mass and damping of floor i, respectively. The restoring force of floor i is expressed by the following equation:

$$f_i = k_i (x_i - x_{i-1}), (i = 1, 2 \cdots n), \qquad (3)$$

where k_i is the stiffness of floor i, and x_i is the relative displacement of floor i.

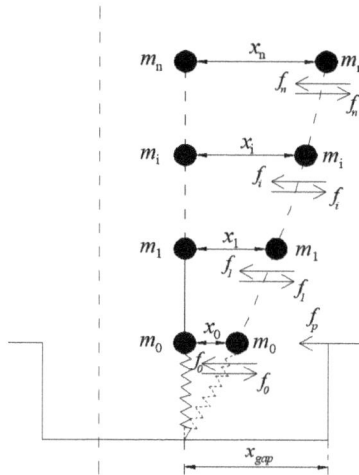

Figure 3. Model of the base-isolated structure with lead-rubber bearing.

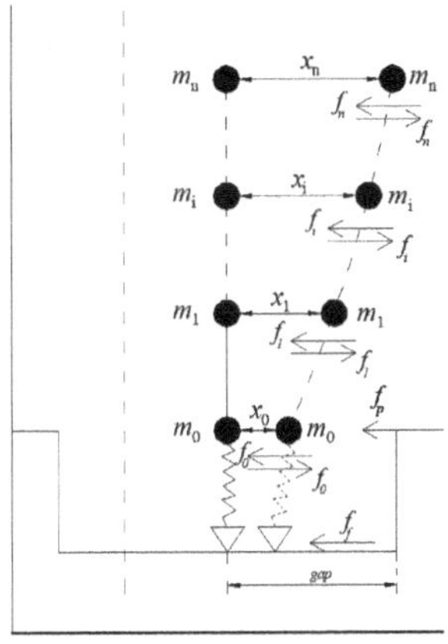

Figure 4. Model of the base-isolated structure with friction pendulum isolation bearing.

Rayleigh Damping is calculated by the following equation:

$$c_i = \alpha m_i + \beta k_i\,, (i = 1, 2 \cdots n),\tag{4}$$

where α, β are calculated by Equation (5) if the damping ratios ξ_i and ξ_j associated with specific frequencies ω_i, ω_j are known.

$$\begin{cases} \alpha = 2\omega_i\xi_j(\omega_j\xi_i - \omega_i\omega_j)/(\omega_j^2 - \omega_i^2) \\ \beta = 2(\omega_j\xi_j - \omega_i\xi_i)/(\omega_j^2 - \omega_i^2) \end{cases},\tag{5}$$

Equations of motion for the isolation layer (the base-isolated structure with lead-rubber bearing) are given as,

$$m_0\ddot{x}_0 + (c_0 + c_1)\dot{x}_0 - c_1\dot{x}_1 + f_0 - f_1 + f_p = -m_0\ddot{x}_g,\tag{6}$$

where m_0 is the mass of isolation layer, $\dot{x}_0$, $\ddot{x}_0$ are relative the velocities and accelerations of the isolation layer, respectively, c_0 is the damping coefficient of the isolation layer, f_0 and f_p are the restoring and the impact force the of isolation layer, respectively.

Equations for the restoring force have been built using the Bouc-wen model:

$$f_0 = \alpha_0 k_0 x_0 + (1 - \alpha_0)k_0 x_y z_0,\tag{7}$$

where k_0 is the isolation layer's initial stiffness, α_0 is the ratio of the yield stiffness to the pre-yield stiffness of bearing, x_0 is the displacement of the isolation layer, z_0 is the hysteretic displacement of the isolation system, and x_y is the yield displacement.

The first order differential equation of the hysteretic displacement is given as,

$$\dot{z}_0 = (-\gamma_0|\dot{x}_0|z_0|z_0|^{n_0-1} - \beta_0\dot{x}_0|z_0|^{n_0} + A_0\dot{x}_0)/x_y,\tag{8}$$

where β_0, A_0, γ_0, and n_0 are related to the amplitude of hysteretic displacement, initial stiffness, and hysteretic shape.

Equations of motion of the isolation layer (the base-isolated structure with friction pendulum isolation bearing) is given as,

$$m_0\ddot{x}_0 + c_1\dot{x}_0 - c_1\dot{x}_1 + f_0 - f_1 + f_p + f_f = -m_0\ddot{x}_g,\tag{9}$$

Restoring force can be calculated by Equation (10).

$$f_0 = k_0 x_0, \tag{10}$$

where k_0 is the stiffness of bearing, x_0 is the displacement of isolation layer.

Friction can be expressed as,

$$f_f = \mu N z_s \text{sgn}(\dot{x}_0), \tag{11}$$

where μ is the coefficient of sliding friction of bearing, N is the weight of superstructure ($N = \sum_{i=1}^{n} m_i g$), z_s is a parameter related to hysteresis characteristics, and z_s is expressed in Equation (12).

$$\dot{Y}z_s = Au - \gamma |u| z_s |z_s|^{\eta-1} - \beta u |z_s|^{\eta} \tag{12}$$

In Equation (12), Y is the elastic shear deformation of bearing before sliding, u is the ground velocity of bearing, and β, A, γ, and n are related to amplitude of hysteretic displacement, initial stiffness, and hysteretic shape.

3. Engineering Case and Numerical Simulation

As mentioned previously, two finite element models were developed. Finite element model A is modeled after a building in Tibetan Qiang Autonomous Prefecture of Ngawa, Sichuan Province, China. The structure of the building is the base-isolated frame structure with lead-rubber bearing. Model A consists of 40 lead-rubber bearings of the same type. The mass of the isolation layer is 2490.55 tons. The equivalent horizontal stiffness is 4.418×10^5 N/mm. The damping ratio of the isolation layer is 0.23. Figure 5 presents the arrangement of the bearings. Figure 6 presents the arrangement of the beams and pillars. Figures 7 and 8 present the structure's front elevation and side elevations, respectively. Table 1 presents the parameters of each story.

Table 1. Parameters of story.

Story	Story Height (mm)	Mass of Story (ton)	Stiffness of Story (10^6 N/mm)
6	4100	231.6	0.231
5	4500	2079.3	1.266
4	3900	2170.5	1.494
3	3900	2184.2	1.676
2	3900	2515.3	1.740
1	4200	2006.1	1576

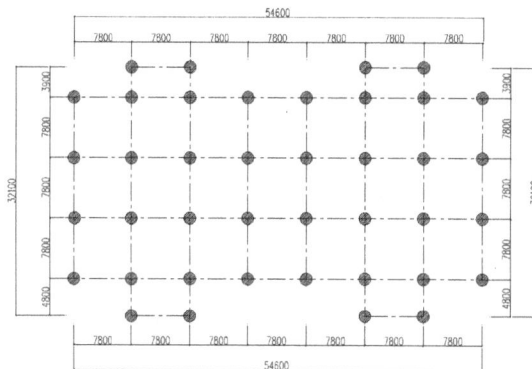

Figure 5. Arrangement of the bearings (mm).

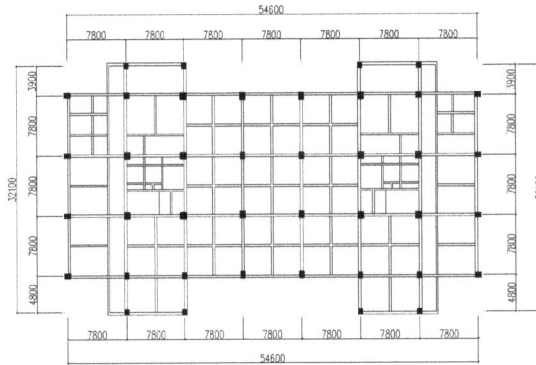

Figure 6. Arrangement of the beams and pillars (mm).

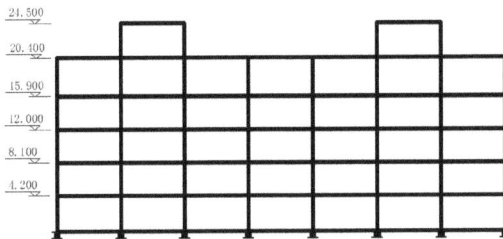

Figure 7. Front elevation of building (m).

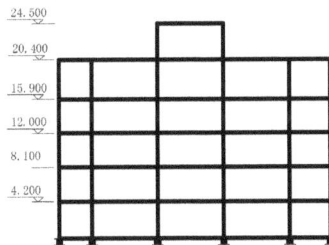

Figure 8. Side elevation of building (m).

Finite element model B is modeled after model A. The superstructure of model B is the same as that of model A. However, the lead-rubber bearings in model A are replaced with friction pendulum isolation bearings. For model B, the mass of the isolation layer is 2490.55 tons. The equivalent horizontal stiffness is 1.6×10^5 N/mm, and the damping ratio of the isolation layer is 0.

3.1. Most Unfavorable Clearance Width

In order to study the pounding effects with different clearance width, a parametric study was conducted. Two sets of strong earthquake records (El Centro (NS) and Taft (EW)), and a set of artificial acceleration time-history curves are used as excitations in the simulations. According to the Code for Seismic Design of Buildings of China (built on Site-class four, intensity 8) [35], 400 gal is adopted as the peak ground acceleration for rare earthquakes. In order to analyze the tendency of absolute acceleration with different clearance widths, the acceleration value (node 858) in the top story is extracted for both models A and B.

In Figures 9 and 10, the tendencies of absolute acceleration are similar while varying the different clearance widths. First, the maximum value of acceleration increased with an increasing clearance width, and then it decreased with continued increase in clearance width, and finally leveled off. When the clearance width was approximately 20 mm, the value of acceleration was the highest.

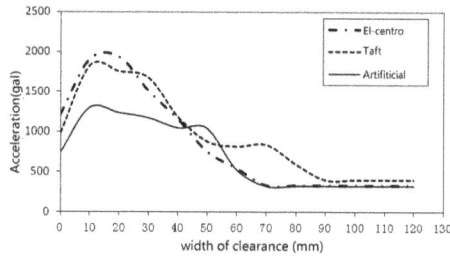

Figure 9. Maximum acceleration value changes with clearance width in model A.

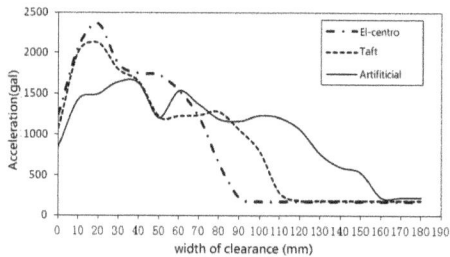

Figure 10. Maximum acceleration value changes with clearance width in model B.

3.2. Effects of Pounding

The effects of the structure pounding against adjacent structures are studied from the perspective of time-history of impact force, story shear-force, velocity, and acceleration. In order to obtain the maximum response of the structure, the clearance width for both models A and B are set to 20 mm.

3.2.1. Impact Force

Figures 11 and 12 show the time-history curve of the impact force under El Centro earthquake excitation. Every peak in the curve represents a pounding. As shown, pounding did not happen just once, but repeatedly during the earthquake.

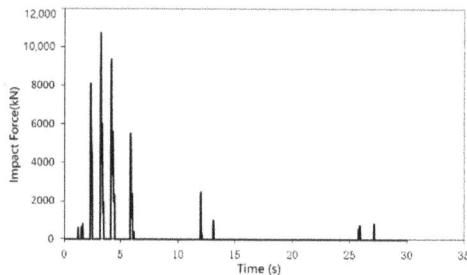

Figure 11. Time-history curve of impact force for model A.

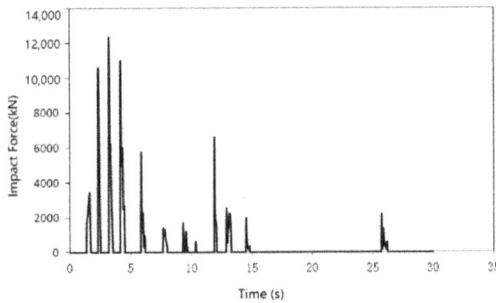

Figure 12. Time-history curve of impact force for model B.

On the basis of models A and B, models with 7, 9, and 12 stories were built to study the effects of story and on pounding. Figure 13 shows the curve of impact force variation with the number of stories under the El Centro earthquake excitation.

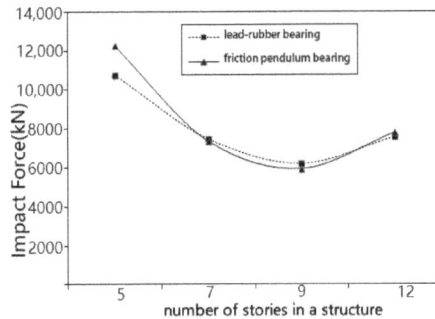

Figure 13. Curves of impact force changes with story.

In Figure 13, it can be seen that the impact force first decreases with an increasing number of stories, but then decreases. This pattern also can be observed in under the other two excitations. Therefore, the number of stories in a building is not a good standard to estimate the magnitude of the impact force. More factors such as the type of structure, and material characteristics should be considered.

3.2.2. Story Shear-Force

Figures 14 and 15 compare story shear-force with and without pounding. Tables 2 and 3 present the maximum values of story shear-force for models A and B under different earthquakes. In Table 2, for the base-isolated structure with lead-rubber bearing, it can be seen that there is a 3.59 to 5.06 times growth of story shear-force for the El Centro earthquake, 2.04 to 3.13 times growth for the Taft earthquake, and 1.03 to 2.63 times for the Artificial earthquake. In Table 3, for the base-isolated structure with friction pendulum isolation bearing, 1.59 to 12.60 times growth of story shear-force can be observed for the El Centro earthquake, as well as 1.30 to 10.93 times growth for the Taft earthquake, and −0.18 to 3.24 times growth for the Artificial earthquake.

It can be inferred that there is a considerable amplification of the story shear-force under pounding for both types of isolated structures. In particular, for the structure with friction pendulum isolation bearing, the amplification of the shear-force in the first story is larger than that in other stories.

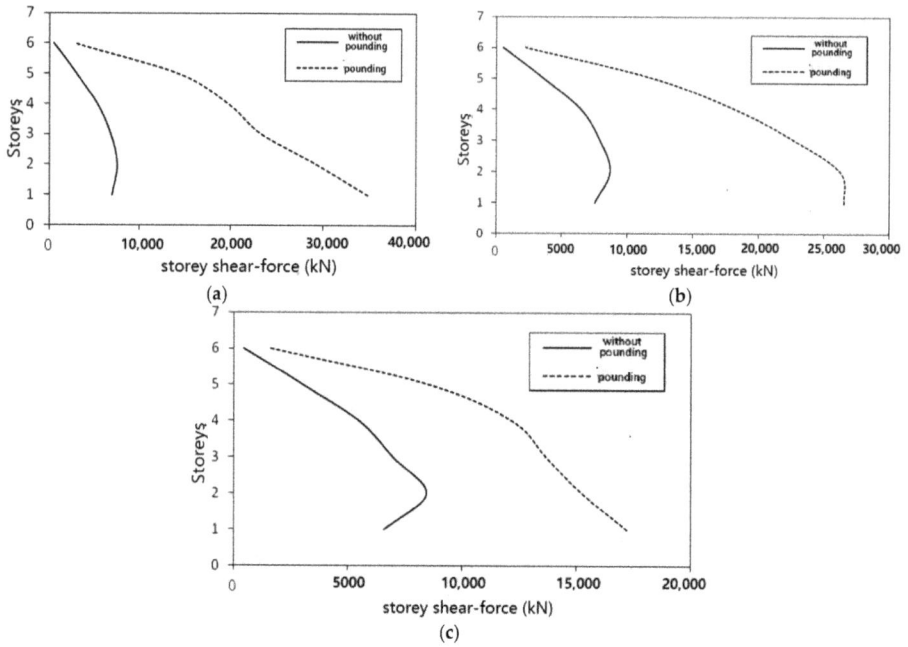

Figure 14. Story shear-force under different earthquakes (model A). (**a**) El Centro earthquake; (**b**) Taft earthquake; (**c**) Artificial excitation.

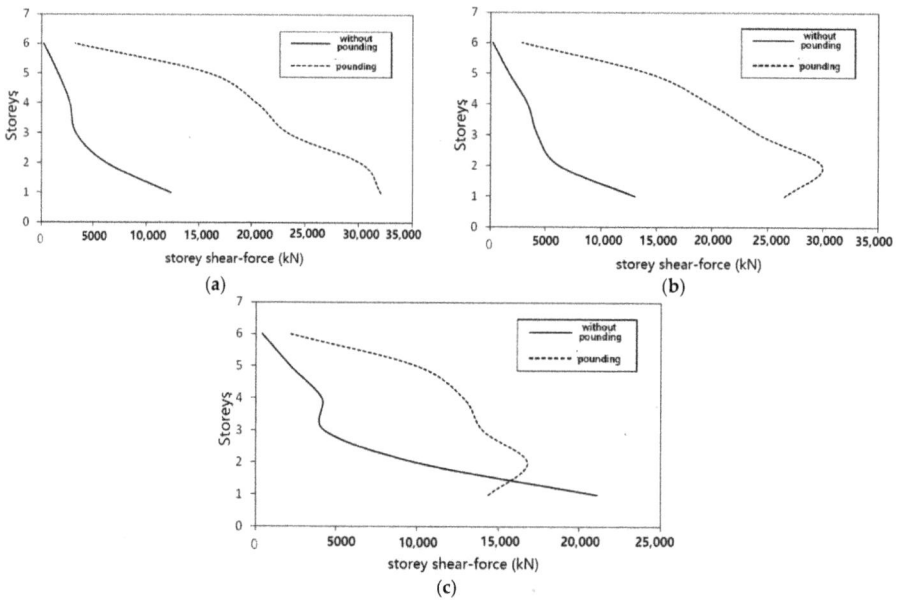

Figure 15. Story shear-force under different earthquakes (model B). (**a**) El Centro earthquake; (**b**) Taft earthquake; (**c**) Artificial excitation.

Table 2. Maximum values of story shear-force for model A.

Earthquake	Range of Maximum Value	Without Pounding (kN)	With Pounding (kN)	Times of Growth
El Centro	low limit	466	2828	5.06
	upper limit	7573	34,770	3.59
Taft	low limit	539	2223	3.13
	upper limit	8715	26,546	2.04
Artificial	low limit	450	1636	2.63
	upper limit	8450	17,213	1.03

Table 3. Maximum values of story shear-force for model B.

Earthquake	Range of Maximum Value	Without Pounding (kN)	With Pounding (kN)	Times of Growth
El Centro	low limit	238	3238	12.60
	upper limit	12,350	32,000	1.59
Taft	low limit	240	2864	10.93
	upper limit	13,080	29,950	1.30
Artificial	low limit	386	1636	3.24
	upper limit	21,114	17,213	−0.18

3.2.3. Acceleration

Figures 16 and 17 show the acceleration time-history curves of node 858, where the maximum values of acceleration were observed, under different earthquake excitations. The maximum accelerations that were obtained in models A and B are presented in Tables 4 and 5.

When compared to cases without pounding, a large amplification can be observed in both models A and B under pounding condition, according to Figures 16 and 17. Furthermore, the maximum values of acceleration appear when excitations are strong.

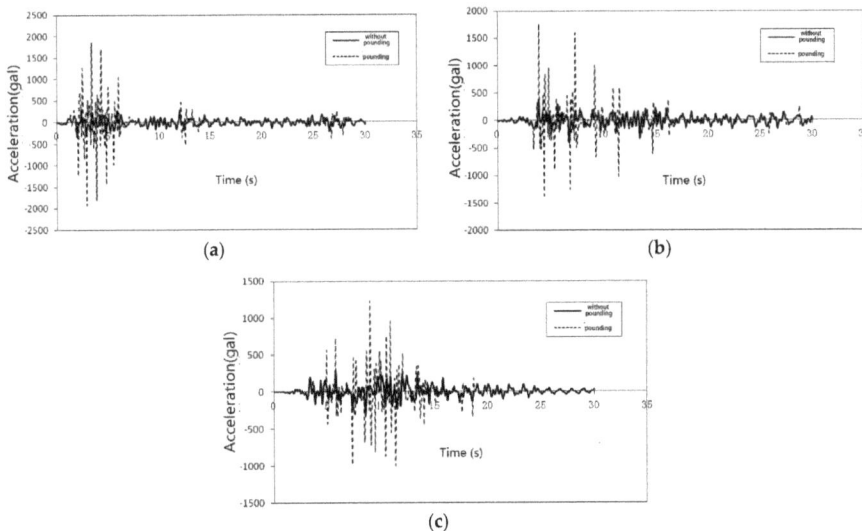

Figure 16. Structural acceleration under different earthquakes (model A). (**a**) El Centro earthquake; (**b**) Taft earthquake; and, (**c**) Artificial excitation.

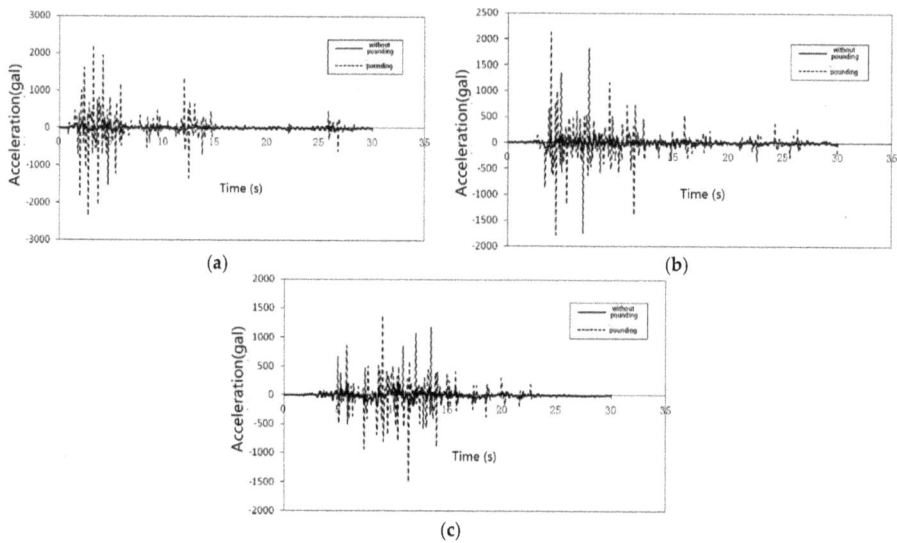

Figure 17. Structural acceleration under different earthquakes (model B). (**a**) El Centro earthquake; (**b**) Taft earthquake; (**c**) Artificial excitation.

Table 4. Maximum acceleration values for model A.

Earthquake	Maximum Acceleration (gal)		
	Without Pounding	With Pounding	Amplification
El Centro	329.10	1934.98	4.88
Taft	398.47	1761.14	3.42
Artificial	323.13	1233.14	2.82

Table 5. Maximum acceleration values for model B.

Earthquake	Maximum Acceleration (gal)		
	Without Pounding	With Pounding	Amplification
El Centro	172.46	2356.69	12.67
Taft	180.09	2119.20	10.77
Artificial	−199.02	−1496.98	6.52

According to Tables 4 and 5, for model A under pounding conditions, there is a 4.88 times growth in acceleration under the El Centro earthquake, a 3.42 times growth under the Taft earthquake, and a 2.82 times growth under the Artificial earthquake. The amplification of model A is larger than that of model B, which was 12.67 times growth under the El Centro earthquake, 10.77 times growth under the Taft earthquake and 6.52 times growth under the Artificial earthquake.

There are great pounding effects on acceleration on top story acceleration of both the structure with lead-rubber bearing and the structure with friction pendulum isolation bearing. However, the acceleration amplification of the structure with friction pendulum isolation bearing is larger.

3.2.4. Velocity

Figures 18 and 19 show the velocity time-history curves of node 858 under different earthquakes. The maximum value of velocity on node 858 of models A and B can be found in Tables 6 and 7.

The amplification of velocity under pounding in model A (1.07 times growth under the El Centro earthquake, 0.47 times growth under the Taft earthquake and 0.31 times growth under the Artificial earthquake) can be obtained in Table 6. The amplification can also be observed in model B (2.00 times growth under the El Centro earthquake, 1.19 times growth under the Taft earthquake and 0.25 times growth under the Artificial earthquake, Table 7).

There are some effects of pounding on velocity in the top story of both types of isolated structure. However, the amplification of acceleration is larger than that of velocity. In addition, the maximum values of velocity appear when the excitations are strong.

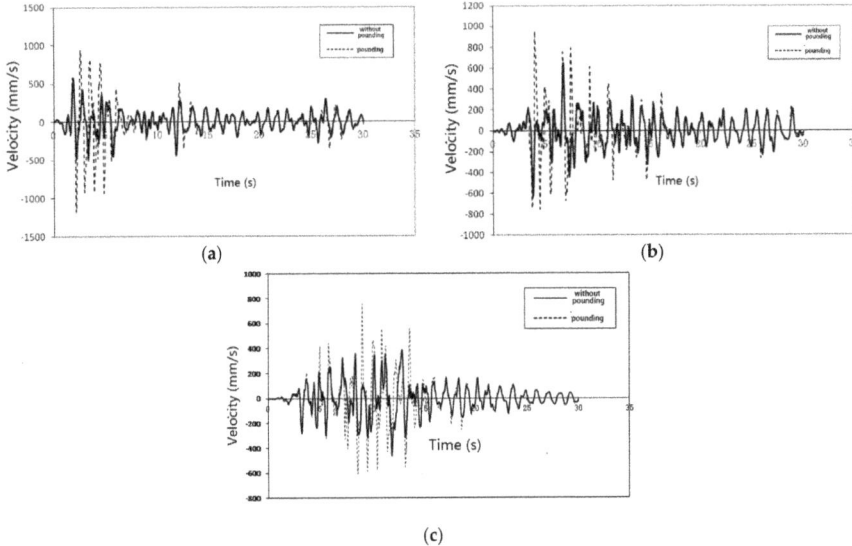

Figure 18. Structural velocity under different earthquakes (model A). (**a**) El Centro earthquakes; (**b**) Taft earthquake; and (**c**) Artificial excitation.

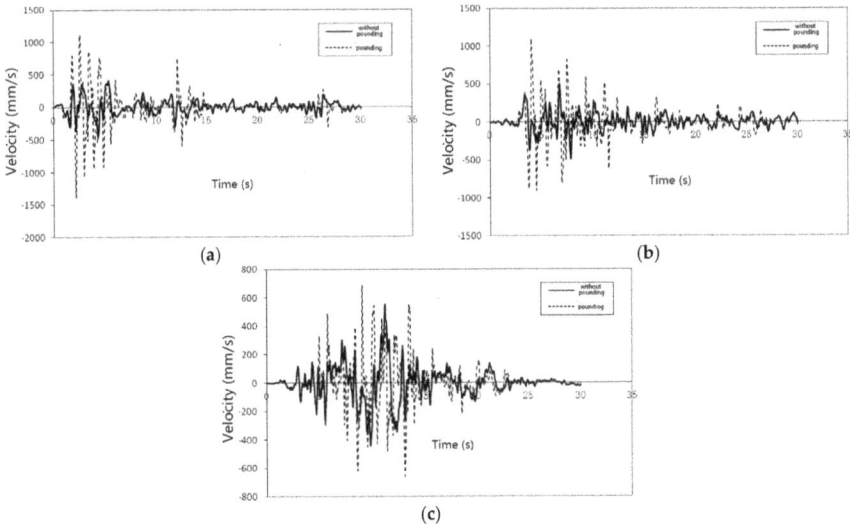

Figure 19. Structural velocity under different earthquakes (model B). (**a**) El Centro earthquakes; (**b**) Taft earthquake; and, (**c**) Artificial excitation.

Table 6. Maximum velocity values for model A.

Earthquake	Maximum Velocity (mm/s)		
	Without Pounding	With Pounding	Amplification
El Centro	568.41	1176.45	1.07
Taft	645.80	951.15	0.47
Artificial	460.34	602.81	0.31

Table 7. Maximum velocity values for model B.

Earthquake	Maximum Velocity (mm/s)		
	Without Pounding	With Pounding	Amplification
El Centro	465.05	1398.14	2.00
Taft	501.82	1100.94	1.19
Artificial	553.26	690.54	0.25

3.2.5. Displacement

Figures 20 and 21 show the displacement time-history curves of node 858 under different earthquake excitations. It can be inferred that there is little amplification of displacement, while the structure was undergoing pounding under the El Centro and the Taft earthquake excitations. Furthermore, the displacement decreased while the structure was undergoing pounding under the artificial earthquake. For model A (Table 8), 0.55 times growth was observed for the El Centro earthquake, 0.09 times growth for the Taft earthquake, and 0.19 times decrease for the Artificial earthquake). For model B (Table 9), 0.21 times growth was observed for the El Centro earthquake, 0.03 times growth for the Taft earthquake, and 0.52 times decrease for the Artificial earthquake).

There is little effect of pounding on displacement due to the restriction of adjacent structures.

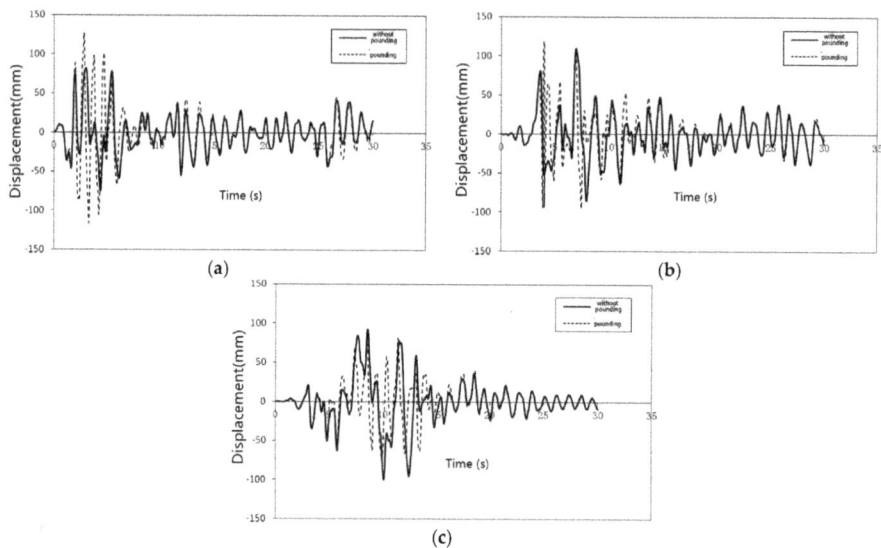

Figure 20. Structural displacement under different earthquakes (model A). (**a**) El Centro earthquake; (**b**) Taft earthquake; and, (**c**) Artificial excitation.

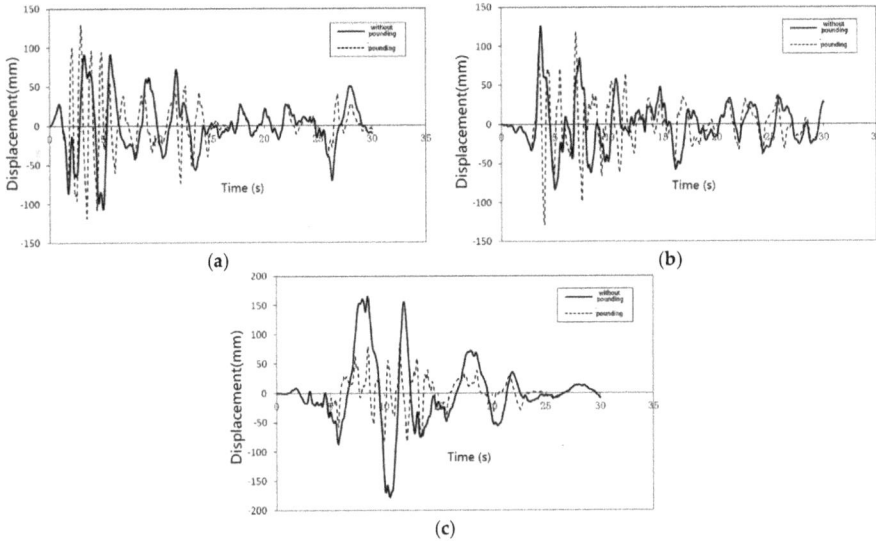

Figure 21. Structural displacement under different earthquakes (model B). (**a**) El Centro earthquake; (**b**) Taft earthquake; and, (**c**) Artificial excitation.

Table 8. Maximum structural displacements for model A.

Earthquake	Maximum Displacement (mm)		
	Without Pounding	With Pounding	Amplification
El Centro	81.80	126.98	0.55
Taft	109.34	118.73	0.09
Artificial	100.04	81.19	−0.19

Table 9. Maximum structural displacements for model B.

Earthquake	Maximum Displacement (mm)		
	Without Pounding	With Pounding	Amplification
El Centro	106.82	129.20	0.21
Taft	125.73	129.58	0.03
Artificial	177.13	85.39	−0.52

4. Conclusions

A base pounding theoretical model with linear spring-gap element is proposed in this paper. On the basis of this theoretical model, the finite element models of a structure with lead-rubber bearing and friction pendulum isolation bearing are built to analyze their seismic response. Some meaningful conclusions obtained are as follows:

(1) The base pounding theoretical model proposed in this paper can be applied easily and efficiently to analyze base-isolated structures when considering base pounding.

(2) There is a most unfavorable clearance width between adjacent structures and the response of base-isolated structures increases in pounding.

(3)   The number of stories in a building should not be uniquely considered to estimate the magnitude of impact force. More considerable factors should be considered, such as the type of structure and the material characteristics

(4)   Significant amplification of the story hear-force, velocity, and acceleration were observed in the analysis, which can bring many risks to base-isolated structures. Therefore, it is necessary to consider base pounding in the seismic design of base-isolated structure.

Acknowledgments: The authors would like to express their gratitude to the support of Natural Science Foundation of China (51778538).

Author Contributions: Chengqing Liu proposed the simulation method and wrote the paper; Wei Yang performed the calculation and analyzed the data; Zhengxi Yan and Nan Luo helped analyzing the data; Zheng Lu conceived the idea, provided valuable discussions and revised the paper.

Conflicts of Interest: The authors declare no conflict of interest

References

1.   Lu, Z.; Wang, Z.X.; Masri, S.F.; Lu, X.L. Particle Impact Dampers: Past, Present, and Future. *Struct. Control Health Monit.* **2017**. [CrossRef]

2.   Lu, Z.; Chen, X.Y.; Zhou, Y. An equivalent method for optimization of particle tuned mass damper based on experimental parametric study. *J. Sound Vib.* **2017**. [CrossRef]

3.   Lu, Z.; Huang, B.; Zhou, Y. Theoretical study and experimental validation on the energy dissipation mechanism of particle dampers. *Struct. Control Health Monit.* **2017**. [CrossRef]

4.   Dai, K.S.; Wang, J.Z.; Mao, R.F.; Lu, Z.; Chen, S.E. Experimental investigation on dynamic characterization and seismic control performance of a TLPD system. *Struct. Des. Tall Spec. Build.* **2017**, *26*, e1350. [CrossRef]

5.   Lu, Z.; Chen, X.Y.; Lu, X.L.; Yang, Z. Shaking table test and numerical simulation of an RC frame-core tube structure for earthquake-induced collapse. *Earthq. Eng. Struct. Dyn.* **2016**, *45*, 1537–1556. [CrossRef]

6.   Kelly, J.M. A seismic base isolation: Review and bibliography. *Soil Dyn. Earthq. Eng.* **1986**, *5*, 202–216. [CrossRef]

7.   Skinner, R.I.; Robinson, W.H.; Mcverry, G.H. *An Introduction to Seismic Isolation;* Wiley: Chichester, UK; New York, NY, USA, 1993.

8.   Housner, G.; Bergman, L.A.; Caughey, T.K.; Chassiakos, A.G.; Claus, R.O.; Masri, S.F.; Skelton, R.E.; Soong, T.T.; Spencer, B.F.; Yao, J.T. Structural Control: Past, Present, and Future. *J. Eng. Mech.* **1997**, *123*, 897–971. [CrossRef]

9.   Lu, Z.; Yang, Y.L.; Lu, X.L.; Liu, C.Q. Preliminary study on the damping effect of a lateral damping buffer under a debris flow load. *Appl. Sci.* **2017**, *7*, 201. [CrossRef]

10.   Kanyilmaz, A.; Castiglioni, C.A. Reducing the seismic vulnerability of existing elevated silos by means of base isolation devices. *Eng. Struct.* **2017**, *143*, 477–497. [CrossRef]

11.   Lu, Z.; Lu, X.L.; Jiang, H.J.; Masri, S.F. Discrete element method simulation and experimental validation of particle damper system. *Eng. Comput.* **2014**, *31*, 810–823. [CrossRef]

12.   Lu, Z.; Chen, X.Y.; Zhang, D.C.; Dai, K.S. Experimental and analytical study on the performance of particle tuned mass dampers under seismic excitation. *Earthq. Eng. Struct. Dyn.* **2017**, *46*, 697–714. [CrossRef]

13.   Lu, Z.; Wang, D.C.; Masri, S.F.; Lu, X.L. An experimental study of vibration control of wind-excited high-rise buildings using particle tuned mass dampers. *Smart Struct. Syst.* **2016**, *18*, 93–115. [CrossRef]

14.   Park, S.W. Simulation of the seismic performance of the bolu viaduct subjected to near-fault ground motions. *Earthq. Eng. Struct. Dyn.* **2004**, *33*, 1249–1270. [CrossRef]

15.   Yen, W.P.; Yashinski, M.; Hashash, Y.; Holub, C. Lessons in bridge damage learned from the wenchuan earthquake. *Earthq. Eng. Eng. Vib.* **2009**, *8*, 275–285. [CrossRef]

16.   Tsai, H.C. Dynamic analysis of base-isolated shear beams bumping against stops. *Earthq. Eng. Struct. Dyn.* **1997**, *26*, 515–528. [CrossRef]

17.   Malhotra, P.K. Dynamics of seismic impacts in base isolated buildings. *Earthq. Eng. Struct. Dyn.* **1997**, *26*, 797–813. [CrossRef]

18.   Komodromos, P. Simulation of the earthquake-induced pounding of seismically isolated buildings. *Comput. Struct.* **2008**, *86*, 618–626. [CrossRef]

19. Polycarpou, P.C.; Papaloizou, L.; Mavronicola, E.; Komodromos, K.; Phocas, M.C. Earthquake induced poundings of seismically isolated buildings: The effect of the vertical location of impacts. In Proceedings of the Tenth Pan American Congress of Applied Mechanics, Cancun, Mexico, 7–11 January 2008.
20. Komodromos, P.; Polycarpou, P.C.; Papaloizou, L.; Phocas, M.C. Response of seismically isolated buildings considering poundings. *Earthq. Eng. Struct. Dyn.* **2010**, *36*, 1605–1622. [CrossRef]
21. Polycarpou, P.C.; Komodromos, P. On poundings of a seismically isolated building with adjacent structures during strong earthquakes. *Earthq. Eng. Struct. Dyn.* **2009**, *39*, 933–940. [CrossRef]
22. Polycarpou, P.C.; Komodromos, P. Simulating the use of rubber shock absorbers for mitigating poundings of seismically isolated buildings during strong earthquakes. In Proceedings of the International Conference on Computational Methods in Structural Dynamics and Earthquake Engineering, Rhodes, Greece, 22–24 June 2009; pp. 22–24.
23. Polycarpou, P.C.; Petros, K. Numerical investigation of potential mitigation measures for poundings of seismically isolated buildings. *Earthq. Struct.* **2011**, *2*, 1–24. [CrossRef]
24. Polycarpou, P.C.; Komodromos, P.; Polycarpou, A.C. A nonlinear impact model for simulating the use of rubber shock absorbers for mitigating the effects of structural pounding during earthquakes. *Earthq. Eng. Struct. Dyn.* **2013**, *42*, 81–100. [CrossRef]
25. Polycarpou, P.C.; Komodromos, P. Earthquake-induced poundings of a seismically isolated building with adjacent structures. *Eng. Struct.* **2010**, *32*, 1937–1951. [CrossRef]
26. Polycarpou, P.C.; Papaloizou, L.; Komodromos, P. An efficient methodology for simulating earthquake-induced 3D pounding of buildings. *Earthq. Eng. Struct. Dyn.* **2014**, *43*, 985–1003. [CrossRef]
27. Mavronicola, E.A.; Polycarpou, P.C.; Komodromos, P. The Effect of modified linear viscoelastic impact models on the pounding response of a base-isolated building with adjacent structures. In Proceedings of the Eccomas Thematic Conference on Computational Methods in Structural Dynamics and Earthquake Engineering, Crete Island, Greece, 25–27 May 2015.
28. Anagnostopoulos, S.A. Pounding of buildings in series during earthquakes. *Earthq. Eng. Struct. Dyn.* **1988**, *16*, 443–456. [CrossRef]
29. Pant, D.R.; Wijeyewickrema, A.C. Equivalent viscous damping for modeling inelastic impacts in earthquake pounding problems. *Earthq. Eng. Struct. Dyn.* **2004**, *33*, 897–902.
30. Fan, J.; Lin, T.; Wei, J.J. Response and protection of the impact of base-friction-isolated structures and displacement-constraint devices under near-fault earthquake. *China Civ. Eng. J.* **2007**, *5*, 10–16.
31. Ye, X.G.; Xie, Y.K.; Li, K.N. Earthquake response analysis of base-isolated structure considering side impact. *J. Earthq. Eng. Eng. Vib.* **2008**, *4*, 161–167.
32. Goldsmith, W. *Impact: The Theory and Physical Behaviour of Colliding Solids*; Edward Arnold: London, UK, 1960; pp. 182–184.
33. Lankarani, H.M.; Nikravesh, P.E. A contact force model with hysteresis damping for impact analysis of multibody systems. *J. Mech. Des.* **1990**, *112*, 369–376. [CrossRef]
34. Susendar, M.; Reginald, D.A. Hertz contact model with non-linear damping for pounding simulation. *Earthq. Eng. Struct. Dyn.* **2006**, *35*, 811–828.
35. *National Standard of People's Republic of China, Code for Seismic Design of Building*; China Architecture & Building Press: Beijing, China, 2010; GB 50011–2010.

*applied
sciences*

MDPI

Article

Development of a Self-Powered Magnetorheological Damper System for Cable Vibration Control

Zhihao Wang [1,*]**, Zhengqing Chen** [2]**, Hui Gao** [1] **and Hao Wang** [3]

[1] School of Civil Engineering and Communication, North China University of Water Resources and Electric Power, Zhengzhou 450045, China; 2587759412@qq.com
[2] Key Laboratory for Wind and Bridge Engineering of Hunan Province, Hunan University, Changsha 410082, China; zqchen@hnu.edu.cn
[3] Key Laboratory of Concrete and Prestressed Concrete Structure of Ministry of Education, Southeast University, Nanjing 210096, China; wanghao1980@seu.edu.cn
* Correspondence: wangzhihao@ncwu.edu.cn; Tel.: +86-150-9340-8299

Received: 22 November 2017; Accepted: 10 January 2018; Published: 15 January 2018

Abstract: A new self-powered magnetorheological (MR) damper control system was developed to mitigate cable vibration. The power source of the MR damper is directly harvested from vibration energy through a rotary permanent magnet direct current (DC) generator. The generator itself can also serve as an electromagnetic damper. The proposed smart passive system also incorporates a roller chain and sprocket, transforming the linear motion of the cable into the rotational motion of the DC generator. The vibration mitigation performance of the presented self-powered MR damper system was evaluated by model tests with a 21.6 m long cable. A series of free vibration tests of the cable with a passively operated MR damper with constant voltage, an electromagnetic damper alone, and a self-powered MR damper system were performed. Finally, the vibration control mechanisms of the self-powered MR damper system were investigated. The experimental results indicate that the supplemental modal damping ratios of the cable in the first four modes can be significantly enhanced by the self-powered MR damper system, demonstrating the feasibility and effectiveness of the new smart passive system. The results also show that both the self-powered MR damper and the generator are quite similar to a combination of a traditional linear viscous damper and a negative stiffness device, and the negative stiffness can enhance the mitigation efficiency against cable vibration.

Keywords: cable vibration mitigation; MR damper; vibration energy harvesting; rotary DC generator; modal damping ratio; negative stiffness

1. Introduction

Due to high flexibility, low inherent damping, and relative small mass, long stay cables are often susceptible to excessive vibrations under various environmental excitations. Large oscillations may result in undue stresses or fatigue failure in cables or connections, which is detrimental to the serviceability and safety of the entire cable-stayed bridge. Hence, it is of great importance to mitigate cable vibrations. Transversely attached dampers near the anchorage of the cable are one of the most common solutions to this problem. However, the performance of passive viscous dampers in mitigating cable vibration is greatly restricted by the small ratio of the distance from the cable anchorage to the damper over the length of the cable [1]. This may cause the supplemental damping induced by a passive damper to be insufficient to eliminate the problematic vibrations of long stay cables without significantly detracting from the aesthetics of the bridge.

As a more promising solution, semi-active control has been proposed to enhance performance since it offers the capability of active control devices without the requirement of large power resources [2]. In particular, magnetorheological (MR) dampers have attracted extensive attention from the community because of their excellent performance in both lab tests and engineering practice [3–10]. To date, MR dampers have been implemented for full-scale applications, including the stay cables on the Dongting Lake Bridge [11], Binzhou Bridge [12], and Sutong Bridge [13]. Nevertheless, these control systems need external power supplies and/or sensors/controllers, which seem to be too costly and complex.

Energy harvesting from structural vibrations is a potential solution to the power supply problem of MR dampers [14–16]. In civil engineering, electromagnetic energy harvesting shows superiority over other mechanisms [17,18], such as piezoelectricity [19], electrostatic generation [20], and dielectric elastomers [21]. Accordingly, a smart damping system, consisting of an MR damper and a linear electromagnetic generator, has been well developed and investigated by a number of researchers [22–25]. However, the energy harvesting capability of a linear generator is quite limited [26]. In view of energy harvesting efficiency, self-powered MR dampers based on rotary generators have also been proposed to enhance the energy harvesting efficiency [27,28].

In this paper, a self-powered MR damper system based on a rotary DC generator for cable vibration control is developed and experimentally investigated. It has been shown that two control devices (i.e., an MR damper and a rotary DC generator in this paper) at the same location of the stay cable will deteriorate the vibration mitigation performance [29,30]. Therefore, unlike previous compact self-powered MR dampers [23,27,28], the proposed system focuses on improving the vibration mitigation efficiency of the stay cable with little consideration to device compactness, where the MR damper and the rotary DC generator are separated rather than integrated into one compact device. In addition, a flexible chain–sprocket is selected as the linear-to-rotation conversion mechanism instead of rigid transmission mechanisms, such as the ball–screw [27] or the rack–pinion [31]. Consequently, the rotary DC generator can be installed at a higher location of the cable to provide more power to the MR damper.

The paper is organized as follows. First, a new self-powered MR damper system is proposed and constructed to suppress cable vibrations, and the energy harvesting performance of the rotary DC generator is highlighted. Next, vibration mitigation tests of a model cable attached with different kinds of dampers are conducted, including a passively operated MR damper with constant voltage, the electromagnetic damping alone due to the generator, and a self-powered MR damper system. The corresponding identified modal damping ratios in the first four modes of the cable are then compared and analyzed. The vibration mitigation mechanisms of the proposed self-powered MR damper system are also investigated.

2. Description of the Self-Powered MR Damper System for Cable Vibration Control

2.1. Configuration and Principle

The proposed self-powered MR damper system consists of a current-adjusted MR damper and an energy harvesting device. Figure 1 depicts a schematic diagram of the system used in a real environmental setting for cable vibration control. The energy harvesting device includes a rotary DC generator and a linear-to-rotational conversion mechanism. The MR damper and the generator are attached to the cable at different locations rather than being integrated into one device. To improve the energy harvesting capability, the generator is attached at a higher location relative to the MR damper. Accordingly, the roller chain–sprocket is selected as the linear-to-rotational conversion mechanism since it can be conveniently installed at a higher location through a flexible connection without detracting from the aesthetics of the bridge.

Under environmental excitation, the reciprocating linear motion of the stay cable can be converted into the rotational motion of the rotary DC generator through the linear-to-rotational conversion mechanism and a pre-tensioned spring. Accordingly, the rotary DC generator will produce electromotive force (EMF), and the induced EMF/current is then used as an input to the MR damper. Consequently, cable vibration can be mitigated by the self-powered MR damper system. It is worth noting that there are two sources of damping in the self-powered MR damper system for suppressing cable vibration. The major one is the self-powered MR damper, and the other one is the electromagnetic damping due to the generator itself.

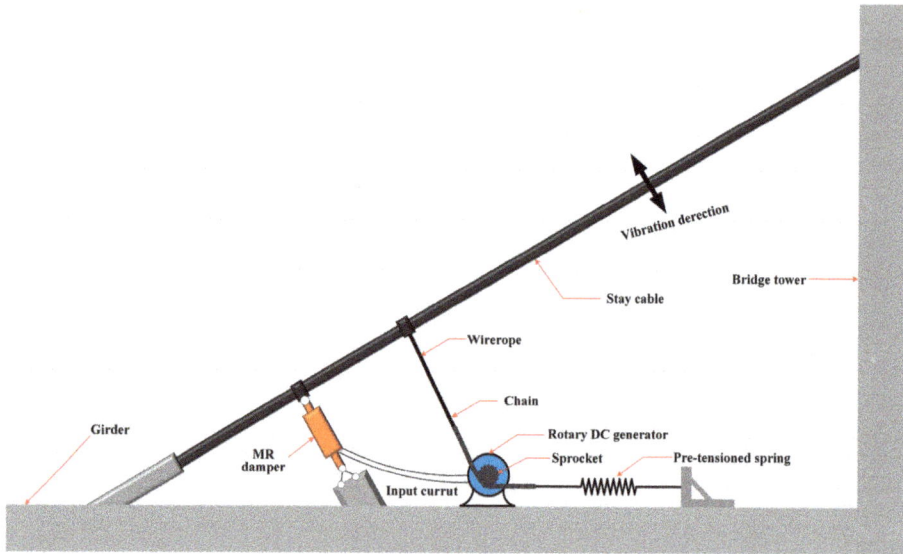

Figure 1. Schematic diagram of the self-powered MR damper system for cable vibration control.

In addition, the EMF induced by the generator is proportional to the velocity of the stay cable [32], which implies that both the supply EMF/current and the damping force will increase with the increase in cable vibration. Consequently, the proposed self-powered MR damper system is expected to be capable of adaptively attenuating cable vibration without an external power supply or any controller. Nonetheless, the system is actually passive.

2.2. Linear-to-Rotational Conversion

To ensure that the reciprocating linear motion of the stay cable can be converted into the bidirectional rotational motion of the sprocket and the rotary DC generator, a pre-tensioned spring is connected to the chain to guarantee that the chain is able to move back and forth with the cable (Figure 1). The stiffness coefficient of the spring should be suitable so that the chain can always keep straight without providing excessive extra tension force to the stay cable. In addition, the initial elongation of the spring should be larger than the predicted amplitude of the cable at the location of the generator. Thus, the linear motion of the cable is successfully translated into the rotational motion of the rotary DC generator. The relationship between the angular velocity of the generator $\dot{\theta}$ and the linear velocity of the stay cable $\dot{u}$ is given as

$$\dot{\theta} = \eta \dot{u} \tag{1}$$

where η is the transmission efficiency of the linear-to-rotational conversion, expressed as

$$\eta = \frac{1}{r}\dot{u} \tag{2}$$

where r is the effective transmission radius of the sprocket.

2.3. Model and Test of a Rotary DC Generator

The equivalent circuit model of a rotary DC generator is depicted in Figure 2. The back EMF of the generator V_g is expressed as

$$V_g = K_e \eta \dot{\theta} = K_e \dot{u} / r \tag{3}$$

where K_e denotes the back EMF constant.

According to Kirchhoff's law, the voltage in the circuit should be in balance as follows:

$$V_g = L_g \dot{I} + R_g I + R_L I \tag{4}$$

where I denotes the current flow in the circuit; L_g and R_g represent the inductance and resistance of the generator, respectively; and R_L s the load resistance in the circuit.

Applying the Laplace transform to Equations (3) and (4), the transfer function of the current I can be obtained using

$$I(s) = \frac{K_e s U(s)/r}{L_g s + R_g + R_L} \tag{5}$$

where $U(s)$ and $I(s)$ are the Laplace transform of $u(t)$ and $I(t)$, respectively.

Equations (3)–(5) indicate that the amplitude of the current flow I is proportional to the back EMF coefficient K_e and the cable velocity at the generator location (i.e., displacement amplitude and frequency of the cable), while it is inversely proportional to the effective transmission radius of the sprocket.

Consequently, the output voltage U_{output} and the output power P_{output} of the generator can be respectively given as

$$U_{output} = R_L I, \; P_{output} = R_L I^2. \tag{6}$$

Equation (6) indicates that the output power of the generator is proportional to the square of the current flow in the circuit, which mainly relates to the square of the cable's displacement amplitude and frequency at the location of the generator.

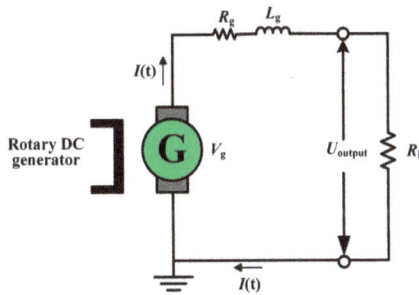

Figure 2. The equivalent circuit model for a rotary DC generator.

Experimental study on the rotary DC generator was conducted to illustrate its potential to provide the power supply for an MR damper. The rotary DC generator adopted in the test is a commercial CFX-04 speed-measuring permanent magnet DC motor, as shown in Figure 3a. The mass of the generator is 3.7 kg, and the maximum angular velocity can reach as much as $2800\pi/\text{min}$. The internal resistance of the generator is 6.4 Ω.

To acquire the back EMF constant of the generator, the shaft of the generator was forced to rotate synchronously with a servomotor at different rotational speeds, as shown in Figure 3b. Figure 4 shows the relationship between the back EMF and the rotational speed of the generator. Finally, the back EMF constant K_e of the generator was identified as 0.0594 v/(r/min).

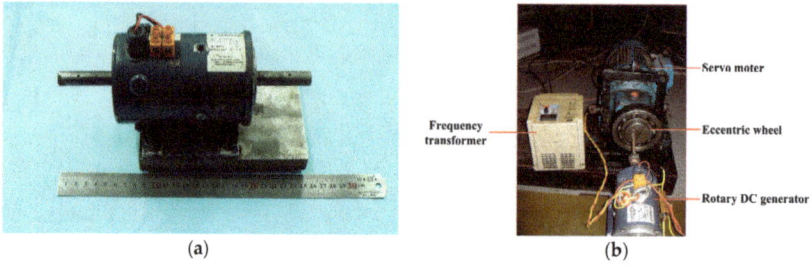

(a)   (b)

Figure 3. Experimental setup for identifying the back electromotive force constant of the generator. (a) The rotary DC generator; (b) Experimental setup.

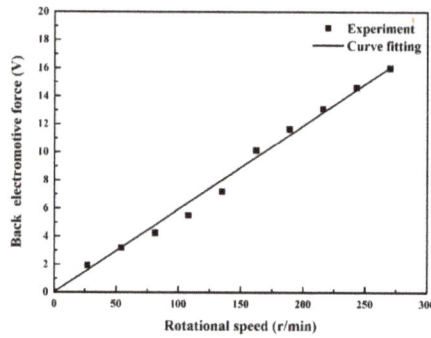

Figure 4. The relationship between the back electromotive force and the rotational speed of the generator.

Next, the chain in the linear-to-rotational conversion mechanism was subjected to sinusoidal excitations with ten different frequencies ranging from 0.5 Hz to 5 Hz at amplitudes of 5 mm and 10 mm, respectively. The external load in the circuit was 15 Ω. The output voltage of the generator was measured. A laser vibrometer was used to monitor the linear displacement of the excitation. The corresponding experimental setup is shown in Figure 5, and all the data were recorded using the Donghua Data Acquisition System.

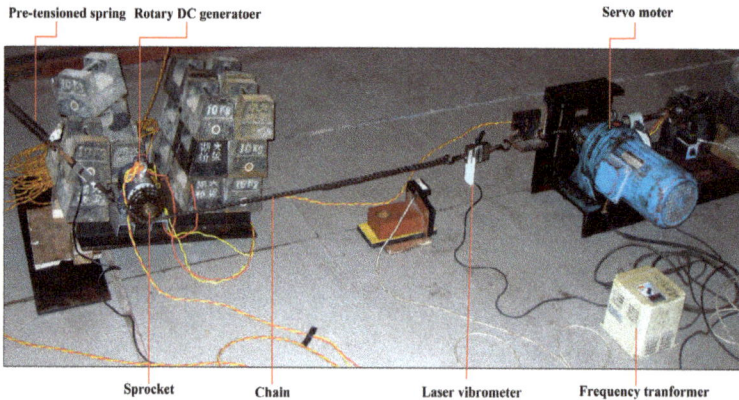

Figure 5. Experimental setup for the energy harvesting capability of the generator.

Figure 6a,b illustrate the relationship between the output voltage and the vibration frequency, and the relationship between the harvested electrical power and the vibration frequency, respectively. It is shown that both the output voltage and the output power are quite dependent on the frequency and the amplitude of the vibration source. Hence, it can be inferred that energy harvesting in civil engineering structures vibrating with low amplitudes and frequencies is quite challenging. However, the electrical energy harvested by the rotary DC generator in this paper is enough to provide a power supply to a small-scale MR damper with a low power requirement.

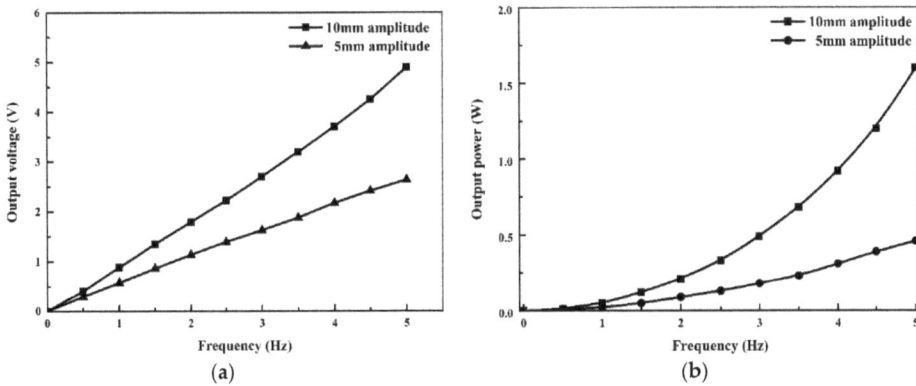

Figure 6. The output voltage and output power of the generator versus the excitation frequency ($R_L = 15\,\Omega$). (**a**) The output voltage versus the excitation frequency; (**b**) The harvested electrical energy versus the excitation frequency.

3. Experimental Setup

3.1. Experimental Platform

To evaluate the vibration control performance of the presented self-powered MR damper system, an experimental platform with a 21.6 m long scaled cable was established in the laboratory. Figure 7 illustrates the general layout of the experimental setup, while corresponding photos are shown in Figure 8. The main properties of the model cable are listed in Table 1. Lumped masses at 160 mm intervals were arranged along the cable to ensure that the model cable had similar dynamic characteristics with those on real cable-stayed bridges. The RD-1005-3 MR damper, manufactured by the Lord Corporation (Cary, NC, USA) (http:www.lord.com), was transversely attached 0.48 m (i.e., 2.2% of the cable length) from the end of the cable, incorporating a load cell and a displacement sensor to monitor the response of the MR damper. This damper type was once applied to suppress the cable vibration in the Dongting Lake Bridge [10]. The rotary DC generator was attached 1.08 m (i.e., 6% of the cable length) from the end of the cable; A laser vibrometer and a load cell were also attached to measure the linear displacement and the total force of the generator, respectively. The elastic force due to the pre-tensioned spring was also measured with a load cell. The output voltage of the DC generator applied to the MR damper was recorded. An accelerator was installed at 12.5% relative to the cable length to measure the cable acceleration, which was then adopted to identify the modal damping ratios of the cable.

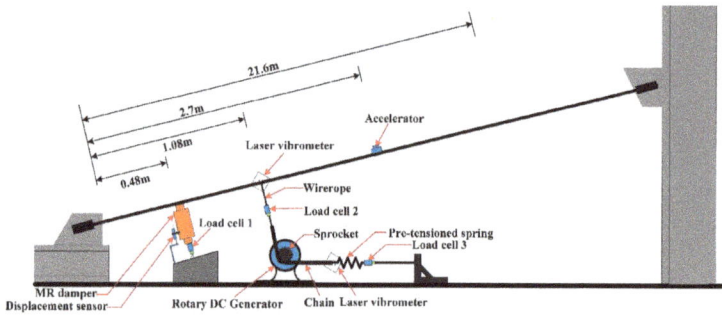

Figure 7. General layout of the experimental setup for cable vibration control.

(a)

(b)

Figure 8. Experimental setup photos for cable vibration control. (**a**) Stay cable; (**b**) Self-powered MR damper installed on the cable.

Table 1. Properties of the model cable.

Parameter	Value
Cable length (l)	21.6 m
Cable cross section area (A)	1.374 cm^2
Mass per unit length (m)	11.01 kg/m
Elastic modulus (E)	200 GPa
Static tension (T)	20.5 kN
Inclination angle (θ)	14.2°
Sag parameter (λ^2)	16.28

3.2. Experimental Procedure and Data Processing

A series of free vibration tests were conducted to assess the performance of the self-powered MR damper system in terms of reducing cable vibration. Four test cases were considered, including an uncontrolled cable, a passively operated MR damper with constant voltage, the electromagnetic damping alone generated by the rotary DC generator, and the self-powered MR damper system.

The cable was excited manually at its natural frequencies in the first four modes, respectively. The excitation rope was not released until the amplitude of the vibration reached a certain level. To get a better acceleration signal with one target mode of the cable, the excitation position for the first two modes was located at $1/4l$ away from the anchorage of the cable, but was moved to $1/10l$ away from the anchorage for the third and the fourth modes. The acceleration signals at the evaluation measurement point were filtered through the band pass filter in MATLAB. Accordingly, the free decay acceleration curve of the cable with almost one single target mode was obtained, and the natural frequencies of the cable were directly identified via the time-domain method.

Generally, the modal damping ratios of structures can be identified using the logarithmic attenuation rate of free decay responses, which works well for exponentially decaying vibrations. However, the modal damping ratios of the cable may be dependent on the amplitude of the cable when the cable is equipped with nonlinear control devices, such as the passively operated MR dampers. The induced problem of amplitude-dependent damping ratio values when a nonlinear damper is attached to the cable has been well described in [33,34]. Nevertheless, except for the case of a passively operated MR damper with constant voltage, cable vibrations decay exponentially. Therefore, a uniform method to identify the modal damping ratios of a cable to which different control devices are attached needs to be defined to compare control performance. In this paper, the logarithmic decrement method was finally adopted, and the first 20 cycles from the same amplitude of free decay acceleration of the cable at the evaluation measurement point were selected to form an estimate of the damping ratios of the cable for each mode. The natural frequencies and modal damping ratios of an uncontrolled cable in the first four modes were identified, and are listed in Table 2.

Table 2. Natural frequencies and modal damping ratios in the first four modes of an uncontrolled cable.

Mode Order	Natural Frequency (Hz)	Modal Damping Ratio (%)
1	1.434	0.59
2	1.969	0.15
3	2.983	0.11
4	3.912	0.12

4. Experimental Results and Discussion

4.1. The Stay Cable with a Passively Operated MR Damper

Figure 9 shows the modal damping ratios in the first four modes of the test cable controlled by a passively operated MR damper with constant voltage. As expected, there is an optimum voltage corresponding to a maximum modal damping ratio for each mode of the cable. Generally, lower voltage input seems to be better for a higher-order mode except for the first mode of the cable. In this study, 0.5 V was finally selected as the optimal constant voltage for the passively operated MR damper since it can perform quite well for all of the first four modes of the cable.

Taking the second mode of the cable as an example, the entire time histories of accelerations at the observation point of the cable with a passively operated MR damper attached are shown in Figure 10. It can be seen that acceleration amplitudes continuously increase with steady external excitation at the first stage, and then decay after the release of excitation. Figure 10 illustrates that the MR damper operated in passive-on mode (i.e., 1 V voltage) can only work well when the amplitude of the cable is relatively large. When the amplitude of the cable falls below one threshold, the MR damper may be locked and lose its energy dissipation capacity. As clearly shown in Figure 10c, the supplemental modal damping due to the MR damper gets quite small after 20 s. Therefore, the performance with a passive-on operated MR damper will be substantially reduced for the case of cable vibration with low amplitude. For the passive-off mode (i.e., 0 V voltage) of the MR damper, the acceleration amplitudes decay exponentially with a low damping ratio in all stages of the vibration, demonstrating amplitude-independent damping behavior. Hence, the MR damper operated in the optimal passive mode (i.e., 0.5 V voltage) may be a better choice as it has benefits from both the passive-on and passive-off modes.

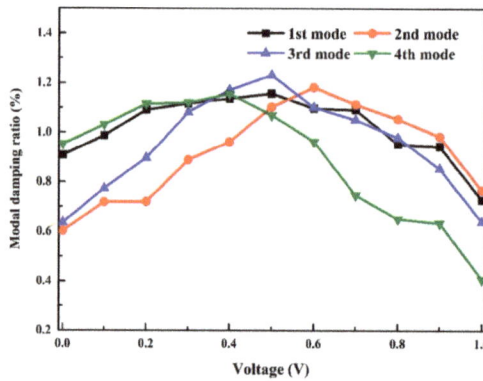

Figure 9. Damping ratios in the first four modes of the cable versus constant voltage of the MR damper.

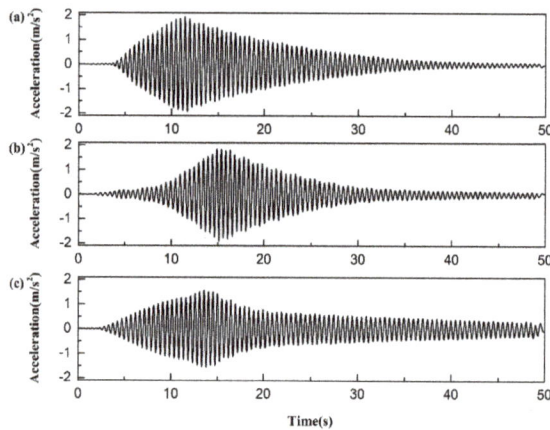

Figure 10. Second mode acceleration time histories of the cable controlled by a passive MR damper with different constant voltages: (a) 0 V; (b) 0.5 V; and (c) 1 V.

4.2. The Stay Cable with Electromagnetic Damping Alone

To investigate the effect of electromagnetic damping alone on the vibration mitigation performance of the cable, two circuits of the generator were considered. One is a short circuit, and the other is loaded with an MR damper. However, the MR damper herein is not attached to the cable. The main difference between them is that the resulting equivalent damping coefficients for the case of the short circuit are larger than those for the other case.

The test results are summarized in Table 3, where the supplemental modal damping ratios are derived from the subtraction between the identified value with the electromagnetic damping and the uncontrolled case. It is noted that the supplemental damping ratio of the cable controlled by the electromagnetic damping alone is pretty limited, and the supplemental modal damping ratios of the same mode in the short circuit case are always larger than those in the circuit loaded with the MR damper. In addition, the supplemental modal damping ratio continuously increases with the increase of the mode order for both of the circuits. However, the supplemental modal damping ratio of the cable with the electromagnetic damping alone is much smaller than the theoretical maximum damping ratio induced by an ideal linear viscous damper (3%). This is mainly because the damping coefficient of the generator is much smaller than the corresponding optimum value. Therefore, the test results above imply that the generator can serve as an alternative damper for cable vibration control, and its efficiency can be further enhanced through optimum design of the generator system.

Table 3. Modal damping ratios of the cable in the first four modes with electromagnetic damping alone.

Mode Order	Short Circuit		Circuit Loaded with an MR Damper	
	Damping Ratio (%)	Supplemental Damping Ratio (%)	Damping Ratio (%)	Supplemental Damping Ratio (%)
1	0.71	0.12	0.68	0.09
2	0.59	0.44	0.54	0.39
3	0.81	0.70	0.65	0.54
4	1.05	0.93	0.74	0.62

The relationship between the damping force and the linear displacement of the generator is shown in Figure 11. The damping characteristics of the generator are very similar to those of a combination of a traditional linear viscous damper and a negative stiffness device in the higher-order mode of the cable. It is noteworthy that the negative stiffness component of the generator is not visible in the 1st mode, which is mainly due to the fact that the negative stiffness of the generator for the 1st mode of the cable is smaller than its internal stiffness.

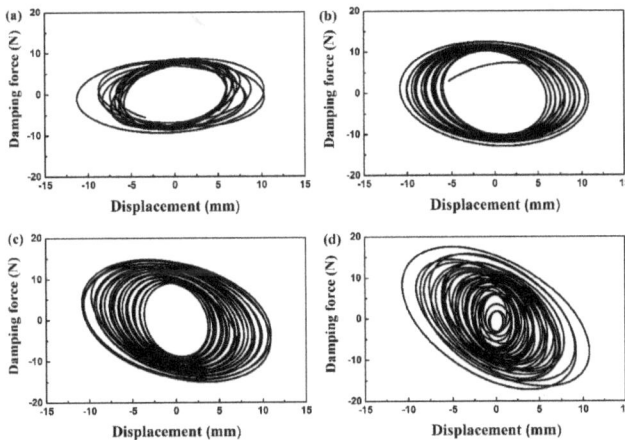

Figure 11. The damping force versus the linear displacement of the DC generator. (**a**) First mode; (**b**) Second mode; (**c**) Third mode; and (**d**) Fourth mode.

In fact, the total force F_g of the rotary DC generator can be divided into the inertial force F_I and the electromagnetic damping force F_d [35]. This relationship can be expressed as

$$F_g = F_I + F_d. \tag{7}$$

When the rotary DC generator is applied to mitigate cable vibration, we have

$$F_I = k_g u = -m_e \omega_i^2 u, \ F_d = c_{eq} \dot{u} \tag{8}$$

where k_g, m_e, and c_{eq} represent the equivalent stiffness, the inertial mass, and the damping coefficient of the generator, respectively; u and ω_i denote the displacement and the natural frequency of the cable in the ith-order mode, respectively. The inertial mass is generated from the rotary part (i.e., rotor) of the generator.

Equation (8) indicates that the negative stiffness of the rotary DC generator is proportional to the natural frequency of the cable, demonstrating frequency-dependent behavior. In other words, a higher mode of the cable corresponds to a larger negative stiffness. Accordingly, experimentally identified equivalent damping coefficients and stiffness coefficients are shown in Figure 12. As expected, the damping coefficient almost remains a constant with the increase of the mode order of the cable for each circuit, while the negative stiffness increases with the increase of the mode order. However, the optimum negative stiffness for a stay cable does not depend on the mode number [36,37]. Hence, the DC generator with frequency-dependent negative stiffness may be better than viscous damping only, but is still suboptimal when compared with optimum viscous damping with negative stiffness for cable vibration control.

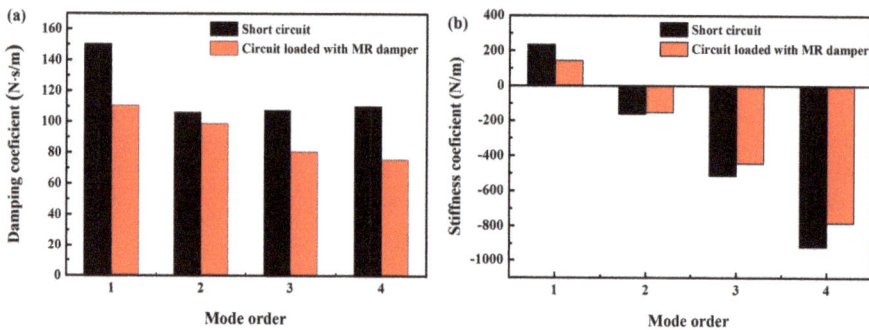

Figure 12. Identified damping coefficients and stiffness coefficients of the DC generator: (**a**) Identified damping coefficients; (**b**) Identified stiffness coefficients.

4.3. The Stay Cable with a Self-Powered MR Damper System

Figures 13 and 14 give the time histories of the acceleration of the cable, the input voltage, and the damping force of the self-powered MR damper in the first and fourth modes, respectively. As expected, the voltage induced by the generator can successfully make the MR damper operate normally, and both the input voltage and damping force increase with the increase of mode order. The shapes of the curves for the acceleration, the input voltage, and the damping force are quite similar. It can be seen that the acceleration amplitudes of the cable decay exponentially, indicating that the damping force of the self-powered MR damper is proportional to the collocated velocity of the cable. In other words, the self-powered MR damper works as a linear viscous damper, and velocity feedback control can be achieved inherently. Moreover, the self-powered MR damper system is always dissipative even at small amplitudes of the cable, in contrast to the passively operated MR damper. In addition, the output voltage signal of the generator seems to be accurate enough to sense the velocity of the cable at the generator's location, demonstrating the good potential of the generator as a velocity sensor for the semi-active control of a self-powered MR damper system.

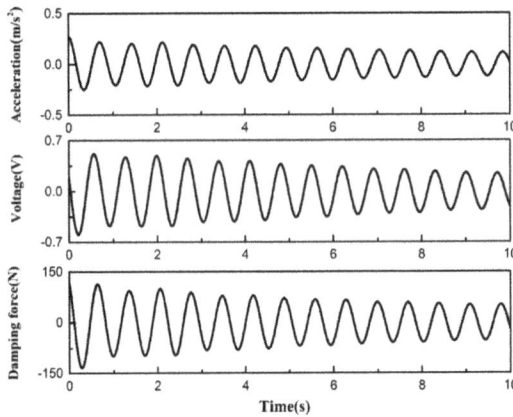

Figure 13. Time histories of acceleration, input voltage, and damping force for the first mode of the cable.

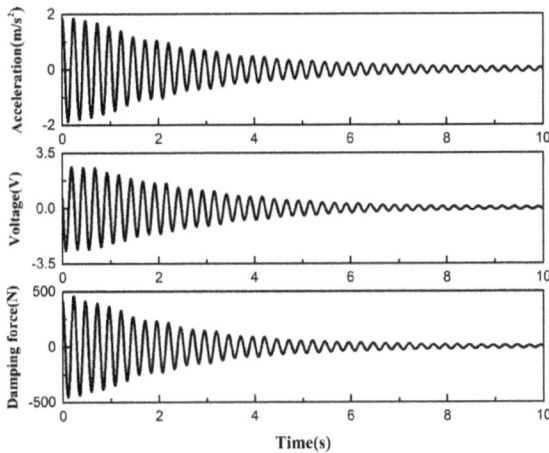

Figure 14. Time histories of acceleration, input voltage, and damping force for the fourth mode of the cable.

The damping ratios in the first four modes of the cable attached with different kinds of dampers are summarized in Table 4. Except for the first mode, the self-powered MR damper system performs the best among the three control strategies for the left three higher-order modes. The poor performance of the self-powered MR damper in the first mode is mainly due to the sag effect of the model cable [38]. As shown in Table 1, the sag parameter (λ^2) of the model cable is 16.28, which is much bigger than practical stay cables. Another reason is that the electrical energy harvested from the cable in the first mode may be too small to have a significant influence on the damping force of the MR damper. The control performance of the system can be further enhanced by decreasing the effective transmission radius of the sprocket or moving the rotary DC generator to a higher location. However, the main purpose of this paper is to demonstrate the feasibility of the self-powered MR damper system in reducing cable vibration, and the system may not show the best performance in the test. Hence, the optimum design of the system should be considered in later research, including the relative locations of the generator and the MR damper, the sizes and parameters of both the generator and the MR damper, and the effective transmission radius of the sprocket.

Table 4. Comparisons of modal damping ratios of the cable with different control devices.

Mode Order	Damping Ratio (%)			
	Uncontrolled	MR Damper in Passive Mode	Electromagnetic Damping Alone (Load MR Damper)	Self-Powered MR Damper System
1	0.59	1.16	0.68	0.87
2	0.15	1.10	0.54	1.45
3	0.11	1.23	0.65	1.69
4	0.12	1.07	0.74	1.53

The hysteretic loops of the self-powered MR damper are shown in Figure 15. The damping characteristics of the self-powered MR damper are quite similar to a combination of a traditional linear viscous damper and a negative stiffness device in the higher mode. Accordingly, the equivalent damping and stiffness coefficients are identified and shown in Figure 16. The equivalent damping coefficients are almost the same for the first four modes, while the stiffness coefficient jumps from positive to negative values after the second mode. Moreover, the negative stiffness coefficient increases with an increase in the mode order of the cable. Hence, the self-powered MR damper is found to be similar to the generator with respect to the negative stiffness characteristics.

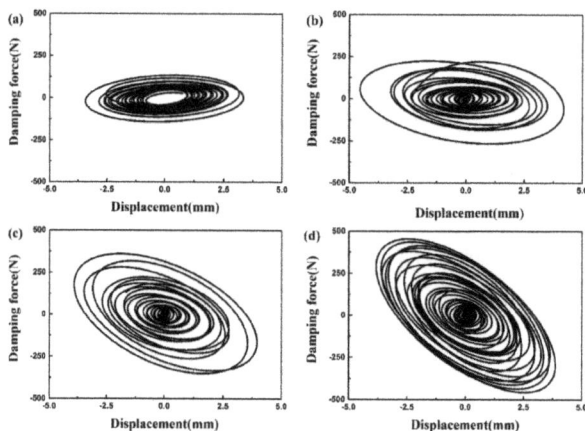

Figure 15. Hysteretic loops of the self-powered MR damper for different modes of the cable. (**a**) First mode; (**b**) Second mode; (**c**) Third mode; and (**d**) Fourth mode.

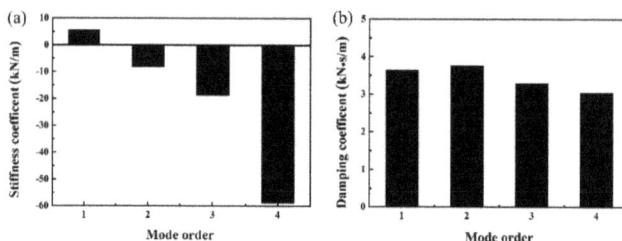

Figure 16. Identified damping coefficients and stiffness coefficients of the self-powered MR damper. (**a**) Identified stiffness coefficients; (**b**) Identified damping coefficients.

Unlike the self-powered MR damper system in this paper, other mentioned self-powered MR dampers, where the MR damper and the generator are at the same location, have not exhibited negative stiffness behavior [27]. The results imply that the negative stiffness may be attributed to the relative positions of the rotary DC generator and the damper in a self-powered MR damper system. Hence, the proposed self-powered MR damper system, where the generator and the MR damper are separated, seems to be more promising than the conventional, compact self-powered MR damper. However, the effects of the relative positions of the rotary DC generator and the damper on the induced negative stiffness of the self-powered MR damper and control performance of the cable still require further investigation.

4.4. Effect of Negative Stiffness of the Damper on the Performance of the Cable

Negative stiffness characteristics of semi-active/active control systems have been found and investigated in the area of structural vibration control [39–42]. The role of negative stiffness in the semi-active control of magnetorheological dampers has been well demonstrated by Høgsberg [43]. According to the research results above, it has been confirmed that the negative stiffness is capable of improving the energy dissipation ability of a conventional damper. In view of such benefits, several passive negative stiffness dampers were also proposed to enhance the vibration control performance of the cable, such as a viscous damper with a negative magnetic stiffness spring [44], an oil damper with two pre-compressed springs [45], and a viscous inertial mass damper [46].

When a negative stiffness viscous damper is quite close to the cable end, the supplemental modal damping ratio of the stay cable ζ_i can be expressed as [36,40]

$$\zeta_i \approx \frac{\eta}{1+\eta^2} \frac{a/l}{1+\bar{k}} \tag{9}$$

where $\bar{k}$ and η respectively denote the dimensionless negative stiffness and the dimensionless damping coefficient, which are given by

$$\bar{k} = ka/T, \ \eta = \frac{c}{\sqrt{Tm}} \frac{i\pi a/l}{1+\bar{k}} \tag{10}$$

where a, k, and c represent the location, the negative stiffness, and the damping coefficient of the damper, respectively; T, m, l, and i denote the tension force, the mass per unit length, the length, and the mode order of the cable, respectively.

When the optimum damping coefficient c^{opt} is determined by

$$c^{\text{opt}} = \frac{T}{a\omega_i}(1+\bar{k}), \tag{11}$$

the maximum modal damping ratio of the cable $\zeta_i^{\max}$ can be obtained as

$$\zeta_i^{\max} = \frac{1}{1+\bar{k}} \frac{a}{2l}. \tag{12}$$

Equation (12) clearly indicates that the incorporation of negative stiffness into a linear viscous damper can enhance the maximum modal damping ratio of the cable. For the new self-powered MR damper system in this paper, the test results shown in Sections 4.2 and 4.3 have demonstrated that both the self-powered MR damper and the generator are able to exhibit negative stiffness behavior for the higher-order mode of the cable. Additionally, the negative stiffness increases as the mode order of the test cable increases. In summary, the proposed self-powered MR damper control system shows superiority over traditional passive MR dampers or viscous dampers.

5. Conclusions

To eliminate the power supply problem of a traditional MR damper, a new self-powered MR damper control system was developed based on electromagnetic energy harvesting technology. Unlike other self-powered MR dampers, the MR damper and therotary DC generator are not at the same location on the stay cable in this paper. The feasibility and effectiveness of the new system for suppressing cable vibration was experimentally evaluated via a 21.6 m inclined cable model. The test results indicate that the developed self-powered MR damper system can provide more cable damping than a passively operated MR damper with constant voltage. The increased supplemental modal damping ratios of the cable in the first four modes are mainly attributed to two factors. One is that the self-powered MR damper can exhibit negative stiffness behavior for the case of the higher-order mode of the cable, which is capable of enhancing the control performance; the other is that the rotary DC generator in the self-powered MR damper system also provides supplemental damping to the cable.

Since the main purpose of this paper is merely to demonstrate the feasibility of the self-powered MR damper system in reducing cable vibration, the system may not show the best possible performance in the test. Hence, the optimum design of the system should be considered in later research, including the relative locations of the generator and the MR damper, the sizes and parameters of both the rotary DC generator and the MR damper, and the effective transmission radius of the sprocket.

Acknowledgments: The authors gratefully acknowledge the financial support from the National Natural Science Foundation of China (Grant No. 51308214 and 51578151) and the National Basic Research Program of China (973 Program) (Grant No. 2015CB057702 and 2015CB060000).

Author Contributions: Zhihao Wang and Zhengqing Chen conceived the idea and designed the experiments; Zhihao Wang performed the experiment and analyzed the data; Zhihao Wang and Hui Gao wrote the paper; Zhengqing Chen and Hao Wang reviewed and revised the paper.

Conflicts of Interest: The authors declare no conflict of interest.

References

1. Krenk, S. Vibration of a taut cable with an external damper. *J. Appl. Mech.* **2000**, *67*, 772–776. [CrossRef]
2. Johnson, E.A.; Christenson, R.E.; Spencer, B.F. Semi-active damping of cables with sag. *Comput.-Aided Civ. Infrastruct. Eng.* **2003**, *18*, 132–146. [CrossRef]
3. Spencer, B.F.; Dyke, S.J.; Sain, M.K.; Carlson, J.D. Phenomenological model of a magnetorheological damper. *J. Eng. Mech.* **1997**, *123*, 230–238. [CrossRef]
4. Duan, Y.F.; Ni, Y.Q.; Ko, J.M. Cable vibration control using magneto-rheological (MR) dampers. *J. Intell. Mater. Syst. Struct.* **2006**, *17*, 321–325. [CrossRef]
5. Christenson, R.E.; Spencer, B.F.; Johnson, E.A. Experimental verification of smart cable damping. *J. Eng. Mech.* **2006**, *132*, 268–278. [CrossRef]
6. Liu, M.; Song, G.B.; Li, H. Non-model-based semi-active vibration suppression of stay cables using magneto-rheological fluid damper. *Smart Mater. Struct.* **2007**, *16*, 1447–1452. [CrossRef]
7. Cai, C.S.; Wu, W.J.; Araujo, M. Cable vibration control with a TMD-MR damper system: Experimental exploration. *J. Struct. Eng.* **2007**, *133*, 629–637. [CrossRef]
8. Huang, H.W.; Sun, L.M.; Jiang, X.L. Vibration mitigation of stay cable using optimally tuned MR damper. *Smart Struct. Syst.* **2012**, *9*, 35–53. [CrossRef]
9. Chang, C.M.; Wang, Z.H.; Spencer, B.F.; Chen, Z.Q. Semi-active damped outriggers for seismic protection of high-rise buildings. *Smart Struct. Syst.* **2013**, *11*, 435–451. [CrossRef]
10. Fu, W.Q.; Zhang, C.W.; Sun, L.; Askari, M.; Samali, B.; Chung, K.L.; Sharafi, P. Experimental investigation of a base isolation system incorporating MR dampers with the high-order single step control algorithm. *Appl. Sci.* **2017**, *7*, 344. [CrossRef]
11. Chen, Z.Q.; Wang, X.Y.; Ko, J.M.; Ni, Y.Q.; Spencer, B.F.; Yang, G.; Hu, J.H. MR damping system for mitigating wind-rain induced vibration on Dongting Lake Cable-Stayed Bridge. *Wind Struct.* **2004**, *7*, 293–304. [CrossRef]

12. Li, H.; Liu, M.; Li, J.H.; Guan, X.C.; Ou, J.P. Vibration control of stay cables of Shandong Binzhou Yellow River Highway Bridge by using magnetorheological fluid dampers. *J. Bridge Eng.* **2007**, *12*, 401–409. [CrossRef]

13. Weber, F.; Distl, H. Amplitude and frequency independent cable damping of Sutong Bridge and Russky Bridge by magnetorheological dampers. *Struct. Control Health Monit.* **2014**, *22*, 237–254. [CrossRef]

14. Scruggs, J.T.; Iwan, W.D. Control of a civil structure using an electric machine with semi-active capability. *J. Struct. Eng.* **2003**, *129*, 951–959. [CrossRef]

15. Wang, Z.H.; Chen, Z.Q.; Spencer, B.F. Self-powered and sensing control system based on MR damper, presentation and application. In Proceedings of the SPIE on Sensors and Smart Structures Technologies for Civil, Mechanical, and Aerospace Systems, San Diego, CA, USA, 8–13 March 2009.

16. Cho, S.W.; Jung, H.J.; Lee, I.W. Smart passive system based on magnetorheological damper. *Smart Mater. Struct.* **2005**, *14*, 707–714. [CrossRef]

17. Cassidy, I.L.; Scruggs, J.T.; Behrens, S.; Gavin, H.P. Design and experimental characterization of an electromagnetic transducer for large-scale vibratory energy harvesting applications. *J. Intell. Mater. Syst. Struct.* **2011**, *22*, 2009–2024. [CrossRef]

18. Shen, W.A.; Zhu, S.Y.; Zhu, H.P. Experimental study on using electromagnetic devices on bridge stay cables for simultaneous energy harvesting and vibration damping. *Smart Mater. Struct.* **2016**, *25*, 065011. [CrossRef]

19. Chin, W.K.; Ong, Z.C.; Kong, K.K.; Khoo, S.Y.; Huang, Y.H.; Chong, W.T. Enhancement of energy harvesting performance by a coupled bluff splitter body and PVEH plate through vortex induced vibration near resonance. *Appl. Sci.* **2017**, *7*, 921. [CrossRef]

20. Mitcheson, P.D.; Miao, P.; Stark, B.H.; Yeatman, E.M.; Holmes, A.S.; Green, T.C. MEMS electrostatic micropower generator for low frequency operation. *Sens. Actuators A-Phys.* **2004**, *115*, 523–529. [CrossRef]

21. Kang, G.; Kim, K.S.; Kim, S. Note: Analysis of the efficiency of a dielectric elastomer generator for energy harvesting. *Rev. Sci. Instrum.* **2011**, *82*, 046101. [CrossRef] [PubMed]

22. Choi, K.M.; Jung, H.J.; Lee, H.J.; Cho, H.W. Feasibility study of an MR damper-based smart passive control system employing an electromagnetic induction device. *Smart Mater. Struct.* **2007**, *16*, 2323–2329. [CrossRef]

23. Choi, Y.T.; Wereley, N.M. Self-powered magnetorheological dampers. *J. Vib. Acoust.* **2009**, *131*, 044501. [CrossRef]

24. Kim, I.H.; Jung, H.J.; Koo, J.H. Experimental evaluation of a self-powered smart damping system in reducing vibration of a full-scale stay cable. *Smart Mater. Struct.* **2010**, *19*, 115027. [CrossRef]

25. Sapiński, B. Energy-harvesting linear MR damper: Prototyping and testing. *Smart Mater. Struct.* **2014**, *23*, 035021. [CrossRef]

26. Gupta, A.; Jendrzejczyk, J.A.; Mulcahy, T.M.; Hull, J.R. Design of electromagnetic shock absorbers. *Int. J. Mech. Mater. Des.* **2006**, *3*, 285–291. [CrossRef]

27. Guan, X.C.; Huang, Y.H.; Ru, Y.; Li, H.; Ou, J.P. A novel self-powered MR damper: Theoretical and experimental analysis. *Smart Mater. Struct.* **2015**, *24*, 105033.

28. Sapiński, B. Experimental investigation of an energy harvesting rotary generator-MR damper System. *J. Theor. Appl. Mech-Pol.* **2016**, *54*, 679–690. [CrossRef]

29. Caracoglia, L.; Jones, N.P. Damping of taut-cable Systems: Two dampers on a single stay. *J. Eng. Mech.* **2007**, *133*, 1050–1060. [CrossRef]

30. Cu, V.H.; Han, B.; Wang, F. Damping of a taut cable with two attached high damping rubber dampers. *Struct. Eng. Mech.* **2015**, *55*, 1261–1278. [CrossRef]

31. Shen, W.A.; Zhu, S.Y. Harvesting energy via electromagnetic damper: Application to bridge stay cables. *J. Intell. Mater. Syst. Struct.* **2015**, *26*, 3–19. [CrossRef]

32. Jung, H.J.; Jang, D.D.; Koo, J.H.; Cho, S.W. Experimental evaluation of a 'self-sensing' capability of an electromagnetic induction system designed for MR dampers. *J. Intell. Mater. Syst. Struct.* **2010**, *21*, 827–835. [CrossRef]

33. Occhiuzzi, A.; Spizzuoco, M.; Serino, G. Experimental analysis of magnetorheological dampers for structural control. *Smart Mater. Struct.* **2003**, *12*, 703–711. [CrossRef]

34. Weber, F.; Høgsberg, J.; Krenk, S. Optimal tuning of amplitude proportional coulomb friction damper for maximum cable damping. *J. Struct. Eng.* **2010**, *136*, 123–134. [CrossRef]

35. Ikago, K.; Saito, K.; Inoue, N. Seismic control of single-degree-of-freedom structure using tuned viscous mass damper. *Earthq. Eng. Struct. Dyn.* **2012**, *41*, 453–474. [CrossRef]

36. Chen, L.; Sun, L.M.; Nagarajaiah, S. Cable with discrete negative stiffness device and viscous damper: Passive realization and general characteristics. *Smart Struct. Syst.* **2015**, *15*, 627–643. [CrossRef]
37. Weber, F.; Distl, H. Semi-active damping with negative stiffness for multi-mode cable vibration mitigation: Approximate collocated control solution. *Smart Mater. Struct.* **2015**, *24*, 115015. [CrossRef]
38. Xu, Y.L.; Yu, Z. Vibration of inclined sag cables with oil dampers in cable-stayed bridges. *J. Bridge Eng.* **1998**, *3*, 194–203. [CrossRef]
39. Iemura, H.; Pradono, M.H. Application of pseudo negative stiffness control to the benchmark cable-stayed bridges. *J. Struct. Control* **2003**, *10*, 187–203. [CrossRef]
40. Li, H.; Liu, M.; Ou, J.P. Negative stiffness characteristics of active and semi-active control systems for stay cables. *Struct. Control Health Monit.* **2008**, *15*, 120–142. [CrossRef]
41. Pradono, M.H.; Iemura, H.; Igarashi, A.; Toyooka, A.; Kalantari, A. Passively controlled MR damper in the benchmark structural control problem for seismically excited highway bridge. *Struct. Control Health Monit.* **2009**, *16*, 626–638. [CrossRef]
42. Weber, F.; Boston, C. Clipped viscous damping with negative stiffness for semi-active cable damping. *Smart Mater. Struct.* **2011**, *20*, 045007. [CrossRef]
43. Høgsberg, J. The role of negative stiffness in semi-active control of magneto-rheological dampers. *Struct. Control Health Monit.* **2011**, *18*, 289–304. [CrossRef]
44. Shi, X.; Zhu, S.Y.; Li, J.Y.; Spencer, B.F. Dynamic behavior of stay cables with passive negative stiffness dampers. *Smart Mater. Struct.* **2016**, *25*, 075044. [CrossRef]
45. Zhou, P.; Li, H. Modeling and control performance of a negative stiffness damper for suppressing stay cable vibrations. *Struct. Control Health Monit.* **2016**, *23*, 764–782. [CrossRef]
46. Lu, L.; Duan, Y.F.; Spencer, B.F.; Lu, X.L.; Zhou, Y. Inertial mass damper for mitigating cable vibration. *Struct. Control Health Monit.* **2017**, *24*, e1986. [CrossRef]

applied
sciences

MDPI

Article

Application of an Artificial Fish Swarm Algorithm in an Optimum Tuned Mass Damper Design for a Pedestrian Bridge

Weixing Shi [1], Liangkun Wang [1], Zheng Lu [1,2,*] and Quanwu Zhang [3]

[1] Research Institute of Structural Engineering and Disaster Reduction, Tongji University, Shanghai 200092, China; swxtgk@tongji.edu.cn (W.S.); 1630634@tongji.edu.cn (L.W.)
[2] State Key Laboratory of Disaster Reduction in Civil Engineering, Tongji University, Shanghai 200092, China
[3] Shanghai Zhili Vibration Technology Co., Ltd., Shanghai 200092, China; 2151xanxus@tongji.edu.cn
* Correspondence: luzheng111@tongji.edu.cn; Tel: +86-21-6598-6186

Received: 26 December 2017; Accepted: 24 January 2018; Published: 25 January 2018

Abstract: Tuned mass damper (TMD) has a wide application in the human-induced vibration control of pedestrian bridges and its parameters have great influence on the control effects, hence it should be well designed. A new optimization method for a TMD system is proposed in this paper, based on the artificial fish swarm algorithm (AFSA), and the primary structural damping is taken into consideration. The optimization goal is to minimize the maximum dynamic amplification factor of the primary structure under external harmonic excitations. As a result, the optimized TMD has a smaller maximum dynamic amplification factor and better robustness. The optimum TMD parameters for a damped primary structure with different damping ratios and different TMD mass ratios are summarized in a table for simple, practical design, and the fitting equation is also provided. The TMD configuration optimized by the proposed method was shown to be superior to that optimized by other classical optimization methods. Finally, the application of an optimized TMD based on AFSA for a pedestrian bridge is proposed as a case study. The results show that the TMD designed based on AFSA has a smaller maximum dynamic amplification factor than the TMD designed based on the classic Den Hartog method and the TMD designed based on the Ioi Toshihiro method, and the optimized TMD has a good effect in controlling human-induced vibrations at different frequencies.

Keywords: tuned mass damper; human-induced vibrations; pedestrian bridge; artificial fish swarm algorithm; passive control

1. Introduction

Pedestrian bridges are more and more common in urban regions. However, with the development of architectural creativeness and structural technologies, they are becoming lighter and more slender, which might lead to serviceability problems under human-induced vibrations [1]. These pedestrian bridges usually have low inherent damping, and large vibrations may be caused by resonance because their natural frequency is usually located within the range of pedestrians' stride frequency. A notable vibration will not only cause serviceability problems but also threaten structural safety.

Tuned mass dampers (TMD) are one of the most traditional vibration control devices [2–6], and it usually consists of a mass, some springs and damping elements [7]. Because TMDs have the advantages of small size and little interference to pedestrian bridges, they are widely used to control human-induced vibrations and solve the serviceability problem [8–10]. The vibration response of a pedestrian bridge is controlled by the inertia force given by the TMD, and the vibration energy of the TMD will be dissipated through its damping element [11–13]. TMD is a single degree of freedom vibration absorber, the three dynamic parameters of TMD, which are mass, stiffness and damping coefficient, have a great influence on its control effect.

TMDs have been researched and used to solve the serviceability problem of pedestrian bridges by many researchers. Casciati et al. [14] proposed a contribution to the modelling of pedestrian-induced excitation of footbridges. Brownjohn et al. [15] used human-induced dynamic excitation to estimate modal mass in pedestrian bridges. Nakhorn et al. [16] proposed the application of non-linear multiple TMDs to suppress human-induced vibrations of a footbridge. Chen et al. [17] introduced the performance enhancement of a long-span bridge using tune mass dampers. Carpineto et al. [18] proposed the application of multiple TMDs to control human-induced vibrations in suspension footbridges. Lu et al. [19] studied the application of TMD to control pedestrian-induced vibration of the Expo Culture Centre. Li et al. [20] studied the pedestrian-induced random vibration of pedestrian bridges and vibration control methods using multiple TMDs. Lievens et al. [21] introduced the robust design and application of a TMD in a footbridge. Caetano et al. [22] studied controlling pedestrian-induced vibrations of a footbridge using tuned mass dampers. Casciati et al. [23] introduced the vibration control effect of multiple TMDs in the towers of bridges.

The mass ratio and damping ratio of a TMD will affect its vibration control effect, and TMD is especially sensitive to the frequency ratio [24–28]. Therefore, the three dynamic parameters of TMD should be well designed. The classic Den Hartog method [29] has a wide application in TMD design. Besides, many new design methods are proposed. In order to search the optimum parameters of TMD for different optimization goals, a natured-inspired computational algorithm has been used for parameters optimization of TMDs by many researchers. Leung et al. [30,31] proposed the particle swarm optimization of TMDs by non-stationary base excitation during earthquake. Bekdas et al. [32] introduced the estimating optimum parameters of tuned mass dampers using harmony search. Mohebbi et al. [33] designed optimal multiple tuned mass dampers using genetic algorithms (GAs) to mitigate the seismic response of structures. Jiménez-Alonso et al. [34] introduced the robust optimum design of TMDs to mitigate pedestrian-induced vibrations using multi-objective genetic algorithms. Therefore, the natured-inspired algorithms play an important role in TMD parameter optimization. Considering the advantages of flexibility, great convergence speed, great accuracy, fault tolerance, etc. [35,36], the artificial fish swarm algorithm (AFSA) is applied to optimize TMD parameters in controlling the human-induced vibrations of footbridges.

The maximum acceleration of a pedestrian bridge under human-induced vibrations is one of the most important indexes to evaluate the serviceability problem [1]. The classic Den Hartog method [29] has a wide application in TMD design; however, it does not consider the damping of the primary system, which may lead to a defective TMD. Currently, practical application analytical proposals that overcome some of the limitations of the Den Hartog proposal, are applied. Ioi and Ikeda [37,38] proposed an improved design method considering structural damping, whose optimization goal is to minimize the maximum acceleration dynamic amplification factor of the primary structure under external harmonic excitations. Butz et al. [39] proposed advanced load models for synchronous pedestrian excitation and optimized design guidelines for steel footbridges. Van et al. [40] introduced the numerical and experimental evaluation of the dynamic performance of a footbridge with tuned mass dampers. Asami et al. [41] studied the analytical solutions to H_∞ and H_2 optimization of dynamic vibration absorbers attached to damped linear systems. In order to depress the maximum acceleration dynamic amplification factor of the primary structure under external excitation and strengthen the robustness of TMD, a novel optimization method based on the artificial fish swarm algorithm (AFSA) is proposed in this paper, considering the structural damping ratio. The optimization goal is the same as the aforementioned.

In this paper, the new design method is first introduced and then it is verified through a case study. The contents are arranged as follows: Section 2 presents the AFSA and introduces the new design method of the TMD. Section 3 presents the optimum TMD parameters table and optimization design equations for the TMD damping ratio and frequency ratio. In Section 4, the accuracy of the proposed design method and the vibration control effect of the optimized TMD is verified through a case study.

2. Optimization Algorithm of TMD

2.1. Artificial Fish Swarm Algorithm

Swarm intelligence algorithms have a wide application in many areas to solve different problems. Swarm intelligence algorithms consist of many algorithms that are developed by imitating the behavior of animals in nature [35,36]. The artificial fish swarm algorithm (AFSA) is one of the swarm intelligence algorithms. The actions of individual fish and the information interactions among them, when they are foraging and preying, are focused by the AFSA [42]. The AFSA is inspired by the individual, group and social behaviors of the fish. In a social form, it is searching for food, immigration, dealing with dangers and interactions between the fish in a swarm that result in intelligent group behavior [43,44]. The AFSA has many advantages such as flexibility, great convergence speed, great accuracy, fault tolerance and so on [45].

Fish do not have the complex logical reasoning ability and comprehensive judgment ability of human beings. Their purpose is achieved through the simple behavior of individuals or groups. It can be found through observing fish activities that they have four basic behaviors which are preying, swarming, following and random behavior. The four behaviors can be interchanged under different conditions. By evaluating their behavior, fish can choose an optimal behavior to achieve higher food concentration.

Among the four behaviors of artificial fish, the preying behavior is a fundamental behavior, which focuses mainly on searching of food. It is usually taken into consideration that artificial fish generally find the tendency to perceive the amount or concentration of food in the water by sight or taste. X_i is the artificial fish current state and in its visual distance, it will select a new state X_j randomly. Y is the food concentration, which means the value of the objective function. The artificial fish will find the global extreme value when the *Visual* is greater, and it will converge more easily. $Rand()$ is a random number between 0 and 1.

$$X_j = X_i + Visual \cdot Rand() \tag{1}$$

In the maximum problem, if $Y_i < Y_j$, the artificial fish will go forward a *Step*, which is a preset number in this direction.

$$X_i^{t+1} = X_i^t + \frac{X_j - X_i^t}{\|X_j - X_i^t\|} \cdot Step \cdot Rand() \tag{2}$$

If not, the artificial fish will choose a new state X_j randomly and judge whether it fits for the progressive condition. If the X_j still cannot fit it, the artificial fish will move a *Step* randomly.

In the process of swarming, fish usually congregate in groups, which is a kind of life habit formed to guarantee the survival of the colony and keep away from danger. Each artificial fish is made as follows: it tries to swarm to the center of the neighboring partner and avoid overcrowding. n_f is the number of the center position artificial fish's companions in its *Visual* distance ($\|X_j - X_i^t\| \leq Visual$) and X_{center} is the center position.

$$X_{center} = \frac{\sum_{j=1}^{n_f} X_j}{n_f} \tag{3}$$

If $Y_{center} > Y_i$ and $\frac{n_f}{n} < \delta$ (n is the number of total fish and δ is the crowding factor which is preset), this means that the center position has more food, which means that the value of fitness function is higher, and it is not very crowded, so the artificial fish will move a *Step* to the center.

$$X_i^{t+1} = X_i^t + \frac{X_{center} - X_i^t}{\|X_{center} - X_i^t\|} \cdot Step \cdot Rand() \tag{4}$$

If not, it will execute the preying behavior. In the swarming factor, the function of the crowd factor is to limit the scale of swarms. Therefore, more artificial fish only cluster at the best area, and a situation in which the artificial fish moves to optimum in a wide field is ensured.

When a fish finds more food and less crowded areas, nearby artificial fish will follow and quickly swim to the food. In the artificial fish's perception range, if it finds a partner in the optimal position, then it will move one *Step* towards it, otherwise, the preying operator is performed. The following operator speeds up the movement of the artificial fish to a better location, which also encourages the artificial fish to move to a better position.

The fish swims freely in the water, which seems to be random. However, actually, they are looking for food or companions in a larger range. The description of random behavior is simple, which is to randomly select a state in the field of view, and then move in that direction. The random behavior is also an unredeemed behavior of preying behavior.

The evaluation of artificial fish behavior is a way to reflect the autonomous behavior of fish. In solving the problem of optimization, the following two kinds of evaluation method can be chosen: one is to choose the optimal execution behavior. Given the current state of the way, the largest behavior in the optimal direction will be chosen. The other is to choose a behavior that can ensure that the artificial fish will go in a better direction.

2.2. Optimization Goal

The following dynamic system is consisted of a single degree of freedom primary structure and a TMD, excited by an external excitation $F(t)$, shown in Figure 1.

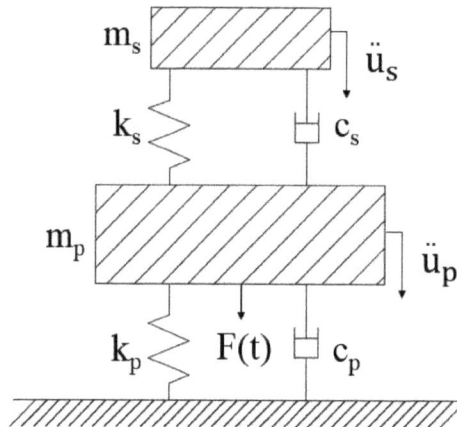

Figure 1. The dynamic system coupled with a tuned mass dampers (TMD) under an external excitation.

When the external excitation is assumed to take the form $F(t) = F_0 e^{i\omega t}$, where F_0 is the forcing amplitude and ω is its frequency. The motion equations of the dynamic system under an external excitation can be written as [1]:

$$\begin{pmatrix} m_p & 0 \\ 0 & m_s \end{pmatrix}\begin{pmatrix} \ddot{u}_p \\ \ddot{u}_s \end{pmatrix} + \begin{pmatrix} c_p + c_s & -c_s \\ -c_s & c_s \end{pmatrix}\begin{pmatrix} \dot{u}_p \\ \dot{u}_s \end{pmatrix} + \begin{pmatrix} k_p + k_s & -k_s \\ -k_s & k_s \end{pmatrix}\begin{pmatrix} u_p \\ u_s \end{pmatrix} = \begin{pmatrix} F_0 e^{i\omega t} \\ 0 \end{pmatrix} \quad (5)$$

where m_p and m_s are the mass of the primary structure and the TMD, respectively. c_p and c_s are the viscous damping coefficient of the primary structure and the TMD, respectively. k_p and k_s denote the stiffness coefficient of the primary structure and the TMD, respectively. u_p and u_s denote the absolute displacement of the primary structure and the TMD, respectively. The raised dot indicates the derivative with respect to time.

Besides, the excitation frequency ratio $\beta = \frac{\omega}{\omega_p}$, the TMD frequency ratio $\gamma = \frac{\omega_s}{\omega_p}$ and the TMD mass ratio $\mu = \frac{m_s}{m_p}$ are utilized.

Under human-induced vibrations, the maximum acceleration of the pedestrian bridge is one of the most important indexes to evaluate the serviceability problem. In order to depress the maximum acceleration dynamic amplification factor of the pedestrian bridge under external excitation and strengthen the robustness of the TMD, in this section, the novel optimization method of TMD is based on the AFSA, considering the structural damping ratio. The optimization goal is to minimize the maximum acceleration dynamic amplification factor of the pedestrian bridge under external harmonic excitations, based on the AFSA.

The acceleration dynamic amplification factor of the pedestrian bridge under a harmonic excitation can be written as:

$$N = \frac{\ddot{X}}{F_0/m_p} = \frac{\beta^2 \sqrt{\left(1 - \frac{\beta^2}{\gamma^2}\right)^2 + 4\left(\frac{\zeta_s \beta}{\gamma}\right)^2}}{\sqrt{\left\{\frac{\beta^4}{\gamma^2} - \frac{\beta^2}{\gamma^2} - \beta^2(1+\mu) - 4\frac{\zeta_s \zeta_p \beta^2}{\gamma} + 1\right\}^2 + \left\{2\frac{\zeta_p \beta^3}{\gamma} + 2\frac{\zeta_s \beta^3}{\gamma}(1+\mu) - 2\frac{\zeta_s \beta}{\gamma} - 2\zeta_p \beta\right\}^2}} \tag{6}$$

where ζ_p and ζ_s are the damping ratio of primary structure and TMD, respectively.

3. Optimum TMD Parameters Based on AFSA

3.1. Optimum TMD Parameters and Fitting Formulas

In this section, the optimum TMD frequency ratio f_{opt}, TMD damping ratio ζ_{opt} and dynamic amplification factor N_{opt} are shown in Table 1.

To illustrate this more clearly, the comparison of the optimum parameters of TMD systems for different mass ratios and damping ratios of the primary structure under external harmonic excitations are presented in Figure 2.

It can be seen from Figure 2a that with the increasing of the TMD mass ratio, the optimum TMD frequency ratio decreases, and with the increasing of the structural damping ratio, the optimum TMD frequency ratio increases too. It is shown in Figure 2b that with the increasing of the mass ratio, the optimum TMD damping ratio also increases, and with the increasing of the structural damping ratio, the optimum TMD damping ratio has a small incensement. It is presented in Figure 2c that with the increasing of mass ratio, the optimum dynamic amplification factor decreases, and with the increasing of structural damping ratio, the optimum dynamic amplification factor decreases steeply for a small mass ratio and smoothly for a large mass ratio.

The explicit expressions for the optimum TMD frequency ratio and optimum TMD damping ratio are given as follows, which result from data fitting through the least square method.

$$f_{opt} = 0.999 - 0.421\mu + 0.0456\zeta_p - 0.588\mu^2 + 1.2477\mu\zeta_p + 1.1325\zeta_p^2 \tag{7}$$

$$\zeta_{opt} = 0.607\sqrt{\mu} + 0.120\mu + 0.129\zeta_p - 3.423\mu^2 - 0.485\zeta_p^2 + 2.312\mu\zeta_p \\ + 22.347\mu^3 + 3.088\zeta_p^3 - 8.366\mu^2\zeta_p + 0.421\mu\zeta_p^2 \tag{8}$$

To verify the accuracy of the fitting Formulas (9) and (10), the fitting error between the formulas and Table 1 are shown in Figure 3.

It is presented in Figure 3a that the fitting error of optimum frequency ratio are all smaller than 0.3%. It is presented in Figure 3b that the fitting error of optimum damping ratio are all smaller than 1.0%. Therefore, the fitting formulas agree well with Table 1. Optimum TMD parameters can be obtained from Table 1 or calculated from Formulas (7) and (8).

Table 1. Optimum TMD parameters for a damped primary structure under external force excitation based on AFSA for a specified mass ratio. TMD: tuned mass dampers; AFSA: artificial fish swarm algorithm.

μ	$\zeta_p = 0.01$			$\zeta_p = 0.02$			$\zeta_p = 0.03$			$\zeta_p = 0.04$			$\zeta_p = 0.05$		
	f_{opt}	ζ_{opt}	N_{opt}	f_{opt}	ζ_{opt}	N_{opt}	f_{opt}	ζ_{opt}	N_{opt}	f_{opt}	ζ_{opt}	N_{opt}	f_{opt}	ζ_{opt}	N_{opt}
0.001	1.000	0.019	24.024	1.001	0.024	17.000	1.002	0.025	12.877	1.003	0.026	10.341	1.004	0.027	8.616
0.002	1.000	0.032	20.227	1.000	0.028	14.782	1.001	0.032	11.562	1.001	0.035	9.502	1.003	0.032	8.054
0.003	0.999	0.030	17.676	1.000	0.034	13.354	1.001	0.038	10.725	1.002	0.037	8.931	1.003	0.038	7.636
0.004	0.999	0.038	15.907	0.999	0.044	12.399	1.000	0.043	10.102	1.001	0.042	8.499	1.003	0.046	7.332
0.005	0.998	0.048	14.698	0.999	0.046	11.665	1.000	0.047	9.592	1.001	0.051	8.151	1.003	0.050	7.078
0.006	0.998	0.049	13.804	0.998	0.048	11.019	0.999	0.053	9.182	1.001	0.050	7.851	1.003	0.051	6.842
0.007	0.998	0.049	12.937	0.998	0.055	10.504	0.999	0.055	8.826	1.001	0.058	7.592	1.002	0.059	6.651
0.008	0.997	0.056	12.258	0.998	0.059	10.076	0.999	0.061	8.531	1.000	0.060	7.373	1.002	0.061	6.496
0.009	0.996	0.061	11.724	0.997	0.065	9.720	0.998	0.060	8.266	0.999	0.065	7.176	1.001	0.067	6.342
0.010	0.996	0.065	11.269	0.997	0.063	9.400	0.998	0.065	8.022	0.999	0.067	7.002	1.001	0.065	6.197
0.015	0.993	0.077	9.533	0.994	0.081	8.161	0.996	0.077	7.121	0.997	0.083	6.304	0.999	0.083	5.656
0.020	0.992	0.086	8.439	0.992	0.090	7.336	0.994	0.093	6.490	0.995	0.095	5.811	0.997	0.097	5.254
0.025	0.988	0.099	7.672	0.990	0.098	6.756	0.991	0.101	6.026	0.994	0.102	5.438	0.995	0.104	4.947
0.030	0.986	0.109	7.072	0.988	0.109	6.285	0.989	0.114	5.655	0.991	0.112	5.134	0.992	0.118	4.699
0.035	0.984	0.116	6.602	0.986	0.113	5.913	0.987	0.119	5.348	0.989	0.120	4.881	0.992	0.121	4.485
0.040	0.982	0.125	6.209	0.983	0.127	5.597	0.984	0.130	5.091	0.987	0.130	4.668	0.988	0.134	4.306
0.045	0.980	0.130	5.885	0.980	0.135	5.333	0.983	0.133	4.870	0.984	0.138	4.481	0.987	0.137	4.146
0.050	0.977	0.138	5.599	0.979	0.141	5.098	0.980	0.143	4.674	0.982	0.146	4.316	0.984	0.148	4.006
0.055	0.974	0.149	5.360	0.976	0.146	4.897	0.977	0.152	4.507	0.980	0.150	4.171	0.982	0.153	3.882
0.060	0.973	0.149	5.141	0.975	0.151	4.715	0.976	0.155	4.350	0.978	0.157	4.038	0.981	0.158	3.766
0.065	0.970	0.157	4.950	0.971	0.163	4.554	0.974	0.162	4.215	0.975	0.168	3.922	0.978	0.167	3.665
0.070	0.967	0.165	4.781	0.970	0.164	4.409	0.972	0.165	4.090	0.974	0.169	3.812	0.977	0.169	3.570
0.075	0.966	0.167	4.621	0.968	0.170	4.273	0.969	0.173	3.973	0.972	0.173	3.713	0.974	0.178	3.481
0.080	0.964	0.173	4.481	0.965	0.176	4.154	0.967	0.180	3.869	0.969	0.180	3.621	0.972	0.183	3.401
0.085	0.961	0.180	4.351	0.964	0.180	4.042	0.965	0.184	3.772	0.968	0.184	3.535	0.970	0.187	3.327
0.090	0.959	0.184	4.230	0.961	0.186	3.936	0.963	0.190	3.680	0.965	0.191	3.454	0.968	0.194	3.255
0.095	0.956	0.192	4.119	0.958	0.192	3.841	0.959	0.198	3.597	0.962	0.200	3.381	0.965	0.200	3.190
0.100	0.955	0.192	4.018	0.956	0.199	3.752	0.958	0.199	3.519	0.961	0.199	3.312	0.963	0.205	3.128

Table 1. Cont.

μ	ζp = 0.06			ζp = 0.07			ζp = 0.08			ζp = 0.09			ζp = 0.10		
	f_{opt}	ζ_{opt}	N_{opt}	f_{opt}	ζ_{opt}	N_{opt}	f_{opt}	ζ_{opt}	N_{opt}	f_{opt}	ζ_{opt}	N_{opt}	f_{opt}	ζ_{opt}	N_{opt}
0.001	1.005	0.023	7.388	1.007	0.029	6.468	1.009	0.023	5.747	1.011	0.024	5.170	1.012	0.032	4.703
0.002	1.005	0.031	6.976	1.006	0.037	6.152	1.008	0.037	5.505	1.011	0.039	4.979	1.013	0.039	4.542
0.003	1.005	0.040	6.670	1.007	0.040	5.922	1.009	0.040	5.318	1.011	0.044	4.831	1.013	0.045	4.425
0.004	1.004	0.049	6.443	1.006	0.046	5.740	1.008	0.050	5.175	1.011	0.052	4.713	1.013	0.051	4.323
0.005	1.004	0.050	6.241	1.006	0.055	5.587	1.008	0.054	5.055	1.010	0.053	4.612	1.013	0.058	4.242
0.006	1.004	0.055	6.070	1.007	0.054	5.451	1.008	0.056	4.941	1.010	0.060	4.523	1.013	0.059	4.168
0.007	1.004	0.061	5.922	1.006	0.061	5.330	1.008	0.062	4.844	1.011	0.062	4.441	1.013	0.063	4.098
0.008	1.003	0.065	5.793	1.005	0.065	5.224	1.008	0.066	4.761	1.011	0.067	4.370	1.013	0.066	4.040
0.009	1.003	0.065	5.671	1.005	0.071	5.130	1.007	0.070	4.682	1.010	0.071	4.303	1.012	0.075	3.983
0.010	1.003	0.070	5.560	1.005	0.070	5.041	1.007	0.071	4.606	1.010	0.075	4.243	1.013	0.073	3.932
0.015	1.001	0.083	5.121	1.003	0.088	4.681	1.006	0.085	4.307	1.008	0.091	3.989	1.011	0.090	3.715
0.020	0.999	0.099	4.797	1.001	0.099	4.407	1.004	0.102	4.078	1.007	0.102	3.792	1.010	0.102	3.546
0.025	0.998	0.105	4.542	1.000	0.106	4.192	1.002	0.110	3.895	1.006	0.109	3.636	1.009	0.113	3.410
0.030	0.995	0.116	4.331	0.997	0.120	4.014	1.001	0.120	3.741	1.003	0.124	3.502	1.006	0.126	3.291
0.035	0.994	0.123	4.151	0.996	0.125	3.860	0.999	0.129	3.608	1.002	0.131	3.386	1.005	0.133	3.190
0.040	0.991	0.133	3.997	0.993	0.139	3.727	0.997	0.137	3.491	0.999	0.142	3.284	1.003	0.141	3.099
0.045	0.989	0.141	3.860	0.992	0.141	3.608	0.995	0.145	3.388	0.998	0.146	3.193	1.001	0.150	3.019
0.050	0.987	0.151	3.738	0.989	0.152	3.501	0.993	0.153	3.294	0.996	0.154	3.110	1.000	0.155	2.945
0.055	0.985	0.155	3.629	0.988	0.155	3.408	0.990	0.161	3.211	0.994	0.160	3.035	0.997	0.166	2.878
0.060	0.983	0.161	3.529	0.986	0.163	3.319	0.989	0.164	3.133	0.993	0.166	2.966	0.996	0.170	2.816
0.065	0.980	0.170	3.440	0.983	0.172	3.240	0.986	0.173	3.062	0.989	0.177	2.903	0.994	0.177	2.759
0.070	0.979	0.174	3.356	0.983	0.173	3.166	0.985	0.179	2.996	0.988	0.180	2.844	0.991	0.185	2.706
0.075	0.977	0.181	3.278	0.980	0.181	3.096	0.983	0.183	2.934	0.986	0.185	2.789	0.990	0.189	2.656
0.080	0.974	0.188	3.207	0.977	0.190	3.033	0.981	0.190	2.877	0.983	0.194	2.737	0.988	0.195	2.609
0.085	0.973	0.190	3.140	0.975	0.195	2.974	0.978	0.197	2.824	0.982	0.197	2.689	0.986	0.200	2.566
0.090	0.971	0.195	3.077	0.974	0.198	2.917	0.978	0.198	2.773	0.979	0.205	2.643	0.983	0.206	2.525
0.095	0.967	0.205	3.018	0.970	0.207	2.864	0.974	0.208	2.725	0.978	0.209	2.599	0.981	0.213	2.485
0.100	0.955	0.192	4.018	0.956	0.199	3.752	0.958	0.199	3.519	0.961	0.199	3.312	0.963	0.205	3.128

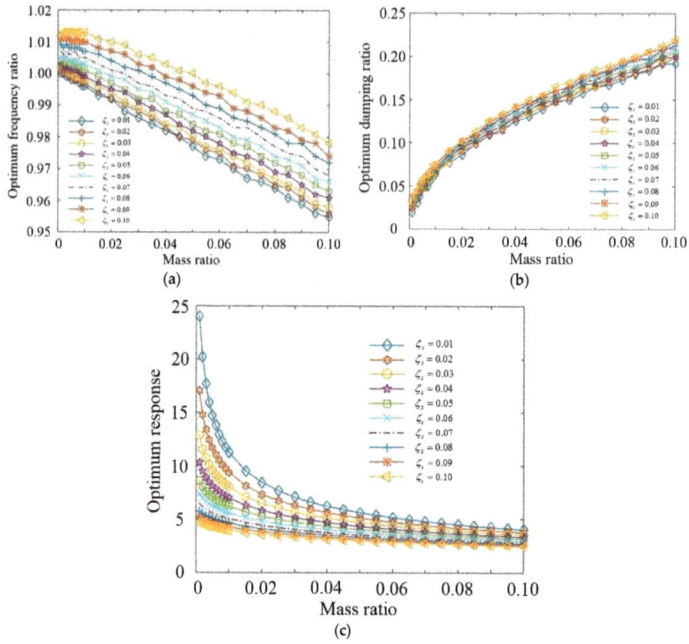

Figure 2. Comparison of the optimum parameters of TMD systems based on artificial fish swarm algorithm (AFSA) for different mass ratios and damping ratios of the primary structure under external excitations. (**a**) Comparison of optimum frequency ratio; (**b**) Comparison of optimum damping ratio; (**c**) Comparison of optimum response.

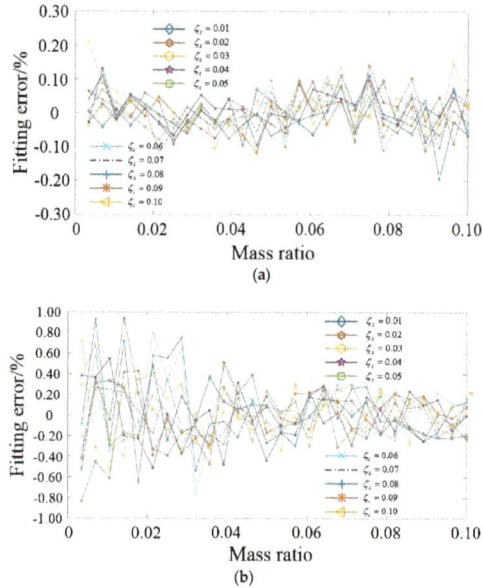

Figure 3. Comparison of fitting error of fitting formulas. (**a**) Comparison of optimum frequency ratio; (**b**) Comparison of optimum damping ratio.

3.2. Comparison Study

The classic Den Hartog method [29] has a wide application in TMD design; however, it does not consider the damping of the primary system, which may lead to a defective TMD. The optimum formulas of the Den Hartog method are shown in following, which is also devoted to minimizing the maximum of the structural acceleration response under external harmonic excitations.

$$f_{opt} = \frac{1}{\sqrt{1+\mu}} \tag{9}$$

$$\zeta_{opt} = \sqrt{\frac{3\mu}{8(1+\mu/2)}} \tag{10}$$

Ioi and Ikeda [37,38] proposed an improved design method considering structural damping, whose optimization goal is to minimize the maximum acceleration dynamic amplification factor of primary structure under external harmonic excitations. The optimum formulas of the Ioi and Ikeda method are shown in the following.

$$f_{opt} = \sqrt{\frac{1}{1+\mu}} + (0.096 + 0.88\mu - 1.8\mu^2)\zeta_p + (1.34 - 2.9\mu + 3\mu^2)\zeta_p^2 \tag{11}$$

$$\zeta_{opt} = \sqrt{\frac{3\mu(1+0.49\mu - 0.2\mu^2)}{8(1+\mu)}} + (0.13 + 0.72\mu + 0.2\mu^2)\zeta_p + (0.19 + 1.6\mu - 4\mu^2)\zeta_p^2 \tag{12}$$

The comparison study between the optimization method based on AFSA proposed in this paper and the classic Den Hartog method is shown in Table 2. The comparison study between the optimization method based on AFSA proposed in this paper and the Ioi and Ikeda method is presented in Table 3.

Table 2. Reduction ratio compared to the classical Den Hartog method/%.

μ \ ζ_s	0.010	0.020	0.030	0.040	0.050	0.060	0.070	0.080	0.090	0.100
0.001	1.039	1.006	1.887	1.952	1.989	2.001	2.341	2.548	2.625	2.546
0.002	1.046	1.125	1.328	2.299	2.625	2.799	2.796	2.940	3.150	3.300
0.003	0.503	1.402	2.281	2.553	2.622	3.117	3.366	3.532	3.486	3.537
0.004	0.767	1.843	2.079	2.362	3.016	3.280	3.418	3.597	3.836	4.020
0.005	1.373	1.506	2.004	2.785	3.073	3.262	3.595	3.841	4.022	4.047
0.006	0.869	1.357	2.417	2.891	3.133	3.542	3.843	4.026	4.074	4.325
0.007	1.149	1.896	2.558	2.783	3.360	3.701	3.903	4.097	4.370	4.549
0.008	0.943	2.063	2.343	2.814	3.290	3.628	3.886	4.241	4.473	4.548
0.009	1.232	1.862	2.078	2.979	3.326	3.577	4.048	4.311	4.501	4.719
0.010	1.182	1.601	2.387	2.972	3.354	3.789	4.131	4.380	4.580	4.851
0.015	1.025	1.904	2.246	3.111	3.501	3.954	4.376	4.671	4.970	5.256
0.020	1.121	1.977	2.607	3.074	3.758	4.088	4.587	4.938	5.220	5.554
0.025	0.954	1.740	2.526	3.161	3.737	4.232	4.644	5.059	5.392	5.668
0.030	0.995	1.875	2.528	3.183	3.773	4.241	4.767	5.111	5.554	5.882
0.035	0.937	1.795	2.580	3.235	3.844	4.371	4.821	5.273	5.610	5.996
0.040	1.030	1.802	2.602	3.207	3.865	4.361	4.909	5.307	5.766	6.110
0.045	0.846	1.803	2.543	3.267	3.878	4.456	4.933	5.412	5.797	6.209
0.050	1.057	1.811	2.661	3.239	3.954	4.441	5.044	5.453	5.933	6.291
0.055	0.959	1.818	2.544	3.305	3.883	4.520	4.999	5.528	5.937	6.381
0.060	0.950	1.865	2.632	3.341	3.989	4.556	5.110	5.578	6.027	6.430
0.065	1.010	1.790	2.621	3.263	3.975	4.532	5.117	5.609	6.079	6.512
0.070	0.945	1.810	2.572	3.336	3.940	4.595	5.107	5.646	6.105	6.547
0.075	0.974	1.887	2.622	3.322	4.021	4.619	5.204	5.681	6.169	6.588
0.080	0.954	1.799	2.620	3.290	4.018	4.604	5.199	5.722	6.203	6.670
0.085	0.967	1.782	2.614	3.350	3.990	4.644	5.176	5.738	6.224	6.681
0.090	0.989	1.878	2.617	3.403	4.040	4.670	5.261	5.759	6.268	6.719
0.095	0.956	1.850	2.620	3.359	4.043	4.661	5.270	5.808	6.304	6.781
0.010	0.928	1.798	2.624	3.332	4.039	4.671	5.238	5.818	6.317	6.809

Table 3. Reduction ratio compared to the Ioi Toshihiro method/%.

μ \ ζs	0.010	0.020	0.030	0.040	0.050	0.060	0.070	0.080	0.090	0.100
0.001	3.286	1.543	1.968	1.834	1.783	1.779	1.657	1.554	1.492	1.308
0.002	0.964	1.531	1.914	1.812	1.721	1.746	1.699	1.542	1.434	1.567
0.003	0.989	1.293	1.232	1.391	1.619	1.712	1.715	1.778	1.693	1.580
0.004	0.126	0.817	1.277	1.518	1.552	1.524	1.531	1.531	1.591	1.680
0.005	0.971	1.131	1.546	1.553	1.550	1.671	1.626	1.559	1.553	1.467
0.006	0.651	1.286	1.275	1.365	1.568	1.549	1.575	1.638	1.573	1.529
0.007	0.895	0.893	0.884	1.074	1.286	1.280	1.409	1.520	1.541	1.573
0.008	0.488	0.773	1.021	1.259	1.223	1.323	1.398	1.343	1.380	1.432
0.009	0.615	0.878	1.090	1.260	1.252	1.379	1.388	1.368	1.394	1.326
0.010	0.671	0.786	1.093	1.093	1.260	1.322	1.322	1.399	1.369	1.352
0.015	0.395	0.646	0.764	0.920	0.942	1.074	1.082	1.138	1.154	1.132
0.020	0.144	0.351	0.520	0.655	0.795	0.811	0.897	0.893	0.930	0.981
0.025	0.208	0.277	0.388	0.437	0.593	0.601	0.736	0.761	0.814	0.833
0.030	0.020	0.278	0.342	0.438	0.499	0.557	0.613	0.646	0.674	0.689
0.035	0.101	0.121	0.196	0.224	0.327	0.364	0.458	0.499	0.562	0.587
0.040	0.036	0.151	0.245	0.271	0.343	0.375	0.429	0.456	0.477	0.497
0.045	0.062	0.049	0.118	0.124	0.220	0.243	0.321	0.345	0.405	0.419
0.050	0.075	0.073	0.169	0.187	0.246	0.262	0.313	0.316	0.320	0.309
0.055	0.060	0.083	0.065	0.125	0.132	0.193	0.214	0.260	0.282	0.305
0.060	0.024	0.048	0.029	0.072	0.123	0.134	0.160	0.150	0.178	0.223
0.065	0.061	0.061	0.083	0.085	0.123	0.130	0.170	0.182	0.191	0.188
0.070	0.016	-0.007	-0.005	0.008	0.016	0.063	0.090	0.116	0.146	0.153
0.075	0.066	0.091	0.036	0.008	0.072	0.082	0.098	0.079	0.048	0.086
0.080	0.041	0.030	0.037	0.034	0.062	0.066	0.094	0.097	0.102	0.091
0.085	0.041	0.011	0.043	0.016	0.013	0.014	0.011	0.038	0.056	0.060
0.090	0.078	0.087	0.072	0.091	0.035	0.051	0.046	0.008	0.008	0.015
0.095	0.058	0.077	0.025	0.066	0.039	0.044	0.052	0.045	0.033	0.041
0.010	0.000	0.041	0.046	0.025	0.048	0.027	0.021	0.041	0.040	0.055

It is presented in Table 2 that the novel optimization method based on AFSA proposed in this paper has a smaller maximum acceleration dynamic amplification factor than the classical Den Hartog method, and in general, the reduction ratio increases with the increasing of the TMD mass ratio and the structural damping ratio. As can be seen in Table 3, the novel optimization method proposed in this paper has a smaller maximum acceleration dynamic amplification factor than the Ioi and Ikeda method, and in general, the reduction ratio increases with the decreasing of the TMD mass ratio and structural damping ratio. Therefore, the novel optimization method based on AFSA proposed in this paper has a better effect in controlling the maximum acceleration dynamic amplification factor.

4. Case Study

In this section, an engineering project of a pedestrian bridge, for which TMDs need to be installed, will be introduced in detail as a case study.

The pedestrian bridge is 45 m long and 6 m wide. It is a simply supported pedestrian bridge. Due to the excessive vibration under human-induced excitations, it has a serviceability problem. The finite element model was established in Midas Gen. The structural modes are shown in Table 4.

Table 4. Modes of the pedestrian bridge.

Mode	Frequency/Hz	UX/%	UY/%	UZ/%	RX/%	RY/%	RZ/%
1	1.006	94.445	3.788	0.058	0.033	0.001	0.588
2	1.488	3.564	84.251	0.000	6.749	0.011	0.007
3	1.946	0.063	0.026	75.499	0.208	0.014	0.000
4	2.098	0.055	2.405	0.097	41.152	0.225	0.019
5	2.482	0.133	8.134	0.035	28.389	0.040	0.000
6	2.726	0.594	0.006	0.001	0.029	0.000	94.733
7	4.276	0.000	0.000	0.003	0.000	0.000	0.000
8	4.278	0.000	0.000	0.000	0.000	0.002	0.000
9	4.425	0.026	0.002	0.000	0.002	0.390	0.003
10	4.681	0.000	0.006	0.000	0.005	43.800	0.109

Notes: UX, UY and UZ are modal mass participation coefficient in transverse, longitudinal and vertical direction. RX, RY and RZ are rotation modal mass participation coefficient in transverse, longitudinal and vertical direction.

Because the transverse and longitudinal loads of pedestrians are relatively smaller than the vertical load [21], the effect of transverse and longitudinal loads can usually be ignored. From Table 4, it is clear that the third vertical mode is dominant. The third order frequency is 1.946 Hz, and the vertical modal mass participation coefficient is 75.499%. This pedestrian bridge is about 357,500 kg in total. Therefore, the third vertical modal participation mass is 269,909 kg. The damping ratio of this pedestrian bridge is set to be 2%. The vertical mode shape of the pedestrian bridge is shown in Figure 4.

In order to solve the serviceability problem, a TMD is designed. The TMD mass ratio was chosen to be 1%. According to Table 1, the TMD frequency ratio is 0.997 and the TMD damping ratio is 0.063. According to the Den Hartog method, the TMD frequency ratio is 0.995 and the TMD damping ratio is 0.061. According to the Ioi and Ikeda method, the TMD frequency ratio is 0.998 and the TMD damping ratio is 0.064. The dynamic amplification factors of the three TMDs under different frequency harmonic excitations are compared in Figure 5.

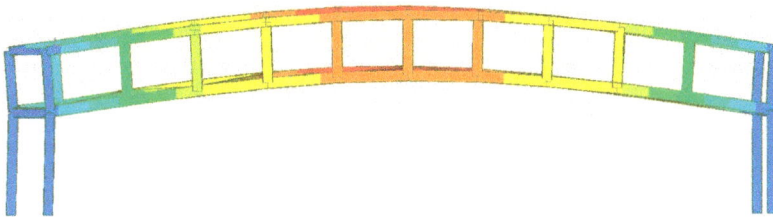

Figure 4. The vertical mode shape of the pedestrian bridge.

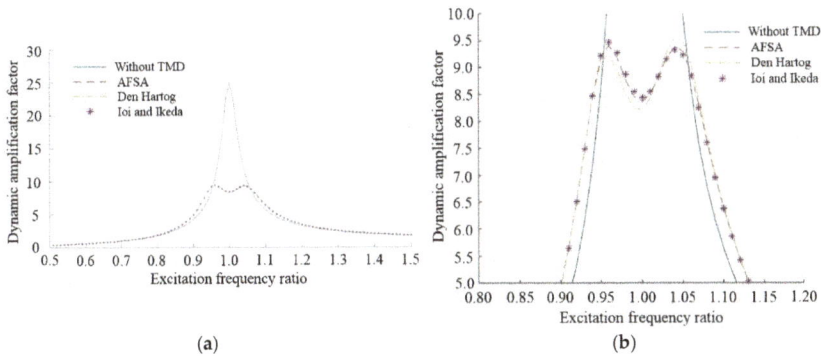

Figure 5. The comparisons of structural dynamic amplification factors under different frequency harmonic excitations. (**a**) Excitation frequency ratio from 0.5 to 1.5; (**b**) Local amplification figure.

It is shown in Figure 5 that the maximum dynamic amplification factor without a TMD is 25.00, while with the TMDs designed based on the AFSA, Den Hartog and Ioi and Ikeda methods it is 9.41, 9.56 and 9.48, respectively. Compared to the situation without a TMD, the TMD designed based on the AFSA, Den Hartog and Ioi and Ikeda methods has a reduction of 62.36%, 61.76% and 62.08%, respectively. It can be seen that the vibration control effect of the TMD designed based on the AFSA method is slightly better than that of the TMD designed based on the remaining two methods.

To further verify the vibration control effect of the optimized TMD designed based on the AFSA, in this section, considering that the frequency of pedestrians is usually around 2.00 Hz, a situation in which a 700 N person walks, runs and jumps on the middle of the bridge at 2.00 Hz will be simulated. According to [46–49], a vertical force can be simulated considering of three different frequencies.

The comparisons of the structural response under 2.00 Hz walking, running and jumping excitations are shown in Figure 6. Note that the unit of acceleration "gal" means "1 cm/s^2".

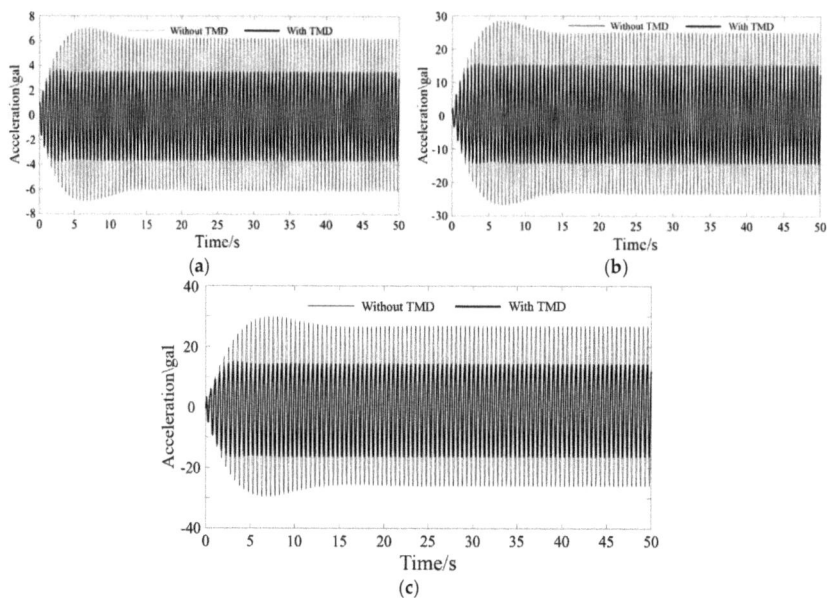

Figure 6. The comparisons of structural responses under 2.00 Hz human-induced excitations. (**a**) Under 2.00 Hz walking excitation; (**b**) Under 2.00 Hz running excitation; (**c**) Under 2.00 Hz jumping excitation.

It is presented in Figure 6 that the peak accelerations of structure under walking, running and jumping excitations are 6.24 gal, 25.17 gal and 26.90 gal, respectively, in the situation without TMD, while the peak accelerations are 3.54 gal, 15.48 gal and 14.39 gal, respectively, in the situation with TMD, and the corresponding peak acceleration reductions of TMD are 43.27%, 38.50% and 46.51%, respectively. Consequently, the vibration control effect of an optimized TMD system based on AFSA is quite obvious. Therefore, under the same design guideline, the maximum permissible pedestrian number is larger in the situation with TMD.

5. Conclusions

TMD is widely applied in the pedestrian-induced vibration control of pedestrian bridges and its parameters are very important. A new TMD optimization method based on the artificial fish swarm algorithm (AFSA) is proposed in this paper, which takes primary structural damping into consideration. The optimization goal is to minimize the maximum dynamic amplification factor of the primary structure under external harmonic excitation, and the resultant optimized TMD has a smaller maximum dynamic amplification factor and better robustness. The optimum TMD parameters table for a damped primary structure with different damping ratios and different TMD mass ratios was presented. At last, as a case study, the application of an optimized TMD based on AFSA for a pedestrian bridge was proposed. The following conclusions can be drawn:

(1) The fitting formulas of the optimum TMD frequency ratio and the optimum TMD damping ratio agree well with Table 1. Optimum TMD parameters can be obtained from Table 1 or calculated from Formulas (3) and (4). Consequently, the design procedure is very simple.

(2) The novel optimization method proposed in this paper has a smaller maximum acceleration dynamic amplification factor than the classic Den Hartog method and the Ioi and Ikeda method.

(3)   The optimized TMD has a good effect in controlling human-induced vibrations at different frequencies for a pedestrian bridge, indicating its good robustness.

Acknowledgments: This work was supported by the National Natural Science Foundation of China (Grant No. 51478361) and the Fundamental Research Funds for the Central Government Supported Universities.

Author Contributions: Weixing Shi proposed the idea. Liangkun Wang carried out the parametric study and wrote the paper. Zheng Lu conceived the analytical method, provided valuable discussions and revised the paper. Quanwu Zhang carried out case study.

Conflicts of Interest: The authors declare no conflict of interest.

References

1.   Shi, W.; Wang, L.; Lu, Z. Study on self-adjustable tuned mass damper with variable mass. *Struct. Control Health Monit.* **2017**, e2114. [CrossRef]
2.   Lu, Z.; Wang, D.; Masri, S.; Lu, X. An experimental study of vibration control of wind-excited high-rise buildings using particle tuned mass dampers. *Smart Struct. Syst.* **2016**, *25*, 1–7. [CrossRef]
3.   Lu, Z.; Chen, X.; Zhang, D.; Dai, K. Experimental and analytical study on the performance of particle tuned mass dampers under seismic excitation. *Earthq. Eng. Struct. Dyn.* **2017**, *46*, 697–714. [CrossRef]
4.   Wang, W.X.; Hua, X.G.; Wang, X.Y.; Chen, Z.Q.; Song, G.B. Optimum design of a novel pounding tuned mass damper under harmonic excitation. *Smart Mater. Struct.* **2017**, *26*, 055024. [CrossRef]
5.   Song, G.B.; Zhang, P.; Li, L.Y.; Singla, M.; Patil, D.; Li, H.N.; Mo, Y.L. Vibration control of a pipeline structure using pounding tuned mass damper. *J. Eng. Mech.* **2016**, *142*, 04016031. [CrossRef]
6.   Zhang, P.; Song, G.B.; Li, H.N.; Lin, Y.X. Seismic control of power transmission tower using pounding TMD. *J. Eng. Mech.* **2013**, *139*, 1395–1406. [CrossRef]
7.   Lu, Z.; Lu, X.; Jiang, H.; Masri, S. Discrete element method simulation and experimental validation of particle damper system. *Eng. Comput.* **2014**, *31*, 810–823. [CrossRef]
8.   Lu, Z.; Chen, X.; Lu, X.; Yang, Z. Shaking table test and numerical simulation of an RC frame-core tube structure for earthquake-induced collapse. *Earthq. Eng. Struct. Dyn.* **2016**, *45*, 1537–1556. [CrossRef]
9.   Lu, Z.; Wang, Z.; Masri, S.F.; Lu, X. Particle Impact Dampers: Past, Present, and Future. *Struct. Control Health Monit.* **2018**, *25*, e2058. [CrossRef]
10.   Lu, Z.; Chen, X.; Zhou, Y. An equivalent method for optimization of particle tuned mass damper based on experimental parametric study. *J. Sound Vib.* **2017**, *9*. [CrossRef]
11.   Lu, Z.; Huang, B.; Zhou, Y. Theoretical study and experimental validation on the energy dissipation mechanism of particle dampers. *Struct. Control Health Monit.* **2017**, *11*. [CrossRef]
12.   Dai, K.; Wang, J.; Mao, R.; Lu, Z.; Chen, S. Experimental investigation on dynamic characterization and seismic control performance of a TLPD system. *Struct. Des. Tall Spec. Build.* **2017**, *26*, e1350. [CrossRef]
13.   Lu, Z.; Lu, X.; Masri, S.F. Studies of the performance of particle dampers under dynamic loads. *J. Sound Vib.* **2010**, *329*, 5415–5433. [CrossRef]
14.   Casciati, F.; Casciati, S.; Faravelli, L. A contribution to the modelling of human induced excitation on pedestrian bridges. *Struct. Saf.* **2017**, *66*, 51–61. [CrossRef]
15.   Brownjohn, J.M.W.; Paivc, A. Experimental methods for estimating modal mass in footbridges using human-induced dynamic excitation. *Eng. Struct.* **2007**, *29*, 2833–2843. [CrossRef]
16.   Nakhorn, P.; Sopak, K.; Pennung, W. Application of non-linear multiple tuned mass dampers to suppress man-induced vibrations of a pedestrian bridge. *Earthq. Eng. Struct. Dyn.* **2010**, *32*, 1117–1131.
17.   Chen, S.R.; Wu, J. Performance enhancement of bridge infrastructure systems: Long-span bridge, moving trucks and wind with tuned mass dampers. *Eng. Struct.* **2008**, *30*, 3316–3324. [CrossRef]
18.   Carpineto, N.; Lacarbonara, W.; Vestroni, F. Mitigation of pedestrian-induced vibrations in suspension footbridges via multiple tuned mass dampers. *J. Vib. Control* **2010**, *16*, 749–776. [CrossRef]
19.   Lu, X.; Ding, K.; Shi, W.; Weng, D. Tuned mass dampers for human-induced vibration control of the Expo Culture Centre at the World Expo 2010 in Shanghai, China. *Struct. Eng. Mech.* **2012**, *43*, 607–621. [CrossRef]
20.   Li, Q.; Fan, J.; Nie, J.; Li, Q.; Chen, Y. Crowd-induced random vibration of footbridge and vibration control using multiple tuned mass dampers. *J. Sound Vib.* **2010**, *329*, 4068–4092. [CrossRef]

21. Lievens, K.; Lombaert, G.; Roeck, G.; Broeck, P. Robust design of a TMD for the vibration serviceability of a footbridge. *Eng. Struct.* **2016**, *123*, 408–418. [CrossRef]
22. Caetano, E.; Cunha, Á.; Moutinho, C.; Magalhães, F. Studies for controlling human-induced vibration of the Pedro e Inês footbridge, Portugal. Part 2: Implementation of tuned mass dampers. *Eng. Struct.* **2010**, *32*, 1082–1091. [CrossRef]
23. Casciati, F.; Giuliano, F. Performance of multi-TMD in the towers of suspension bridges. *J. Vib. Control* **2009**, *15*, 821–847. [CrossRef]
24. Nagarajaiah, S. Adaptive passive, semi-active, smart tuned mass dampers: Identification and control using empirical mode decomposition, Hilbert transform, and short-term Fourier transform. *Struct. Control Health Monit.* **2009**, *16*, 800–841. [CrossRef]
25. Sun, C.; Nagarajaiah, S.; Dick, A.J. Family of smart tuned mass damper with variable frequency under harmonic excitation and ground motions: Closed-form evaluation. *Smart Struct. Syst.* **2014**, *13*, 319–341. [CrossRef]
26. Sun, C.; Nagarajaiah, S. Study on semi-active tuned mass damper with variable damping and stiffness under seismic excitations. *Struct. Control Health Monit.* **2013**, *21*, 890–906. [CrossRef]
27. Lu, Z.; Yang, Y.; Lu, X.; Liu, C. Preliminary study on the damping effect of a lateral damping buffer under a debris flow load. *Appl. Sci.* **2017**, *7*, 201. [CrossRef]
28. Lu, X.; Liu, Z.; Lu, Z. Optimization design and experimental verification of track nonlinear energy sink for vibration control under seismic excitation. *Struct. Control Health Monit.* **2017**, *24*, e2033. [CrossRef]
29. Den Hartog, J.P. *Mechanical Vibration*, 4th ed.; McGraw-Hill: New York, NY, USA, 1956.
30. Leung, A.Y.T.; Zhang, H. Particle swarm optimization of tuned mass dampers. *Eng. Struct.* **2009**, *31*, 715–728. [CrossRef]
31. Leung, A.Y.T.; Zhang, H.; Cheng, C.C.; Lee, Y.Y. Particle swarm optimization of TMD by non-stationary base excitation during earthquake. *Earthq. Eng. Struct. Dyn.* **2008**, *37*, 1223–1246. [CrossRef]
32. Bekdas, G.; Nigdeli, S.M. Estimating optimum parameters of tuned mass dampers using harmony search. *Eng. Struct.* **2011**, *33*, 2716–2723. [CrossRef]
33. Mohebbi, M.; Shakeri, K.; Ghanbarpour, Y.; Majzoub, H. Designing optimal multiple tuned mass dampers using genetic algorithms (GAs) for mitigating the seismic response of structures. *J. Vib. Control* **2012**, *19*, 605–625. [CrossRef]
34. Jiménez-Alonso, J.F.; Sáez, A. Robust optimum design of TMDs to mitigate pedestrian induced vibrations using multi-objective genetic algorithms. *Struct. Eng. Int.* **2017**, *4*, 492–501. [CrossRef]
35. Li, X.; Shao, Z.; Qian, J. An optimizing method based on autonomous animates: Fish swarm algorithm. *Chin. J. Syst. Eng. Theory Pract.* **2002**, *22*, 32–38.
36. Wang, L.; An, L.; Pi, J.; Fei, M.; Pardalos, P. A diverse human learning optimization algorithm. *J. Glob. Optim.* **2017**, *67*, 283–323. [CrossRef]
37. Toi, T.; Ikeda, K. On the dynamic vibration damped absorber of the vibration system. *JSME Int. J.* **1978**, *21*, 64–71.
38. Ikeda, K.; Toi, T. On the houde damper for a damped vibration system. *Trans. Jpn. Soc. Mech. Eng. Ser. C* **1979**, *45*, 663–670. [CrossRef]
39. Butz, C.; Feldmann, M.; Heinemeyer, C. Advanced load models for synchronous pedestrian excitation and optimised design guidelines for steel footbridges. *Agrociencia* **2013**, *47*, 781–794.
40. Van Nimmen, K.; Verbeke, P.; Lombaert, G.; De Roeck, G. Numerical and experimental evaluation of the dynamic performance of a footbridge with tuned mass dampers. *J. Bridg. Eng. ASCE* **2016**, *21*, C4016001. [CrossRef]
41. Asami, T.; Nishihara, O.; Baz, A.M. Analytical solutions to H_∞ and H_2 optimization of dynamic vibration absorbers attached to damped linear systems. *J. Vib. Acoust.* **2002**, *124*, 284–295. [CrossRef]
42. Wang, H.; Fan, C.; Tu, X. AFSAOCP: A novel artificial fish swarm optimization algorithm aided by ocean current power. *Appl. Intell.* **2016**, *45*, 992–1007. [CrossRef]
43. Liu, Q.; Odaka, T.; Kuroiwa, J.; Shirai, H.; Ogura, H. An artificial fish swarm algorithm for the multicast routing problem. *Ieice Trans. Commun.* **2014**, *97*, 996–1011. [CrossRef]
44. Neshat, M.; Sepidnam, G.; Sargolzaei, M.; Toosi, A. Artificial fish swarm algorithm: A survey of the state-of-the-art, hybridization, combinatorial and indicative applications. *Artif. Intell. Rev.* **2014**, *42*, 965–997. [CrossRef]

45. Zhu, X.; Ni, Z.; Cheng, M.; Jin, F.; Li, J.; Weckman, G. Selective ensemble based on extreme learning machine and improved discrete artificial fish swarm algorithm for haze forecast. *Appl. Intell.* **2017**, *3*, 1–19. [CrossRef]
46. Casado, C.; Díaz, I.; Sebastián, J.; Poncela, A.; Lorenzana, A. Implementation of passive and active vibration control on an in-service footbridge. *Struct. Control Health Monit.* **2013**, *20*, 70–87. [CrossRef]
47. Nimmen, K.; Lombaert, G.; Roeck, G.; Broeck, P. Vibration serviceability of footbridges: Evaluation of the current codes of practice. *Eng. Struct.* **2014**, *59*, 448–461. [CrossRef]
48. Occhiuzzi, A.; Spizzuoco, M.; Ricciardelli, F. Loading models and response control of footbridges excited by running pedestrians. *Struct. Control Health Monit.* **2008**, *15*, 349–368. [CrossRef]
49. Shi, W.; Wang, L.; Lu, Z.; Gao, H. Study on adaptive-passive and semi-active eddy current tuned mass damper with variable damping. *Sustainability* **2018**, *10*, 99. [CrossRef]

applied sciences

MDPI

Article

Application of the Hybrid Simulation Method for the Full-Scale Precast Reinforced Concrete Shear Wall Structure

Zaixian Chen [1,*], Huanding Wang [1,2], Hao Wang [3], Hongbin Jiang [2], Xingji Zhu [1,*] and Kun Wang [4]

[1]   Department of Civil Engineering, Harbin Institute of Technology at Weihai, Weihai 264209, China; hdwhrb@hit.edu.cn
[2]   Harbin Institute of Technology, School of Civil Engineering, Harbin 150090, China; 86282081@163.com
[3]   Department of Civil Engineering, Southeast University, Nanjing 210096, China; wanghao1980@seu.edu.cn
[4]   College of Civil Science and Engineering, Yangzhou University, Yangzhou 225127, China; wangkun@yzu.edu.cn
*   Correspondence: zaixian_chen@sina.com (Z.C.); zhuxingji@hit.edu.cn (X.Z.); Tel.: +86-0631-568-7845 (Z.C.)

Received: 2 January 2018; Accepted: 1 February 2018; Published: 7 February 2018

Abstract: The hybrid simulation (HS) testing method combines physical test and numerical simulation, and provides a viable alternative to evaluate the structural seismic performance. Most studies focused on the accuracy, stability and reliability of the HS method in the small-scale tests. It is a challenge to evaluate the seismic performance of a twelve-story pre-cast reinforced concrete shear-wall structure using this HS method which takes the full-scale bottom three-story structural model as the physical substructure and the elastic non-linear model as the numerical substructure. This paper employs an equivalent force control (EFC) method with implicit integration algorithm to deal with the numerical integration of the equation of motion (EOM) and the control of the loading device. Because of the arrangement of the test model, an elastic non-linear numerical model is used to simulate the numerical substructure. And non-subdivision strategy for the displacement inflection point of numerical substructure is used to easily realize the simulation of the numerical substructure and thus reduce the measured error. The parameters of the EFC method are calculated basing on analytical and numerical studies and used to the actual full-scale HS test. Finally, the accuracy and feasibility of the EFC-based HS method is verified experimentally through the substructure HS tests of the pre-cast reinforced concrete shear-wall structure model. And the testing results of the descending stage can be conveniently obtained from the EFC-based HS method.

Keywords: equivalent force control; hybrid simulation; full-scale; nonlinear seismic performance; descent stage

1. Introduction

The seismic testing method is an important way to evaluate the structure nonlinear seismic performance and verify the feasibility of the structure seismic theory. The hybrid simulation (HS) testing method, which is named initially as the pseudo-dynamic substructure testing method, was proposed in 1969 [1]. This method combines the physical test of the critical nonlinear parts of a structure and the numerical simulation of the remainder. It can effectively reduce the size of the test model and is an efficient and versatile testing tool for the large-scale structural model.

The novel technique of the HS method, which mainly includes (1) the robust numerical integration techniques [2–7]; (2) the loading control method [8–14]; (3) the geographically distributed HS [15–20]; and (4) the parameter identification of HS [21–26] for the implement and improvement of HS method.

Especially, the real-time HS plays a powerful role to evaluate the rate dependent behaviors, such as magneto-rheological damper or vibration isolation system [18,27–36]. Additionally, the Network for Earthquake Engineering Simulation (NEES), which featured 14 geographically distributed and shared-use laboratories to test modern structural system [37–40], accelerates the development of the HS platforms (OpenFresco manufactured by dSpace, and MTS (STS and FlexTest systems) National Instruments [41], and UI-Simcor developed by the MUST-SIM facility and the MAE Center at the University of Illinois at Urbana-Champaign [42]). It was widely used to realize the performance assessment of modern building structure systems conveniently.

Although HS method has been advanced continuously in the last four decades, previous studies mainly focused on the accuracy, stability and reliability. And only the small-scale tests were used to examine the feasibility of the HS method [43,44]. A few large-scale structural testing models were examined by the HS method [24,31,45,46]. To accelerate the development of the HS method in the large-scale model, a full-scale three-story precast reinforced concrete shear-wall substructure model is built as the physical substructure to evaluate the seismic performance of the entire twelve-story structure, in this paper.

Note that there are two challenges associated with the implementation of the controller for this full-scale HS method, including: (1) the complex test setup which employed four electro-hydraulic servo actuators as the horizontal loading device; and (2) the complex structural model which is asymmetric due to the asymmetric structural arrangement of the model. In order to solve the above two problems, an equivalent force control (EFC) method with implicit integration algorithm is proposed to deal with the numerical integration of the equation of motion (EOM) and the control of the loading device. The general arrangement of the whole test which includes the test model, the loading device and the testing device are all described. Following that, a discussion of the EFC method for the test model is presented. Then, the parameters for the EFC-based HS method are determined by the numerical simulation. Finally, an experimental validation is carried out to demonstrate the efficacy of the proposed approach.

2. Test Arrangement

2.1. Test Model

The twelve-story pre-cast reinforced concrete shear wall structure was designed based on both the Chinese Code GB50010 for Design of Concrete Structures and the Chinese Code for Seismic Building Design GB50011. The seismic precautionary intensity of the region is 7 degrees for the structural model. As shown in Figure 1, the plan dimension of the test model is 4200 mm × 3000 mm. And the height of each story is 3000 mm. The thickness of the shear wall is 200 mm. The cross-section of the coupling beam connected to the walls is 200 mm × 400 mm. The design value of the concrete compression strength for the test model is 30 MPa. The HRB335 steel is employed for the reinforcements of the beams and the HPB235 for the stirrups and slab. Because of the limitations of the test site and loading device, a full-scale three-story precast concrete shear wall substructure model was built as the physical substructure in the Structural and Seismic Testing Center of Harbin Institute of Technology, as shown in Figure 2. A foundation beam, which is 600 mm high, was constructed to ensure the fixed connection with the laboratory floor. Therefore, the total height of the physical substructure model is 9600 mm.

From Figures 1 and 2, the tension stiffness of the model is not equal with the compression stiffness due to the arrangement of the door and window in the plan. Table 1 shows the inter-story stiffness of the test model. Note that the initial stiffness in positive and negative direction is calculated by the measured drift and shear force at different floors of the physical substructure, which are obtained by loading at the top of the physical substructure. Because of the same structural layouts from the second floor to the sixth floor, here an average value of the stiffness between the second and third floor can be considered as the initial stiffness for over the second floor. The details of material properties are shown in Tables 2 and 3.

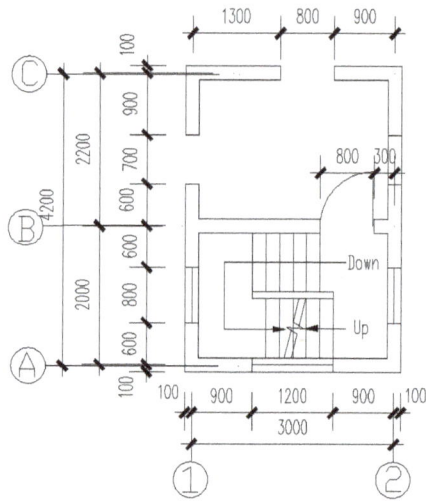

Figure 1. Plan of the test model.

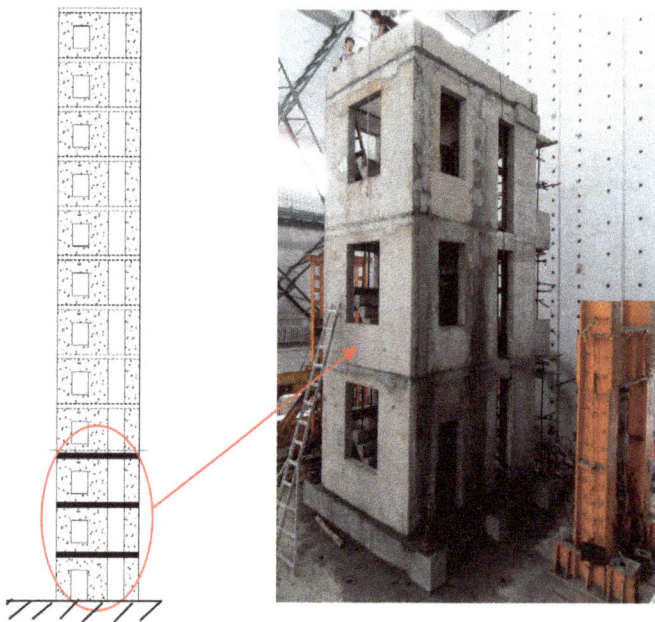

Figure 2. Full-scale three-story precast concrete shear wall substructure model.

Table 1. Calculated parameters for test model.

Floors	1	2	3	4~12
Initial stiffness in positive direction (kN/mm)	417	295	295	295
Initial stiffness in negative direction (kN/mm)	475	359	359	359

Table 2. Material properties of the concrete.

Type	Compress Strength of Concrete	Compress Strength of Grouting Material	Bending Strength of Grouting Material
	(N/mm²)	(N/mm²)	(N/mm²)
Precast parts of each floor	31.6	—	—
Cast-in-place parts of 1st floor	22.9	75.7	6.1
Cast-in-place parts of 2nd floor	29.5	73.8	5.9
Cast-in-place parts of 3rd floor	32.8	80.1	8.0

Table 3. Material properties of the steel reinforcement.

Diameter	Yield Strength	Limit Strength
(mm)	(N/mm²)	(N/mm²)
6	353.1	504.2
8	290.6	433.3
10	308.9	429.9
12	386.6	560.3
14	340.8	483.8
16	283.1	443.7

2.2. Loading Device

Figure 3 shows the horizontal loading devices which include four hydraulic servo actuators. Only the third floor was loaded by two actuators with the capacity of ±630 kN. A same actuator was employed on the middle of the Axes 1 and 2 for the second floor while an actuator with the capacity of ±250 kN was employed for the first floor. The displacement measurement range of all the actuators was ±250 mm. The errors of both force and displacement are the 0.1% corresponding to the capacity. To prevent the local damage of the beams connecting the actuator, two side beams of the third floor, which located at the end of horizontal loading devices, were partly enlarged to the section of 400 mm × 400 mm. For the bottom two floors, to enable a uniform transfer of horizontal forces to each story, as well as to ensure a tight connection among the beam, the actuator and floor slab, the far end was passed through the wall along the B axis and beam-anchored by high-strength bolts, as shown in Figure 4.

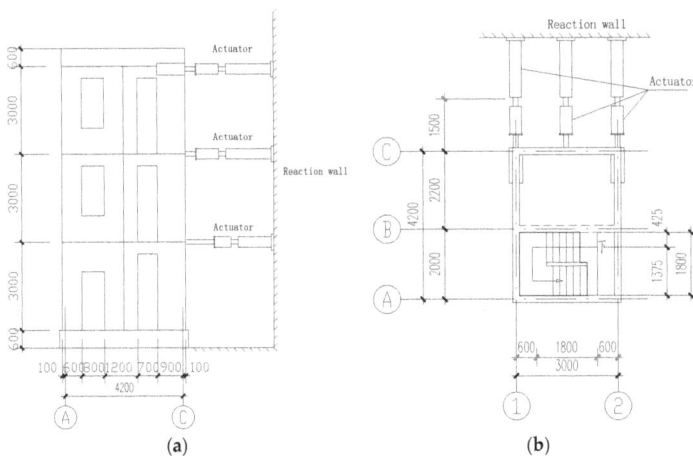

Figure 3. Planned arrangement of horizontal loading devices for the hybrid simulation testing: (a) elevation; (b) Plan view.

Figure 4. Arrangement of horizontal loading devices.

2.3. Arrangement of Measuring Points

During the HS test, a total of ten Linear Variable Differential Transformers (LVDTs) were employed to measure the displacement of the test model. The arrangement of the measuring points is illustrated in Figure 5. Two LVDTs were set up at the top of the foundation beam and each story along Axes 1 and 2, respectively (8 in total); the average value of the pair of LVDTs was used to determine the story drift. Another two LVDTs were arranged at a height of 1 m and 2 m above the ground along the Axe 2.

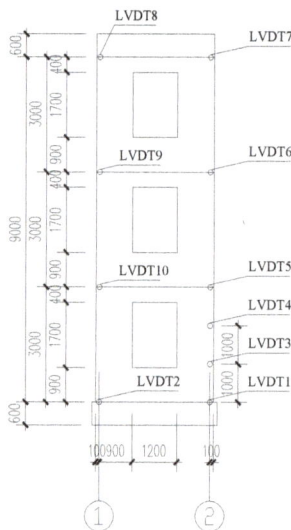

Figure 5. Arrangement of displacement measuring points.

3. Parameters of Equivalent Force Control HS Method

3.1. Parameter of Numerical Substructure Model

The EFC method was used for HS [47] with an implicit integration algorithm, during which the mathematical iteration procedure was improved by using a force feedback control. Numerical and experimental results of simple model verified that, as long as the equivalent force controller was properly designed, proposed EFC method was sufficient. This section introduces the detailed test procedure for the aforementioned sophisticated structures with the elastic non-linear numerical substructure.

Considering the asymmetric stiffness in the tension and compression direction caused by the asymmetric structural arrangement of the test model, an asymmetric restoring force model was used in this paper, as shown in Figure 6. In this model, an inflection point exists at the location of the zero displacement for the asymmetric restoring force model of the numerical substructure. Traditionally, the subdivision technique should be used to reduce the computational error at the location of the displacement inflection point. However, the measured error will be inevitably involved because of the increase of the loading step caused by the subdivision technique for the EFC-based HS test. Furthermore, the structural responses were calculated using the feedback from the previous integration steps to gradually approach the exact response at the current step by using the EFC-based HS test. It is difficult to trace back to the previous finished test steps. To this end, no subdivision of integral time interval was taken into account at the location of the displacement inflection point. To make sure the subdivision influence upon the test results, the numerical simulation comparisons between the subdivision and non-subdivision cases were done.

The model parameters of the numerical simulation were taken from the test measurements, as listed in Table 1. The structural damping was $C = 0.05 K$; the restoring force model of each story can be shown in Figure 6. The flowchart of both the subdivision and non-subdivision method is illustrated in Figure 7. The difference between subdivision and non-subdivision method is the step after checking the inflection point of displacement. For the subdivision method, once variations in the inter-story shear force are observed, the subdivision of integral time interval is processed, meaning that the time integration interval is subdivided one step prior to the displacement inflection point. At the same time, the earthquake record was subdivided accordingly. The stiffness at current step is updated and used in the calculations for the remainder of this time interval. For the non-subdivision method, a re-assembly for the stiffness matrices is following in the next step.

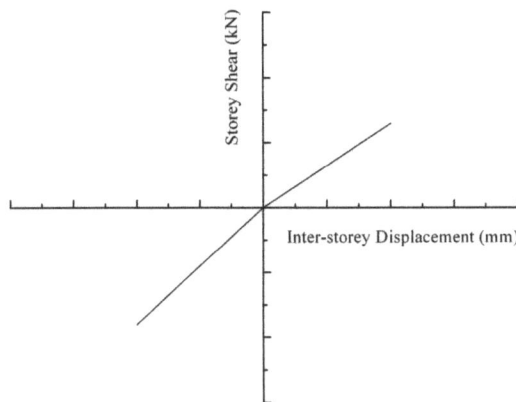

Figure 6. Restoring force model of numerical substructure.

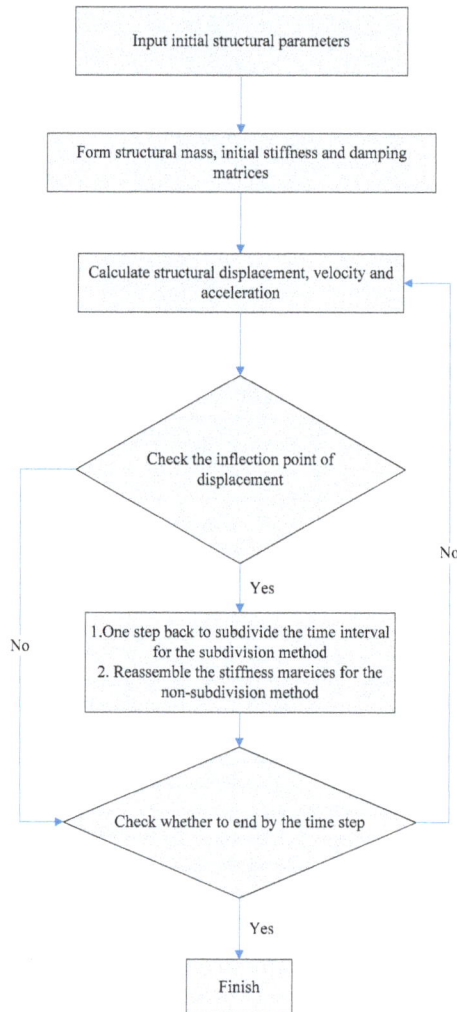

Figure 7. Numerical flowchart using the subdivided integral step and the full integral step.

Figures 8 and 9 show the hysteric curves on the first and second floors under a peak ground acceleration of 110 gal using subdivision and non-subdivision techniques, respectively (other floors are similar to these two floors). Note that both results agree well with the restoring force model as shown in Figure 6. This agreement suggests that the above-mentioned techniques and the corresponding analysis procedures are reliable.

Figure 10 shows the comparisons of the displacement time history curve between the subdivided and non-subdivided integral steps for the first story (the case of the other story is similar and not shown in this research). In order to observe the relationship more clearly, Figure 11 shows the results from 4 s to 8 s. From the Figures 10 and 11, we can see that both the subdivided and non-subdivided results nearly coincide. Figure 12 shows the time histories of the equivalent force (EF) command and feedback of the first story using the strategy of non-subdivided integral steps (the similar trend in the other story). It is easy to see that the EF feedback tracks the EF command well. In summary, it is feasible that the non-subdivision strategy was used to determine the inflection point of displacement.

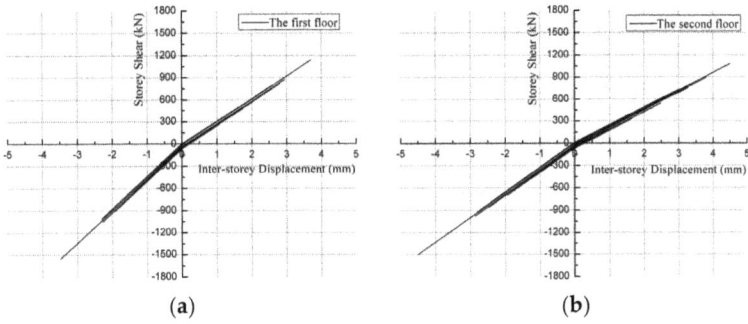

Figure 8. Relationship of the inter-story restoring force and story drift using the subdivided integral step for (**a**) the first floor, and (**b**) the second floor.

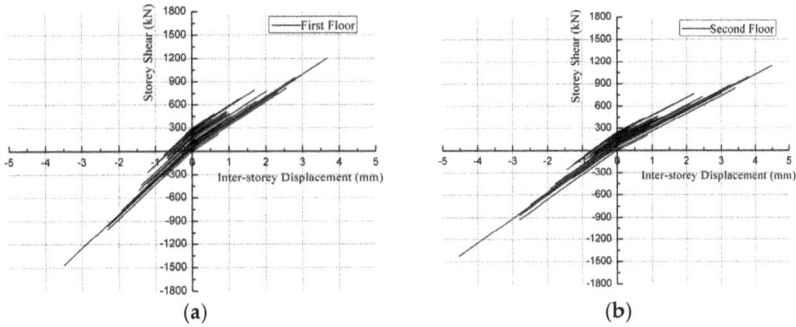

Figure 9. Relationship of the inter-story restoring force and story drift using the non-subdivided integral step for (**a**) the first floor, and (**b**) the second floor.

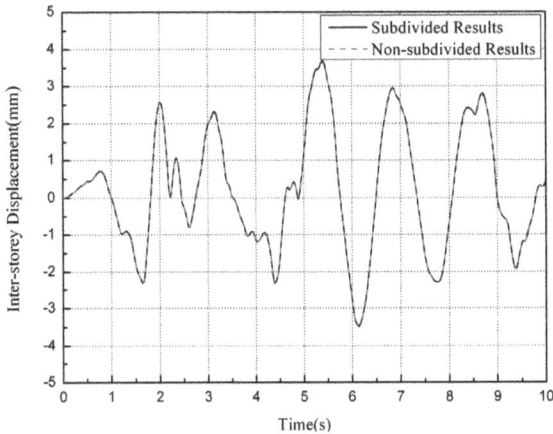

Figure 10. Time history curve of the story drift with different integral steps.

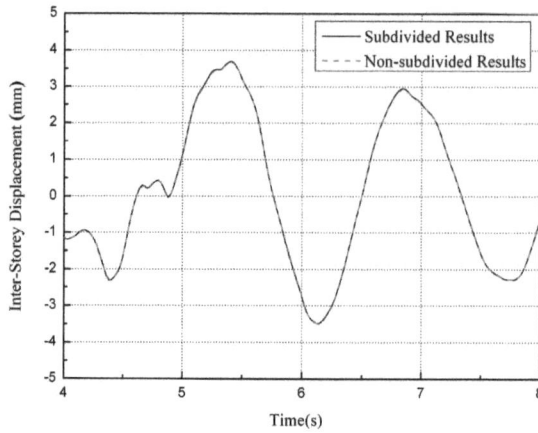

Figure 11. Detailed time history curve of the story drift with different integral steps at 4 s~8 s.

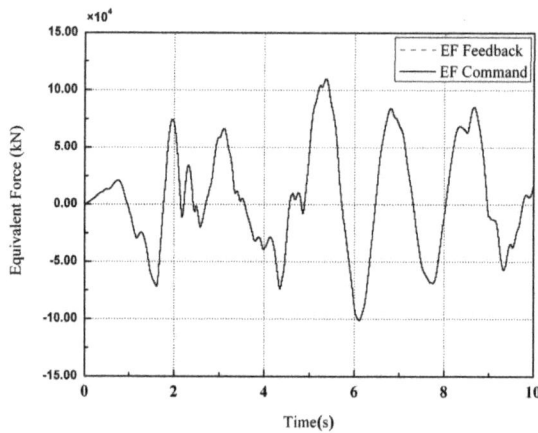

Figure 12. Comparison of the equivalent force (EF) feedback and EF command on the first floor using the non-subdivision method.

3.2. EFC Parameters

The previous sections focused on the parameters of the structural restoring force model for the numerical substructure, which was applied with asymmetric tension and compression stiffness, as shown in Figure 6, and the non-subdivision strategy was used to determine the inflection point of displacement. This section presents the determination of the EFC parameters such as force-displacement conversion factor C_F, proportional gain K_P, and integral gain K_I.

Generally, the force-displacement conversion factor C_F can be expressed as

$$C_F = (K_N + K_{PD} + K_E)^{-1} \tag{1}$$

$$K_{PD} = \frac{4M}{\Delta t^2} + \frac{2C}{\Delta t} \tag{2}$$

where K_N, K_E are the stiffness matrices of the numerical and physical substructures, respectively; K_{PD} is the pseudo-dynamic stiffness; M, C are the mass and damping matrices, respectively, and usually taken to be constant. From Equation (2), given a relatively small time interval Δt generally,

K_{PD} is considered to be a relatively large component compared to K_N and K_E. In the above discussion, the proportional gain K_P, and integral gain K_I are the control parameters, while the force-displacement conversion factor C_F calculated by the Equations (1) or (2) is used to transform the equivalent force to the displacement of the model.

Note that it is difficult to calculate the stiffness of the physical substructure due to the test error. For the sake of convenience, the factor C_F is always calculated in terms of the initial stiffness matrices of the numerical and experimental substructures. Then, Equation (1) can be adjusted to

$$C_F = (K_{N,0} + K_{PD} + K_{E,0})^{-1} \tag{3}$$

where $K_{N,0}$, $K_{E,0}$ are the initial stiffness matrices of the numerical and experimental substructures, respectively. Because of the asymmetry in the tension and compression stiffness, the following method was used to determine the force-displacement conversion factor C_F: the $K_{N,0}$ was used as the practical value while $K_{E,0}$ was used as the mean value of the tension and compression stiffness matrices. For the nonlinear test models, the calculation of C_F was based on how the above method impacted the test results. But the effect of the EFC parameters upon the response performance can be adequately reduced through a reasonable design of the EF controller using a self-adaptive control, sliding mode control [48,49], etc. The PI controller was used in this study, as shown in Figure 13. The following section presents a detailed determination of the proportional gain K_P and integral gain K_I.

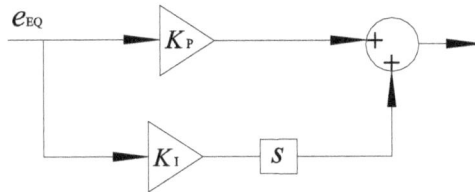

Figure 13. Equivalent force controller.

In order to guarantee that the force-displacement conversion factor C_F in the numerical simulation identically corresponds with the correspondent in the hybrid simulation test, the stiffness of the physical substructure using for the numerical simulation is taken to be the average value of the tested stiffness in the positive and negative directions listed in Table 1. The updated stiffness was listed in Table 4. The elastic nonlinear restoring force model, shown in Figure 6, was adopted for the numerical substructure; at the same time, an elastic linear model was used for the physical substructure.

Table 4. Calculated parameters for test model.

Floors	1	2	3	4~12
Initial stiffness in the positive direction (kN/mm)	446	327	327	295
Initial stiffness in the negative direction (kN/mm)	446	327	327	359

Figure 14 shows the time histories of the inter-story displacement with a varying K_I of 0.001, 0.0065, and 0.01, and a fixed K_P of 0.05 (only the result of the first floor is shown here, because there were similar results for the other floors). The reference results are obtained using the fourth-order single-step integration method and these can serve as exact solutions (W). Note that with increase of the K_I, the results based on the PI controller begin to approach the theoretical solution. This indicates that a moderate increase in K_I can enhance the accuracy of the numerical integration method. However, the comparison between the EF feedback and EF command, shown in Figure 15, suggests that the overshoot (Figure 15c) or undershoot (Figure 15a) phenomenon occurs when K_I when is too large or too small. In other words, a moderate integral gain can enhance both the EF response speed and the accuracy of the integration algorithm, while an excessive increment will lead to an overestimation and thereby compromise the results. It is important to point out that increasing the frequency of the EF feedback can relieve some of these accuracy issues. However, this will also dramatically increase the testing duration and potentially lead to a greater rate of test error.

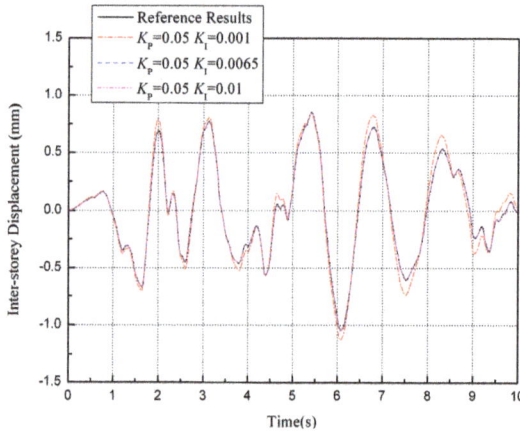

Figure 14. Time history curves of story drift with different integral gains.

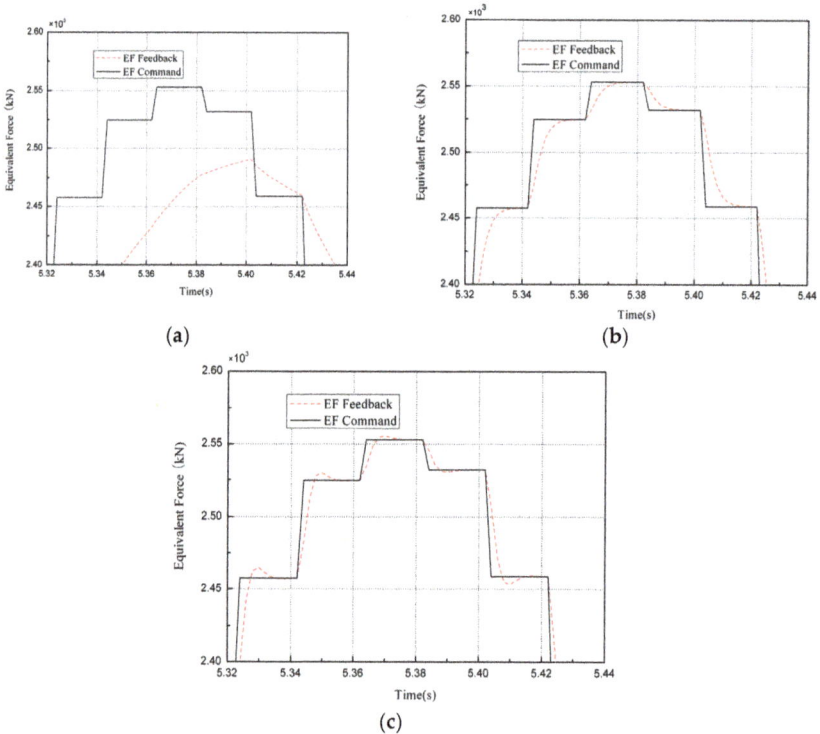

(a)

(b)

(c)

Figure 15. Comparison of the EF feedback and EF command: (a) $K_P = 0.05$, $K_I = 0.001$; (b) $K_P = 0.05$, $K_I = 0.0065$; (c) $K_P = 0.05$, $K_I = 0.01$.

The time histories of the inter-storey displacement with a varying K_P of 0.05, 0.01, 0.90 and 1.00, and a fixed K_I of 0.0065 are shown in Figure 16. As the proportional gain K_I increases, the results from the PI controller deviates from the theoretical results—something similar can be seen in the comparison of the EF feedback and EF command as shown in Figure 17. The excessive actuating displacement may arise due to an excessive proportional gain—which can cause the overshooting and oscillation of the EF feedback that produces unrealistic hysteretic responses. Wu et al. [49] analyzed the EFC method for real-time substructure testing and found that an extensively small proportional gain often led to a time-lag problem of actuator response. This, in turn, introduced negative damping into the numerical substructure model and affected the test results. Increasing the proportional gain can accelerate the EF response, which reduces the time lag in the EF response as well as the negative damping. In contrast, an oversized proportional gain can easily lead to oscillation phenomenon in the EF feedback, and distorted results. In conclusion, the $K_P = 0.05$, $K_I = 0.0065$ was adopted in this study.

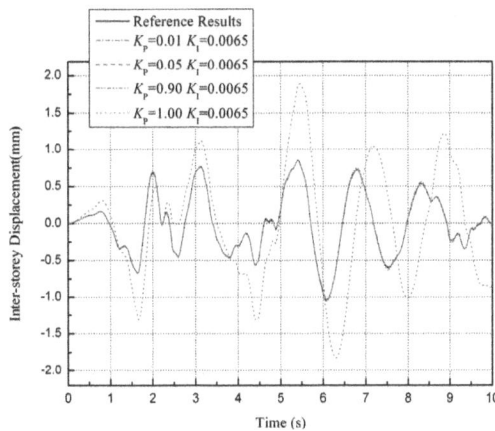

Figure 16. Time history curves of story drift with different proportional gains.

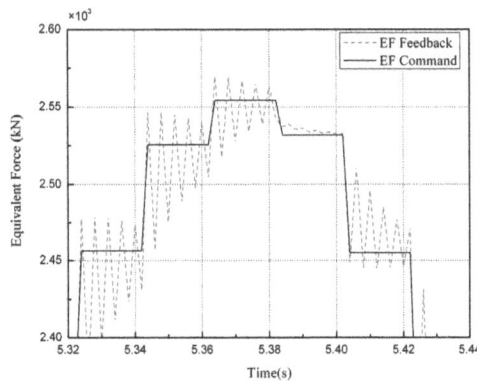

Figure 17. Comparison of the EF feedback and EF command when $K_P = 0.9$, $K_I = 0.0065$.

4. Experimental Validation

Using the previously mentioned method, a series of hybrid simulation tests were carried out on the test model under the following peak ground accelerations: 35 gal→70 gal→110 gal→220 gal. Figure 18 shows the time histories of the inter-story displacement under different peak ground accelerations.

Note that in Figure 18d, only the results for the duration of 6.26 s are present. This was because after the peak displacement, the test model entered the descending stage in the hysteretic curve and the test was artificially terminated when the restoring force recovered to zero at t = 6.26 s. From Figure 18, we can see that the structural responses are controlled by the first mode. As the peak ground acceleration increases, the whole displacement time history becomes increasingly sparse. This suggests that the stiffness of each story, as well as the whole structure, has been degraded. Figures 19 and 20 show the experimental results with a peak ground acceleration of 220 gal. More experimental results will be shown in other study on the structural seismic performance. Figure 19 shows the comparison of the displacement time histories between the EF feedback and EF command. The comparison suggested a good agreement, which indicating that the proposed EFC method performed well. As seen in Figure 20, when the substructure response enters the descending path of the hysteretic curve, the ultimate load of the substructure under the pulling force of the first and second floor is reached. This demonstrates that the EFC can effectively simulate the degrading behavior of the substructure following the ultimate load in the hybrid simulation test.

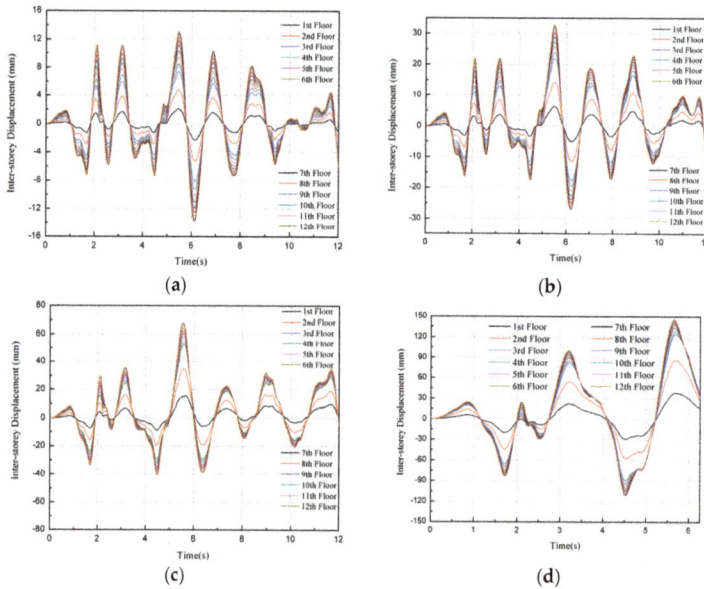

Figure 18. Time history curves of story drift with different peak accelerations: (**a**) 35 gal; (**b**) 70 gal; (**c**) 110 gal; (**d**) 220 gal.

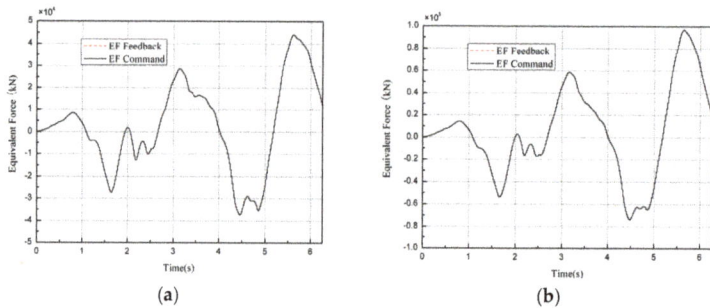

Figure 19. *Cont.*

(c)

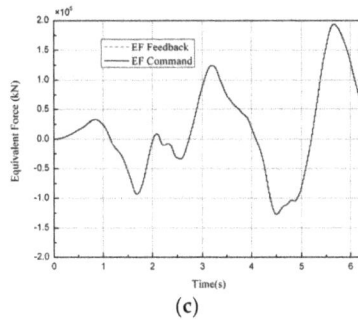

Figure 19. Comparison of the EF feedback and EF command for (**a**) the first floor; (**b**) the second floor; (**c**) the third floor.

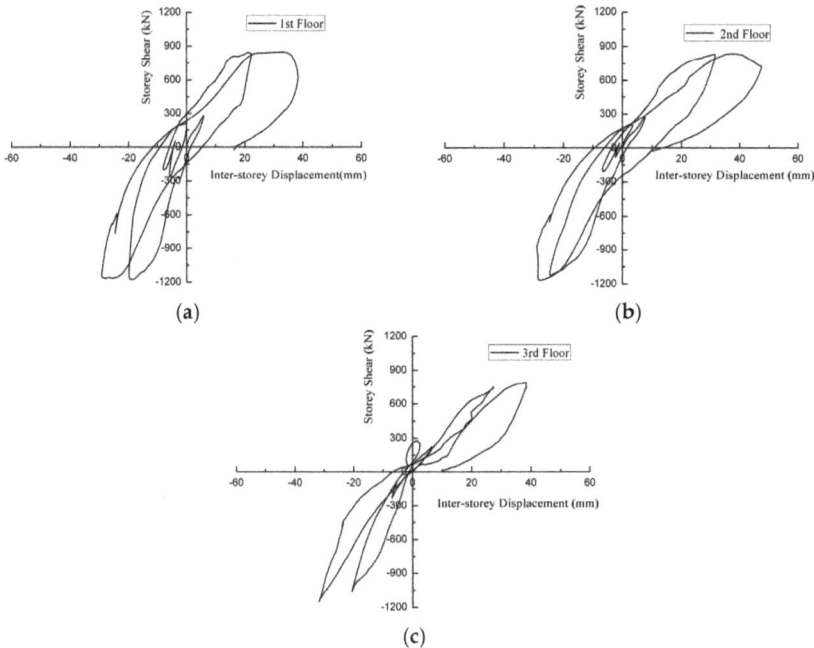

(a)

(b)

(c)

Figure 20. Relationship of the inter-story restoring force and story drift in the physical substructure under the peak ground acceleration of 220 gal for (**a**) 1st floor; (**b**) 2nd floor; (**c**) 3rd floor.

5. Conclusions

This paper presents an application of the EFC for the multy degree of freedom hybrid simulation test combined with the elastic nonlinear numerical substructure model. The following conclusions may be drawn from the results:

(1) The EFC method with implicit integration algorithm was successfully used to the HS test of a twelve-story pre-cast reinforced concrete shear wall structure model. And non-subdivision strategy for the displacement inflection point of numerical substructure is used to easily realize the simulation of the numerical substructure and thus reduce the measured error. The testing results of the descending stage can be conveniently obtained by the EFC based HS method.

The physical test results show that the EFC based HS method has the good performance in both reliability and accuracy.

(2)   A moderate increment in the integral gain can enhance the EF feedback speed as well as the accuracy of the integration. However, an excessively large value may lead to problems of overshooting, and compromise the results. Increasing the proportional gain can accelerate the EF response, reduce the time-leg effect, and decrease damping. Likewise, an excessive value can cause oscillation in the EF feedback while also distorting the results. The displacement control problem can be effectively addressed through reasonable arrangements of the equivalent force controllers.

Acknowledgments: The research is financially supported by the National Science Foundation of China (51678199, 51508495, 51161120360). The authors also gratefully acknowledge the financial support provided by the China Scholarship Council (CSC) at UIUC (201606125079).

Author Contributions: All authors have made a substantial contribution to this study. Z.C. provided the concept and design of the study. X.Z. wrote and revised the manuscript. H.D.W. and H.J. performed the experiment and collected the data. H.W. and K.W analyzed the experimental data.

Conflicts of Interest: The authors declare no conflict of interest.

References

1.   Hakuno, M.; Shidowara, M.; Haa, T. Dynamic destructive test of a cantilevers beam, Controlled by an analog-computer. *Trans. Jpn. Soc. Civ. Eng.* **1969**, *1969*, 1–9. [CrossRef]
2.   Chang, S.Y. Explicit pseudo-dynamic algorithm with unconditional stability. *J. Eng. Mech. (ASCE)* **2002**, *128*, 935–947. [CrossRef]
3.   Chen, C.; Ricles, J.M.; Marullo, T.; Mercan, O. Real-time hybrid testing using the unconditionally stable explicit CR integration algorithm. *Earthq. Eng. Struct. Dyn.* **2009**, *38*, 23–44. [CrossRef]
4.   Phillips, B.M.; Spencer, B.F. Model-Based Feedforward-Feedback Actuator Control for Real-Time Hybrid Simulation. *J. Struct. Eng.* **2013**, *139*, 1205–1214. [CrossRef]
5.   Liu, Y.; Goorts, K.; Ashasi-Sorkhabi, A.; Mercan, O.; Narasimhan, S. A state space-based explicit integration method for real-time hybrid simulation. *Struct. Control Health Monit.* **2016**, *23*, 641–658. [CrossRef]
6.   Kolay, C.; Ricles, J.M.; Marullo, T.M.; Mahvashmohammadi, A.; Sause, R. Implementation and application of the unconditionally stable explicit parametrically dissipative KR-αmethod for real-time hybrid simulation. *Earthq. Eng. Struct. Dyn.* **2015**, *44*, 735–755. [CrossRef]
7.   Kolay, C.; Ricles, J.M. Force-Based Frame Element Implementation for Real-Time Hybrid Simulation Using Explicit Direct Integration Algorithms. *J. Struct. Eng.* **2018**, *144*, 04017191. [CrossRef]
8.   Dimig, J.; Shield, C.; French, C.; Bailey, F.; Clark, A. Effective force testing: A method of seismic simulation for structural testing. *J. Struct. Eng. (ASCE)* **1999**, *125*, 1028–1037. [CrossRef]
9.   Pan, P.; Nakashima, M.; Tomofuji, H. Online test using displacement-force mixed control. *Earthq. Eng. Struct. Dyn.* **2005**, *34*, 869–888. [CrossRef]
10. Nakata, N. Effective force testing using a robust loop shaping controller. *Earthq. Eng. Struct. Dyn.* **2013**, *42*, 261–275. [CrossRef]
11. Günay, S.; Mosalam, K.M. Enhancement of real-time hybrid simulation on a shaking table configuration with implementation of an advanced control method. *Earthq. Eng. Struct. Dyn.* **2015**, *44*, 657–675. [CrossRef]
12. Ou, G.; Ozdagli, A.I.; Dyke, S.J.; Wu, B. Robust integrated actuator control experimental verification and real-time hybrid simulation implementation. *Earthq. Eng. Struct. Dyn.* **2015**, *44*, 441–460. [CrossRef]
13. Yang, T.Y.; Tung, D.P.; Li, Y.; Lin, J.Y.; Li, K.; Guo, W. Theory and implementation of switch-based hybrid simulation technology for earthquake engineering applications. *Earthq. Eng. Struct. Dyn.* **2017**, *46*, 2603–2617. [CrossRef]
14. Wu, B.; Ning, X.; Xu, G.; Wang, Z.; Mei, Z.; Yang, G. Online numerical simulation: A hybrid simulation method for in complete boundary conditions. *Earthq. Eng. Struct. Dyn.* **2017**. [CrossRef]
15. Campbell, S.; Stojadinovic, B. A system for simultaneous pseudodynamic testing of multiple substructures. In Proceedings of the 6th U.S. National Conference on Earthquake Engineering, Earthquake Engineering Research Institute, Washington, DC, USA, 31 May–4 June 1998.

16. Watanabe, E.; Yun, C.B.; Sugiura, K.; Park, D.U.; Nagata, K. On-line interactive testing between KAIST and Kyoto University. In Proceedings of the 14th KKNN Symposiumon Civil Engineering, KKNN Symposium, Kyoto, Japan, 5–7 November 2001.

17. Tsai, K.C.; Yeh, C.C.; Yang, Y.C.; Wang, K.J.; Chen, P.C. Seismic hazard mitigation: Internet-based hybrid testing framework and examples. In *Colloquium on Natural Hazard Mitigation: Methods and Applications*; Université Pierre et Marie CURIE: Paris, France, 2003.

18. Spencer, B.F.; Carrion, J.E.; Phillips, B.M. Real-time hybrid testing of semi-actively controlled structure with MR damper. In Proceedings of the Second International Conference on Advances in Experimental Structural Engineering, St. Louis, MO, USA, 10–12 June 2009.

19. Kim, S.; Christenson, R.; Phillips, B.; Spencer, B. Geographically distributed real-time hybrid simulation of MR dampers for seismic hazard mitigation. In *Proceedings of the 20th Analysis and Computation Specialty Conference, Chicago, IL, USA, 29–31 March 2012*; ASCE: Reston, VA, USA, 2012.

20. Li, X.; Ozdagli, A.I.; Dyke, S.J.; Lu, X. Christenson Richard. Development and Verification of Distributed Real-Time Hybrid Simulation Methods. *J. Comput. Civ. Eng.* **2017**, *31*, 04017014. [CrossRef]

21. Kwon, O.; Kammula, V. Model updating method for substructure pseudo-dynamic hybrid simulation. *Earthq. Eng. Struct. Dyn.* **2013**, *42*, 1971–1984. [CrossRef]

22. Hashemi, M.J.; Masroor, A.; Mosqueda, G. Implementation of online model updating in hybrid simulation. *Earthq. Eng. Struct. Dyn.* **2014**, *43*, 395–412. [CrossRef]

23. Wu, B.; Chen, Y.; Xu, G.; Mei, Z.; Pan, T.; Zeng, C. Hybrid simulation of steel frame structures with sectional model updating. *Earthq. Eng. Struct. Dyn.* **2016**, *45*, 1251–1269. [CrossRef]

24. Shao, X.; Pang, W.; Griffith, C.; Ziaei, E.; Lindt, J. Development of a hybrid simulation controller for full-scale experimental investigation of seismic retrofits for soft-story woodframe buildings. *Earthq. Eng. Struct. Dyn.* **2016**, *45*, 1233–1249. [CrossRef]

25. Chuang, M.; Hsieh, S.; Tsai, K.; Li, C.; Wang, K.; Wu, A. Parameter identification for on-line model updating in hybrid simulations using a gradient-based method. *Earthq. Eng. Struct Dyn.* **2017**. [CrossRef]

26. Mei, Z.; Wu, B.; Bursi, O.S.; Yang, G.; Wang, Z. Hybrid simulation of structural systems with online updating of concrete constitutive law parameters by unscented Kalman filter. *Struct. Control Health Monit.* **2017**, *25*, e2069. [CrossRef]

27. Nakashima, M. Development, potential, and limitation of real-time on-line (pseudo-dynamic) testing. *Math. Phys. Eng. Sci.* **2001**, *359*, 1851–1867. [CrossRef]

28. Chen, C.; Ricles, J. Improving the inverse compensation method for real-time hybrid simulation through a dual compensation scheme. *Earthq. Eng. Struct. Dyn.* **2009**, *38*, 815–832. [CrossRef]

29. Wu, B.; Wang, Z.; Bursi, O. Actuator dynamics compensation based on upper bound delay for real-time hybrid simulation. *Earthq. Eng. Struct. Dyn.* **2013**, *42*, 1749–1765. [CrossRef]

30. Gao, X.; Castaneda, N.; Dyke, S. Real-time hybrid simulation: From dynamic system, motion control to experimental error. *Earthq. Eng. Struct. Dyn.* **2013**, *42*, 815–832. [CrossRef]

31. Chae, Y.; Kazemibidokhti, K.; Ricles, J. Adaptive time series compensator for delay compensation of servo-hydraulic actuator systems for real-time hybrid simulation. *Earthq. Eng. Struct. Dyn.* **2013**, *42*, 1697–1715. [CrossRef]

32. Asai, T.; Chang, C.; Spencer, B. Real-Time Hybrid Simulation of a Smart Base-Isolated Building. *J. Eng. Mech.* **2015**, *141*, 04014128. [CrossRef]

33. Wang, J.; Gui, Y.; Zhu, F.; Jin, F.; Zhou, M. Real-time hybrid simulation of multi-story structures installed with tuned liquid damper. *Struct. Control Health Monit.* **2016**, *23*, 1015–1031. [CrossRef]

34. Maghareh, A.; Waldbjørn, J.; Dyke, S.; Arun, P.; Ali, I. Adaptive multi-rate interface: Development and experimental verification for real-time hybrid simulation. *Earthq. Eng. Struct. Dyn.* **2016**, *45*, 1411–1425. [CrossRef]

35. Zhang, R.; Phillips, B.; Taniguchi, S.; Ikenaga, M.; Ikago, K. Shake table real-time hybrid simulation techniques for the performance evaluation of buildings with inter-story isolation. *Struct. Control Health Monit.* **2017**, *24*, e1971. [CrossRef]

36. Zhou, H.; Wagg, D.; Li, M. Equivalent force control combined with adaptive polynomialbased forward prediction for real-time hybrid simulation. *Struct. Control Health Monit.* **2017**, *24*, e2018. [CrossRef]

37. Buckle, I.; Reitherman, R. The Consortium for the George E. Brown, J. Network for Earthquake Engineering Simulation. In Proceedings of the 13th World Conference on Earthquake Engineering, Vancouver, BC, Canada, 1–6 August 2004; Paper No. 4016.

38. Stojadinovic, B.; Mosqueda, G.; Mahin, S.A. Event-driven control system for geographically distributed hybrid simulation. *J. Struct. Eng.* **2006**, *132*, 68–77. [CrossRef]

39. Kim, J.K. KOCED collaboratory program. In Proceedings of the ANCER Annual Meeting: Networking of Young Earthquake Engineering Researchers and Professionals, Honolulu, HI, USA, 28–30 July 2004.

40. Ohtani, K.; Ogawa, N.; Katayama, T.; Shibata, H. Project 'E-Defense' (3-D Full-Scale Earthqyake Testing Facility). In Proceedings of the Joint NCREE/JRC Workshop on International Collaboration on Earthquake Disaster Mitigation Research, Taipei, Taiwan, 17–18 November 2003.

41. Takahashi, Y.; Fenves, G. Software framework for distributed experimental-computational simulation of structural systems. *Earthq. Eng. Struct. Dyn.* **2006**, *35*, 267–291. [CrossRef]

42. Kwon, O.S.; Elnashai, A.S.; Spencer, B.F. UI-SIMCOR: A global platform for hybrid distributed simulation. In *Hybrid Simulation: Theory, Implementation and Applications*; CRC Press: Boca Raton, FL, USA, 2008; p. 157.

43. Mahmoud, H.N.; Elnashai, A.S.; Spencer, B.F.; Kwon, O.S.; Bennier, D.J. Hybrid simulation for earthquake response of semirigid partial-strength steel frames. *J. Struct. Eng. (U.S.)* **2013**, *139*, 1134–1148. [CrossRef]

44. Murray Justin, A.; Sasani, M. Near-collapse response of existing RC building under severe pulse-type ground motion using hybrid simulation. *Earthq. Eng. Struct. Dyn.* **2016**, *45*, 1109–1127. [CrossRef]

45. Abbiati, G.; Bursi, O.S.; Caperan, P.; Sarno, L.D.; Molina, F.J.; Paolacci, F.; Pegon, P. Hybrid simulation of a multi-span RC viaduct with plain bars and sliding bearings. *Earthq. Eng. Struct. Dyn.* **2015**, *44*, 2221–2240. [CrossRef]

46. Dong, B.; Sause, R.; Ricles, J.M. Accurate real-time hybrid earthquake simulations on large-scale MDOF steel structure with nonlinear viscous dampers. *Earthq. Eng. Struct. Dyn.* **2015**, *44*, 2035–2055. [CrossRef]

47. Wu, B.; Wang, Q.; Shing, P.B.; Ou, J. Equivalent force method for generalized real-time substructure testing with implicit integration. *Earthq. Eng. Struct. Dyn.* **2007**, *36*, 1127–1149. [CrossRef]

48. Wagg, D.J.; Stoten, D.P. Substructuring of Dynamical Systems via the Adaptive Minnimal Control Synthesis Algorithm. *Earthq. Eng. Struct. Dyn.* **2001**, *30*, 865–877. [CrossRef]

49. Wu, B.; Wang, X.; Wang, Q. Sliding Mode Control of Servohydraulic Testing System and Its Application to Real-time Substructure Testing. In Proceedings of the 4th National Conference on Structural Control, Dalian, China, 11–13 July 2004.

MDPI

St. Alban-Anlage 66

4052 Basel

Switzerland

Tel. +41 61 683 77 34

Fax +41 61 302 89 18

www.mdpi.com

Applied Sciences Editorial Office

E-mail: applsci@mdpi.com

www.mdpi.com/journal/applsci

www.ingramcontent.com/pod-product-compliance
Lightning Source LLC
Chambersburg PA
CBHW051729210326
41597CB00032B/5662